A TOM LUONG NOVEL

THE TOMORROW COLLECTION

# MARS OUTPOST
## Surviving Tharsis Montes

## JULIAN PHILLIPS

## TOM LUONG

ISBN-13: 978-1499578416
ISBN-10: 1499578415

Inside art "Mars Colony" by David Schleinkofer
Cover art "Mars Outpost" (BookBaby.com for Luong Films)

## DEDICATION

The writer wishes to dedicate this novel with co-author Tom Luong, in memory of beloved friend Karen Thomason, bed-ridden in a paralytic-coma at a care-home in Orange, California, from the turn-of-the-century, year 2000, to 2014.

Tom Luong would like to dedicate this novel to his parents, Mike Manh Van Luong and Nancy Ngat Thi Le, both from Vietnam.

# CONTENTS

# CONTENTS

# ACKNOWLEDGMENTS

Tom Luong Films, and author Julian Phillips, would like to acknowledge all of those who provided support and real assistance in the accomplishment of this novel, and other works in the Tom Luong Tomorrow Collection series, including: The Luong Family; the Tom Luong Films group of fans, talent, crew and volunteers; Mr. Phillips' wife Carol Lynn, son Preston Laverne, and other family members; the National Aeronautics and Space Association (NASA), the U.S. Mars program and Mars probe researchers; the Los Angeles Times; Wikipedia; the Mars One online community; Edgar Rice Burroughs, author of the John Carter on Mars 'Barsoom' books (1912-1964); author Robert Heinlein (1907-1988), and his heirs; the Orion Institute of Santa Margarita, California; the research, work, science and efforts of the Pasadena Jet Propulsion Laboratory, the International Space Station program, the Hubble Space Telescope program, and the National Space Society; the 'Star Trek' family of writers and talented performers; as well as dreamers, and forward-looking, positive-thinking readers everywhere.

## CHAPTER 1: Flight Path Error

*"Space---the final frontier?  Kiss my ass, Captain Kirk---get me the hell out of here!"*

*---Guy Reisling, US space-program transport pilot, 2075*

*Nothing like a good shake-down and brutal emotional drubbing by your commanding officer,* Guy was thinking. *Makes me feel so important.* The man now under the threat of career-ruin and complete professional and personal humiliation was Guy Askilav Reisling, a standard-issue muscle boy airman for duty in the abyss. Not un-important, but far from leadership or super-star status among his peers, which included astronauts, off-world walkers, Nobel-Prize scientists, space-pilots, and folks who knew how to get safely from Point-A to Point-B, when either spot was a few million miles apart, separated by vast regions of nothingness, and hard to find in the dark. His type were known as 'vacuum-cleaners', maybe because it's hard to draw a breath in the deep-deep-deep, without a few preparations, at least, and about three million years of human evolution and science to help out. Not to mention progress in anti-gravity toilets *("To boldly go where no man has gone before--- ").*

Guy was seated now for his punishment, back on Earth, in a briefing room at the Vandenberg, California Space-Port Center. He was generally a joyful type of human being, as men go. But not at the moment. Solid and strong, only about 40 years old, with a wide jaw, big shoulders and overall big-build. He had more-than-just-curly, slightly reddish hair, and tended to sport an unruly beard. *Take it like a man, Guy,* he told himself. *It wasn't your fault.* If Guy had any actual freckles, on his face or shoulders, few would mock him for it. He was, after all, a space-ship pilot. The 'right stuff' for that, in the year 2075, definitely included kicking ass if he felt like it, maybe on leave, at a bar, or at the race-track, where the ground-level types seemed to get a thrill pretending that what he did for a living was a mere fly-boy luxury, or vanity for over-educated college-types who signed up only so they could brag about their true understanding of Elton John's 1970's-era pop-song, 'Rocket Man'---ancient history, true, but the romance never really fades, when you've earned the privilege to work in space. But for the vast populace of humanity, work in space had no redeeming value.

The briefing room where Guy was reaping what he had sowed, now months behind him, but oh-so present, was like any such office or conference room the Earth had known for any time-period after 1975 or so. The Vandenberg Space-Port included many facilities that were that old, often a source of complaint from workers. Chairs, fluorescent-tube lights, walls, tables, plants, windows with thick bars or screens, a computer, a wall-map or two, chalk-board, plenty of good old-fashioned gravity---and Commander Okman, the Transport Crew head-honcho for the Western-region space-program. Although Okman had actually done 'real' space-flight work in the past, these days he only told program sub-ordinates the way it is, will be, and was, and should be, and cannot-be-otherwise, without instant death as a consequence. Okman was about 60 years old, but presented a formidable form and content of opinion, standing on the other side of the conference table, across from Guy, pretty much raging. Both men were dressed in street-clothes, with nametags and base-passes, and had known each other about five years.

"What were you thinking, Guy?" Okman spouted in his high-sounding vocal pattern. "What were you thinking? Every flight plan we provide for you to navigate from Earth to Mars on your run, is the product of about ten years of work, you know this, right? Did you forget or something? What were you going to do if you missed you re-entry corridor and ended up floating around in empty space for a year waiting for planet Earth to circle the Sun and come back up to meet your lifeless ship, if you even survived that long? Is that even possible? Have you done the math on that?? C'mon! You blew it!"

"I told you what happened, Commander," Guy replied. "I was forced to make a flight-plan adjustment and I may have miscalculated in the rush. But I was able to correct the mistake later. It all worked out. It will never happen again, okay?"

Okman flexed his shoulders like his body was an old coat he had dragged from the closet that morning, wrinkled, dusty and unkempt. The look on his face could only be called disgust---something he probably practiced in private, for just such an occasion.

"No course-correction was necessary in the first place, Guy. You never needed to change course the first time. Those solar gas-cloud flares pass through the corridor all the time. They're harmless, unless you take them at more than 100K caloric. You were less than 5K. That's nothing. The energy-levels were acceptable, there was no danger."

"My information at the time was otherwise, sir."

"Do you have the luxury of being wrong, less than halfway home from Mars, in a ship worth more than entire nations, with a crew of seven highly skilled workers, transporting life-or-death supplies, in a deep-space environment? Give me a break, Guy. You were drinking or something. And why did you make alterations in your fuel consumption? Why did you switch to reserve-engines on hydrogen-only? What was the purpose of that?" Okman fumed.

Guy heaved a bitter sigh. "I've made this run more than 20 times, both ways, Commander," he said. "We had the report from Molinari on the solar gas-flare---"

Okman huffed at mention of the deep-space Earth-Mars corridor mid-point dock-station. "Ha! Yeah! Molinari! Your girl-friend!"

"What's that got to do with anything? Yeah, Lila works there, sure. But it didn't come from her anyway. So my corridor had the solar-heat---the caloric wasn't specific. If I get the heat, I either have to divert, slow-and-wait, shield, or just fucking go around it. So that's what I did. Afterwards I went to hydrogen to make up for lost time, and match the previous flight-plan. The hydrogen is faster in deep space than the peroxide. So that's it. It all worked out. No one was hurt. No big loss."

"Not really," Okman said. "I had to bring in a team of five astro-navigators on triple-time for two weeks to figure out how to land you and coordinate docking. If they hadn't been able to do it, you and your crew would be dead. I'm serious. You could have survived another six weeks, but planet Earth would have left you so far behind you'd never get back. Because we would have been about 100,000 miles ahead in orbit around the Sun, while you went spinning off the other way. I don't care about the ship, Guy, and I don't care about the cargo. I don't even care you went to hydrogen engines without authorization. But I sure as hell care about my people. Including you."

"Yes, sir," Guy said, muffling resentment in the tone of his voice. Okman walked twice across the room, as Guy squirmed.

"So, okay. That's it. You're grounded, space-man. Your last flight is up for committee review for illegal operations and maneuvers. If the review finds you significantly at fault, your career is over. You'll never go up again. And that's okay; it's for the best. We all make mistakes. We all get tired. But I can't have any of my pilots making those kinds of errors, because lives are on the line, every single time."

"Yes, sir. I realize that, sir. I care about my people, too."

"Of course you do, Guy. I'm not---I'm not saying it was you---it's not your fault, of course. And I know you give a damn. You were just tired. Burnt-out. Shit happens. You're dismissed. Get out of here. I'll message you on the next step and orders. But you're grounded and so is your ship and your entire crew. At least for now. Now get out."

Guy rose from his chair, looking sheepish. He waited long enough for Okman to look him in the eye.

"Go on, fly-boy," Okman said tersely.

As he left the room, Guy knew Okman was basically right. The truth on the solar-heat flare needed to be confirmed prior to his course-correction, with more science to back up his decisions. At 5K caloric-energy, the heat was really minimal. At 100K caloric, his transport ship could fry. But at the same time, a flight-path alteration mid-way between worlds could easily have resulted in exactly the deadly situation Okman had described. And that's a heck of a way to die, comparable to an Earthside nuclear submarine that sinks to the bottom of the North Sea with all hands, losing power, and no way back, the crew left inside, only to count the hours until they suffocate. Plenty of time for the ship's captain to apologize to the dying men, something no one should ever have to do. And their only job, as transport-crew, was to haul needed items to the base on Mars, and then other items back to Earth. Not very exciting unless something went wrong.

It was because of the base on Mars that guy's Condrum 21 Deep-Space Local Planetary Cruiser (his ship, known affectionately as 'Penelope') had been commissioned for service at all. The Penelope was designed only for the journey she had been intended for, the deep-space run between the two planets. She had a standard crew of nine highly skilled persons. The pilot (Guy), his Second, two staff on nothing but life-support systems and ship-integrity, two tech-science staff with variable roles, also inter-changeable, two on navigation and astro-physics, and one to control and maintain propulsion-fuel and high-energy thruster mechanisms. Okman had said seven---true enough, one of the crew on life-support, and one of the navigators, had been dismissed at that time, which was fairly common, and not an alarm.

Guy made his way across the Vandenberg Space-Port campus, exiting from the hallways of the building where he was brought under orders for his meeting with Commander Okman. California, as always, even for hundreds of years, was the best the Earth had to offer, in many ways. The brisk winter air and high clouds, beneath the partridge blue hen's egg sky, it was home forever, Paradise, so different than his 'workplace' as to bring to mind an entirely different realm of being---another planet, you might say. And one he certainly favored. Like a military installation, the Vandenberg facilities were filled with men and women in uniform, large concrete structures of imposing nature, huge equipment and machinery, various weapons. As he walked, he met a few people who recognized him, but not many. No super-star. A transport-pilot was minor-league around here. But that was okay with Guy. A smile and a wave, and any friends or co-workers were on their way. His big screw-up was maybe only painful to himself, and not really any kind of big news that would embarrass him forever.

Why Vandenberg? Why California? Why Mars? Relieved after his meeting with Okman, Guy suddenly remembered that he was 'grounded' from space-travel for a while. After about 20 round-trips in deep space---this was also a relief. He wondered how he would spend his time until the big-shots decided if he was fit for service or not. His training had cost billions of dollars and more than 15 years of his life, and years of many other lives. All so the folks at the Mars Base could have a steady supply of toilet paper. And a few other essentials--- like fresh O2 they could breathe. Or fresh H20 they could drink. Or methods to fabricate both. Or food they could eat.

Guy paused on one of the walkways, beneath a US flag, blustering high aloft in the winter air. He took a deep breath, just for himself. Then another, then counting, four-counts in, two-counts out, one-count hold.

He glanced at a nearby trash-receptacle. "We recycle everything, but we've never developed a system that's totally free of any kind of trash at all, or any wrappers or any packaging whatsoever," he said to himself. "Look at this."

He picked up a few bits of trash that had fallen near the receptacle. *This is 2075*, he thought to himself. *We can do better.*

"Plastic, cellophane, super-light metals, wood-fiber, plant-fiber, coal-and-oil by-products. What a waste! We could easily buy and sell, or share the same consumer goods, with complete convenience, using no packaging whatsoever! This is insane!"

The space-transport pilot dropped the loose trash-items into the receptacle and pressed them down, alone in his thoughts. Somewhere, far away, yet nearer-than-near, Mars loomed and glared back at him, a distant star now, as he would see with his own eyes when night fell again. He didn't miss the place. He didn't miss the vast 'space' in-between, cold, empty, indigo, deadly. Being grounded for a while never looked so good.

## CHAPTER 2: The Margie Effect

*"Yet still the lacy spires of Truth sing beauties madrigal. And she herself will ever dwell along the grand-canal." From 'Along the Grand-Canal' by Robert Heinlein*

Reisling, the transport space-pilot, had a place when in Vandenberg, where he could stay. Guy wasn't married, but knew a few women who were 'more than friends'. After months in space, therefore, and on recommendation of numerous Space Program doctors, it was for his own health and enjoyment, and psychological well-being, when he returned, to get 'down to business' and have a good time for a bit. His home in the Vandenberg-area was very nice, but perhaps not extravagant by many standards, and on the lazy California afternoons in the heat beneath those high clouds, among the tender green leaves on the bushes in his very own backyard, where birds would play, and grass underfoot, short-cut by the handyman who watched the place---fresh air, quality gravity, food and friends---it was always good to be back, every single time. Earth is home forever, as any space-pilot knows.

The women he knew were mostly of the same variety that were impressed by race-car drivers, jet-pilots, bull-rider rodeo boys, surfers, heavy-equipment operators, and police or emergency workers. Space-travel was not yet truly a 'normal job', in 2075. But there were many runs these days, far more than in the past, and the Earth's Space-Program back-and-forth to Mars, in particular, was now almost like any other very expensive and high-tech industry. So for Guy, it was a 'cool gig', and the ladies loved it. Guy was attractive and healthy, athletic, and he loved them right back. So any returning was a celebration, and barbecues in his backyard were a favorite mode of enjoyment---easy, private, outdoors, with music and friends.

Guy's fast-grill wood barbecue lofted pale-blue wood-smoke into the air and sunlight. He was making hamburgers. Margie, one of the cafeteria-girls at the base, who had a wonderfully pronounced and appealing body for a woman of about age 35-years, had joined him, and for today it was just the two of them, rather than a 'big deal' party or anything. *The better to eat you with, my dear*, Guy thought to himself. He was cooking hamburgers, and other food, and they had beer and snacks. The radio was playing popular current tunes. *It's good to be back*, Guy thought. *I wonder if Margie is up for a hot shower with a cold space-man??*

"Why did all the early science-fiction writers 100 years ago think that Mars had big canals, Guy? Have you ever wondered? You know---Arthur C. Clarke and Robert Heinlein, or Edgar Rice Burroughs and Ray Bradbury? They all invented stories where Mars was covered with these big giant ancient canals, like deep canyons. Why was that? Do you know?" Even for a cafeteria-worker, Margie was well read, probably because it was a spaceport, and all that. Everyone at the base got into the fun and glory of space-travel, the excitement and legend.

"Guy, your hamburgers are burning!"

Guy replaced his beer on top of a picnic table, and grabbed his spatula. He started flipping hamburger-patties as quickly as he could---the bluish smoke was getting thick. "Mars never had canals, Margie," he answered her as he worked. "The idea was because the planet might have had plenty of water, in the past, which had created the canals, like deep rivers. There were early photos, and also early artwork, that perpetuated the myth, and caught on as a popular delusion. So, it was only a romantic notion. The whole canal thing had something or other to do with the Western or US common lack of world-travel experience, and the American's desire to travel to places like Venice, Italy, where they had the famous gondolas, another popular cartoonish image."

"Maybe a root canal," Margie said. "Like at the dentist's."

Margie's pronounced and very appealing 35 year-old female body had at this point cleverly maneuvered close by Guy's shoulder, and she skillfully leaned her slightly rounded arm on his shoulder, and another on his hip. She sipped her beer. Guy finished flipping his seven hamburger patties---the fire was a bit too hot, but they were fine. He loved the way the smoke curled into the air. And he also loved the nearby local heavenly bodies, such as Margie's.

In that year, 2075, the Mars program was now at least 20 years old. Much progress had been made. The most significant was the establishment of a functioning base, on Mars. The Snikta-Ridge Volcanic-Basin Mars Base was a wonder. It had been built in about 10 years, for the first part, and really was always expanding with new parts or new sections. The US Space-Program was the lead agency, but other nations had participated---Japanese, Chinese, UK, Brazilian, Russian, and Israeli. Guy was very proud to be a part of it all, and of course his job was merely to move equipment, gear and supplies, and sometimes people, from Earth to Mars, for the base and its various needs. The ships, the base-machinery, the staff and crews, the propulsion-systems, the navigation, and the things learned, were all a stunning victory for the program, at least to an extent.

So, of course, driving a transport back-and-forth, sometimes made beautiful women like Margie desirous of Guy's company, he guessed as a status-thing for the gals on the base, who probably bragged later in their girly-talk about which space-pilot was the best lover. Margie was no favorite, or any long-term relationship or significant-other status for Guy. In terms of simple lust, he was a lot more interested in the hamburgers, at the moment. The in-flight food they ate on the 'Penelope', his Cundrum-21 Deep-Space Local Planetary Cruiser (by DuPont-Monsanto), was really pretty awful. No outdoor barbecues, cold beers, gently lofting smoke from a greasy grill, toasted bread-buns, mustard, ketchup and onions. So he always craved 'that sort of thing' when he was on his own, and back home. The doctors didn't approve of much 'junk food' for his nutritional routine, but he and Margie had other plans.

He told Margie more about why he had been grounded. The story was now common-knowledge at the base, as far as his peers. They lay in bed in their afternoon-delight, later, full of beer, burgers and each other, between the cotton sheets, made from cotton-puff fiber clouds only Mother-Earth could yield, like her white-clouds above. For Margie it was a gossip-plumb she could cherish and savor for later---the inside scoop.

It was their fifth run of the season, for Guy and the crew of the Penelope. Because of the relatedness of each planet's position as they would orbit the Sun, transports could only make so many runs to Mars with any real success. The two planets were only near enough during a short window of months. After that, the journey was almost insanely difficult. The so-called Mars Effect, back on Earth, was even calculated as far as gravitational-pull on Earth oceans, and other effects, of course seemingly minor, but others speculated the nearness of the planets could create emotional and psychological upsets among the common populace of Earth, during those seasons. Wars, market crashes, domestic violence, disease---it might just be Mars, or the Mars Effect, though of course it was impossible to really know. But during those seasons, when Mars and Earth were near enough, that was when Guy went to work, and voyaged in his ship, with goods for the Mars Base. *I guess everyone needs a scapegoat,* Guy thought. *In the case of Earth, we need an entire other planet!*

"Just rub my back," Margie moaned. "You have such strong hands."

Otherwise known as the Margie-Effect, as far as guy was concerned. He complied happily.

"*Ohhhhh.* Thanks, Guy. So go ahead. It was your fifth-run on your way back. They had the solar flares. Why didn't you just---you know----just solve the problem like a good astronaut? You're supposed to know how and everything, right? How else could you survive?"

Guy's thoughts drifted back in his memories of the flight. How could he explain so a cafeteria-worker would understand? It was now three months in the past. They had performed their duties flawlessly, another great run. The chores as far as space-men and pilots were concerned, were now fairly routine. But, like any work of this sort, even a small irregularity, or unforeseen event, could literally create sudden terror, and death, for Guy and his crew. The Penelope had been required to ferry over a more-than-usual cargo of Condensed Water----things get dry on the Red Planet. Every transport run was essential, and there was no turning back after the Halfway-Point Docking at Molinari, the long-term orbiting deep-space rest stop.

Memories of the abyss are only darkness, except aboard the ship, where they all lived every moment, the mother, the womb. Great place for Buddhists---emptiness its only feature, with the stars so distant, as to appear as non-real. So, he and his team did their thing, the ship was fine. But the solar flares, or what they call heated emissions---any movement there on the local solar tracking and monitoring, and Guy, as the pilot, had to take care. If he didn't handle it properly, based on the ship's position, the movement and speed of the hot gases, and the heat-levels of the hot gases---they could fry. Not a happy outcome. And in this case, apparently, even though they had a successful return to Earth, Guy had not handled it properly. Flight-path error. It means, 'we don't do things that way'. Space flight is like running a huge stock-market fund. If you screw up, it can be like a very long row of very carefully stacked dominoes, which happen to be very expensive as well. One error leads to another and another. So the pilots are always on the spot. If he and his crew fry, that's one thing. If he makes an in-flight decision, it's more about flight-training and procedural methods. As anyone would guess, when making the choice that had landed him in bed with Margie that peaceful afternoon, incineration was the first-up motivating-factor, compared with being grounded.

Now Margie's pink-white butt was in his hands, which he was kneading like bread-dough. "It's obvious, Guy," she was commenting. "You were just sick of space-travel and being out there so long in the dark. You wanted to come home, and be with me. So you sub-consciously chose to mess up the flight-plan, when maybe you could have done it another way."

"Sure Margie," Guy said. "But I also didn't want to be burned to death along with my crew. Besides, you and I are NSA, you know? No-strings-attached. "

She smirked back at his talk. And here Margie thought it was all about her. What a heart breaker!! "So you used the hydrogen-fuel instead of the regular fuel. Big deal. Or you changed course when maybe it wasn't planned out. You still got home. You still saved their lives. Isn't that enough?"

"Yeah, for me," Guy said. He now had also had enough, and gave up on Margie's massage. Later, as the afternoon rolled by in a bliss of warmth and welcome, they kissed and said good-bye. *Great gal*, Guy thought. *Smart, too.*

Burn, or----take a vacation back home. Not a difficult set of options. Maybe it was true, Guy thought. Maybe Mars wasn't worth it after all. There was nothing there of much value. Someone at the base was talking about a Big Meeting that was ahead for Earth's Space-Program leadership---something about the Snikta-base on Mars. There had been a lot of rumors, as usual. Whether Guy would still be flying any transports, whatever future was ahead, was 'up-in-the-air' for now. But he still wanted to be a part of it all. He still wanted the thrill of his job, the daring-do kind of tough-guy self-confidence, and the beauty and wonder. This was why being grounded meant so much. Because it's not just a job. Oh no. Not just a driver's spot on a transit bus. So much more. Maybe even---too much.

"There are no hamburgers in space," Guy said to himself, watching the sunset over the Pacific after Margie was safely on her way home.

## CHAPTER 3: Roll Call: Vandenberg

*"Mars, the ancient Roman deity of war, a god in olden times. Why not then to tread it underfoot, and end all wars forever?"*

*--Ibrahim Mehudi, Ph.D., US-Earth Mars Command Second-Director of Sciences, Vandenberg, CA Spaceport, 2075*

ROLL CALL: these are the men and women of the US-Earth Mars-Mission Space Command. Skilled, trained, brave, fearless, unwavering, and sick to death of crappy 'McAbyss Happy Spaceman Meals', among other discomforts.

Talk on the Earth-side US continental West-coast center for space-flight stuff had turned to an announced and scheduled all-hands conference and meeting, even for weeks prior to Guy Reisling's ill-fated grounding and return to planet-side, from his last Mars-run. The space-program was considered both important and even vital, and a total waste of time, even among its long-term workers and students. The cost was outlandish, but given that by 2075, old-style money or wealth had become virtually meaningless, it was thought wise to remove the blockage of typical funding approaches, and 'just do it'. Damn the financial torpedoes, we're going to Mars, they all said. Much as kings in the ancient world could build the Great Pyramids of Egypt without regard to mathematical money, the intellectual community had revealed over many years, to the powers and authorities, that common Earth-technicians really could do almost anything, within our grasp of science and technology, with or without actual money. And this was liberation, to be sure. But the goals, the 'value', the payback, the harvest or wealth, or return-on-investment, for the citizenry---this had always been a huge public relations stumbling block for the space-program, even in the old days. Without the discovery of mountains of solid gold or oceans of high-quality petro-fuel on Mars, that would enrich the Earth somehow---other goals, such as long-term species survival, seemed rather empty. Many other issues occupied the various crews and staff at the Vandenberg base from time to time, like an unusual anthill of uniquely endowed ants, always busy. But the current alarm, regarding this up-coming 'big-deal' meeting, was much a mystery.


The Mars-Mission Fleet Commander at this juncture was a man named Winton 'Kick' Berle. Winton was a large-faced man, about age 52-years, with a huge jaw and ruddy complexion, seeming always to smile, but a harder smile than most. His role was to oversee all the US space ships and vessels, related to the Mars missions and travel. All flight-travel, all traffic and coordinated navigations and plans, all equipment, gear, readiness, all the pilots and crews, and much more. It was a lot to handle, but really not unlimited. Winton was nick-named 'Kick' by the pilots, for a story about one of his own space-flights, as a pilot, early in the Mars program, when a small inner-ship doorway hatch was stuck, mid-way in their deep-space run, blocking the crew's use of important life-sustaining gear, and he had to kick in the door with his foot to save their lives. He still enjoyed complaining about the consequences of a tiny metatarsal bone fracture deep within his right foot, and occasional pain, when chatting up the newbies. Winton was no longer a pilot, and was very much beloved and respected by all.

"Every damn meeting like this in the last ten years, I end up selling Mars rocks and free hot-dogs to the Japanese rocket fuel-merchants. They love the hot dogs, and re-sell the rocks for paperweights. Go figure," Winton complained to his secretary, two weeks prior to the event. His secretary chuckled as she always did at his colorful asides, which were never quite true.

Ibrahim Mehudi was the Israeli-born Ph.D., Second-Director of Sciences, for the entire Mars program. Male about age 45-years, thin and dark-skinned, muscular and lanky. He had a heavy-looking brow-forehead and deep eyes, but watery and Buddha-like within. Obviously, the vast arena of sciences related to space-travel encompassed enormous knowledge and technique, from metallurgy and rocketry, to astro-physics, and including the hypercritical area of life-support and human biology. Within the hierarchy of the program, a few select men needed to pull it all together into a cohesive, standardized whole that all the various teams could understand and relate to in a common language focused on the task at hand. Ibrahim's feet never left the ground, and he had never even traveled in space at all. But his mind and thoughts soared infinitely. The actual science they used was not really 'new'. Ibrahim's job was to channel all that info-data and technique, to the widely different critical moments that made up even a single space flight---the launch, the trajectory, the deep-space movement, re-entry, crew-safety and well being, and also much of the work on the actual Snikta-base on Mars itself. Ibrahim's knowledge was truly awesome, god-like, and very valuable to every person working on the Mars program. He was also a great tennis-player.

"There is no God in space-travel. No Jesus, no Spirit. No Space-Ghost. You understand this? It's a no-no," Ibrahim said to his wife, at their beachfront California home one afternoon. He was also making preparations for the meeting planned for the base. Ibrahim was one of those who actually knew what the meeting was really about, and the important announcement they were releasing for the entire Mars program staff and crews. His wife was not ignorant of all this as well.

"But Ibrahim, my husband," she said over their wine at dinner, in complete privacy. "The destruction of the planet, the end of life on Earth, a huge meteor that will wipe us all out. Surely the Supreme Being should at least be consulted? A thing like that!"

"Not for many years, Golda. Not for many years."

Branson Porter had a pretzel-logic job or role among the higher-up's in the Mars-mission Space-Command. Porter was male, age 45, a Texan. Not much to look at, tough-as-hell. The program really had no military aspect. Space-travel was considered to be research-oriented and pure science, and the militarization of the space-program had been a hot debate many years past. If the program went to military purposes, it would be exploited for typical Earth wars, deaths, powers, wealth, conquest and control, with potentially gigantic damage, because of the advantages. Bombs on orbiting platforms, or weapons launched from the moon, would be a ruin, and a waste. On the other hand, the program included very high-tech, controversial, costly, and safety-specific bases, gear, equipment, satellites, communications, ships, and people---very similar to any modern military effort. And all that had to be protected. This was Branson's main gig: Mars Mission Security Command. And any attacking alien armadas of trans-light speed space ships who wanted to swipe a Mars transport ship for study or to melt down

22

for cheap fuel would have to go through Branson and his security crews first; and Earth-types too.

"I hope they're going to war with China from the Molinari Space Dock so I can try out my ship-side EMP emitter-arrays. 50-million watts of radio-energy at a high enough frequency to toast every PC in Hong Kong from orbit in about an hour. And I never even get to play with 'em!" Porter was bragging that same night, at a beer-bar in Santa Barbara, near Vandenberg, concerning the up-coming meeting at the base Conference Center. "Not that I have anything against the Chinese, of course." He and his Security-type buddies kicked back a few Millers, with sports shows on TV at the bar, speculating about the Mars program and the meeting. But no one really knew anything.

Karen Tutturo was on her way to Mars within only two weeks, on a completely different job assignment, that had very little to do with any big-deal announcements or meetings. Or so she thought. A science type, or high-tech worker-bee, Karen was about age 42, female, very attractive and intelligent, with a somewhat big build physically, or slightly 'butch', appropriate for space-travel in a woman. Karen was a Communications Systems-Specialist, and a citizen of the UK. The base on Mars was designed with a very sophisticated deep-space radio communications system that could link voice-chatter, and tech-commands or tech-instructions, Earth-to-Mars and Mars-to-Earth, far faster than any ship. Unfortunately, now ten years into the Mars Base success story, there were problems with the links and system. So Karen was on her way to the Mars Base, supposedly to do the needed 'fixes'. She had of course studied the system and knew all about it, how it worked and what it was made of. But it was her first trip into the Abyss, and she was pretty nervous about it all. The up-coming meeting at

Vandenberg everyone was talking about meant very little to her.

"It's always the end of the world somewhere," Karen told one of her girl friends, as a joke, driving up the coast highway in the heat and sunshine, on another day. Karen's voyage to Mars was scheduled far in advance, and of course she was preparing. Her road-car purred with hydrogen-fuel efficiency, a blur of motion in the wind and fresh air. But her heart was full of butterflies about the whole space-travel experience. Nevertheless, as part of the Mars program, if her skills could restore the Mars Base communications to full working order, she had to do it, she had to go, she had to be strong, she had to face her fears. And she knew she would.

Transport space-ship pilot Guy Reisling's Second-Pilot was Rob Cowan, an experienced space-flight veteran, age 45, lean, lanky, athletic, Caucasian. The Penelope's Fuel-and-Propulsion Technician was Herbie French, age 35, an Egyptian-born US citizen and student of rocketry, very skilled with that tech. Of course on Guy's runs, their lives depended on their knowledge and hard work and attention. It wasn't just 'one mistake' that would kill them, out there. It was attitude, focus, alertness---and errors, too, if any were made. Errors to space-men on these flights in 2075 were like tiny sparks of static-electricity in a dry forest or in the grassy brown California hillsides in the hot summers, when wild-fires still destroyed homes valued at multi-millions of dollars, and took lives as well. Errors add up, and then---shit happens. Guy trusted both these men and they trusted him. They all knew the up-coming meeting would probably affect them somehow.

"Cut-backs," Herbie chimed in, back at the base, in a locker-room area, where they had run into each other, now 'out-of-service' as a working transport crew, given Guy's punishment and flight-ready demotion by Commander Okman. They hadn't forgotten. But like Guy, it wasn't half-bad at all, and his crew was enjoying the break. They had months between voyages anyway, on their return. Like the rest, they just figured the Mars Program was going through some changes, and the leadership wanted everyone to be informed. The transport guys were not all that important or glamorous.

The US Mars Mission Commander, known to the lower-level types, and other higher-up's, as 'The Queen of the Galaxy', or alternately, 'Commander of Angels', was a formidable woman named Lynn Rodgers-Smith, PhD, about age 60 years. Lynn was a Texan, like Branson Porter, and of course Texas had a long history with the US space-program. If there was a single person who had the final word on choices and decision for the Mars program, it was Lynn Rodgers-Smith. But there really was no single individual pulling all the strings, no Commander-in-Chief approach. Much of a bureaucrat, Lynn was also very experience and knew all the ins-and-outs of the program, having worked with the crews, even on some flights, and deep into the science-tech side as well, in the past, prior to recruitment as a 'desk-jockey', and then to the very top. She was a short woman, busty and brassy, and she could command a room full of astronauts with a hard whisper, much respected as 'knowing'. She loved to cook and also made home-brewed wine and beer. For the Big Meeting they were all anticipating, now only a week away, Lynn was reflective as they all were, intending only to rise-to-the-occasion once again, mother-to-millions in her way, but humble and careful on all points.

"We need to be very clear about the data and info on the meteor," Lynn said to her immediate peer, a man named Willy Atta-Bowman. Bowman was a specialist, brought in for only this meeting, rather as the Explainer-in-Chief, who would bring down-to-Earth for the three-day conference, what they needed to share. The two of them were in the office complex where Lynn held court, private and removed from other chores at the base. Bowman was 52 years-old, an educated science-and-space University teacher, who was in charge of the meeting, and would make the presentations to an assembly of about 300 people at the Conference Center, next Wednesday (including the Japanese rocket-fuel merchants, who were invited out of habit more than need). Of course Bowman also had a staff and assistants for the work involved.

"It's not hard to visualize, just hard to prove to a room full of Ph.D. astro-scientists," Bowman said. "They tend to be picky about fact vs. fiction where giant meteors are concerned. But we know what we know. That's the thing. Given another six years, there will be no doubt whatsoever. Now is the time to bring this out with your Mars program people, so needed changes and decisions can go forward immediately."

"And your researchers still feel the Ukrainian space-program is hostile to our interests?"

"Well---researchers, or you might say, good old-fashioned spies. But yes, in a way. We'll get into all that. Bottom line though, if the Russian Space-Program really does want the Mars Base, in the next year or so, and if they have plans to take action, the new data on the meteor will set things in place for some kind of showdown, either formally or informally. And we feel they do want the Mars Base, as a safe-haven if the meteor hits. We also feel they have the same data we have, and have been making these kinds of plans," Bowman said, with a heavy sound in his voice.

The so-called Queen of the Galaxy, Lynn Rodgers-Smith, present Commander of the US Mars Space Program, for the year 2075, leaned over in the chair at her large oak desk, spitting out her chewing gum unceremoniously into a trash-can, with a tiny 'plink' sound. Her thick pink fingers ran across her face nervously, looking back at Bowman.

"So ten years into the success of the Mars Base, we're now talking about some kind of a conflict for control or occupation of the Mars facility, with Russian or Ukrainian Space-forces, or even military, because of a meteor approaching the Earth, which is not scheduled to even come close to the planet for about another ten years," she said. "Is that correct?"

"That's about it," Bowman responded. "Yeah."

Lynn paused. "Well, we need to be very clear about the data and info on the meteor, that's all," she said. "One damn thing at a time."

## CHAPTER 4: U2753b

*"A trip to Mars was a fantasy, a mere dream, a muse or complete fiction, for thousands of years, or even much longer, here on Earth. Nothing or next-to-nothing was known, and the idea would be such as credited to the mad or lunatics, or sinister wizards. And of course, it's still much the same, with the only exception being that now we are actually doing it."*

*--Lynn Rodgers-Smith, 2075, US Space-Program Mars Mission Commander, overheard at a luncheon speech*

The Japanese rocket-fuel merchants included a posse of about ten Asian men and women, whose services were essential to the Mars program for the US. These were mature science-industry business-people, really not very involved in the global space program, as far as any personal flights or adventures in orbit or beyond. Their education and background made them invaluable participants, however, and as a result they were always invited to the conferences, meetings, seminars, speeches, and governance-boards, for the US program, over many years. They had industrial sources and connections for ready supplies of numerous types of fuel, in particular refined hydrogen, hard to acquire in the amounts, volume, type and purity needed. Launch fuel was different than deep-space fuel, and so on. Their work was under-appreciated and controversial among environmentalists, because of the toxicity of some of the fuels, but it made them very wealthy anyway, and their visits to places like the Vandenberg Space-Port were well-financed, as well as their US lobbyists.

All ten Japanese men and women wandered across the Vandenberg base, about a week after Rodgers-Smith had met with Willy Atta-Bowman, concerning the conference meeting about the Mars program. They all had badges and passes, and were not noticed as being unusual or out-of-place, mostly between the ages of 30 and 60 years-old, dressed variously in modern or somewhat geeky outfits. The Conference-Hall was a large auditorium, on the South side base, surrounded by grassy lawns, walkways, minor security, electronic displays with information and deluxe-pixilated 3D images, and people entering, or hanging about talking. It was about 3 p.m. in the afternoon, somewhat overcast with watery gray clouds, and a bit of a chill.

Kick Berle, the Fleet Commander, found himself waiting just outside the hall entrance, with his Secretary, as the Asian fuel-supply delegation passed by, chatting amongst themselves in Japanese. One man recognized him, and stopped briefly to greet him with only a smile, bowing slightly as was the Japanese custom even to that day in 2075. Berle smiled back at him broadly and shook his hand, American-style. They laughed, then the man moved off with his group into the building.

"I was just joking about the hotdogs" Berle said to his secretary, a bright young woman who had all his notes and books for the conference in a bag. "They eat much better than that, let me assure you. US hot-dogs are not very healthy. They like fish, and raw vegetables."

"I kinda' figured, Commander Berle," his assistant said.

"But they do go for the Mars rocks. They really do," he added. She smiled, and they also went inside to find their seats for the meeting.

The conference-meeting was organized to efficiently inform about 300 people about details and information, related to the topic, which was posted overhead on a large banner, in the front, and elsewhere around the large room, and on stationary and press-releases, reading: MARS-MISSION UPDATE CONFERENCE, SPRING 2075 'THE CHALLENGE OF ASTEROID U2753b'.

At the front of the hall, there was a panel-discussion style set of two long tables, and a podium-table in the center. On either side were an array of computers and projection-gear, and behind were two very large image-screens, and a sound-system. In all, about eight 'experts' were seated at the tables, or just settling into their seats as the room quieted down. Each had name-plaques, their own laptops, assistants, and so on. Before them, the hall was filled with chairs, like any large assembly, numbering about 300, and now filling up with guests. Doorways in the back were still allowing people to enter and find their seats. The room was lighted, and other tables had documentation-material, and food-snacks, coffee, tea. The 300 guests included anyone the program leadership felt needed to be informed about Asteroid U2753b. Pilots, commanders, crew, support tech-staff, engineers, Earth navigation plotters and also in-flight navigators, life-support crews, and also staff and crew from the Molinari Deep-space Dock. Every other sort who was directly involved in the program was in attendance, which was mandatory.

Despite Cargo-Crew Commander Okman's slow pace at settling the final decision on transport-pilot Guy Reisling's 'grounding' or flight-ready status, Guy and his entire crew were at the meeting, with a row of seats just for themselves. Each crew for each of about 20 other active Mars space ships also had a row of seats. Karen Tutturro, Branson Porter, Ibrahim Mehudi, and many other un-named and unknown heroes of the program, were seated here and there, with their own guests and assistants or helpers and crews or seconds. Within about twenty minutes of chatter and blustering around, all 300 people quieted down. A Moderator introduced the program and topics, and each panel-expert was introduced as well, which took another half-hour.

"Pretty boring so far," Guy whispered to Rob Cowan, his Second Pilot, seated next to him.

"Boring is good," Cowan said. "If it's anything really dramatic, we're all sure to suffer."

The so-called Queen of the Galaxy, Commander Rodgers-Smith, took the speaker's podium, after everyone was introduced. Light applause greeted her for a short moment, and a few good-natured hoots.

"Thank you, people, thank you," she said. Her voice echoed into the hall over the sound-system. "I think we're ready for the meat-and-bones of this meeting, and I know many of you are very curious about what's going on. As we get started, be advised that this meeting is rather serious, and no simple matter at all. With Asteroid U2753b, the Mars Program is facing an unprecedented challenge, with vast consequences. It's not just our program. It's not just us. This is truly a global issue, although still ten years away from any real harm. But I'll leave that to the experts to explain. Nevertheless, as your program head-honcho and Mars program bureaucratic guru, please be alert. This is a very serious matter."

Now the room was quiet, a hush.

"Let me introduce Doctor Willy Atta-Bowman, Ph.D., from the University of Berkeley," she said. Bowman, who had already been introduced once, now rose from his seat at the long panel-discussion tables, and took the speaker's podium. Rodgers-

Smith shook his hand and sat down quickly.

Bowman waved his hand at the operator of a computer-projection system. The large screens behind him came to life, with a huge image showing a navigation map of the Earth's Sun (Sol), it's relation to the Earth, and aspects in between (Mercury, Venus, and other). Distances and planetary orbit paths were also indicated.

"Hello, Mars program," Bowman started. "I'm Doctor Bowman, and I think my background has already been sufficiently presented. I'm a planetary-science specialist at UC Berkeley, for the past 15 years, basically. For tonight, I'm the Explainer-in-Chief. So, hope you like my style, I guess."

He paused, a bit nervous. "Okay. So, it's really no mystery, is it? U2753b has been tracked now for many moons--- even years. Who knows where these large rocks come from, I guess the Big Bang. Earth hasn't been hit by a very large meteor in thousands if not millions of years. Hollywood gets a lot of mileage out of this kind of thing, and we've all seen those old movies and read those books. Great stories, right? As we know, the space-program has been tracking large meteors and asteroids for more than 20 years, for obvious reasons. And we do get them, ranging in size from a few feet long, to much larger, maybe the size of a Chevy truck, or a school bus. The chances of Earth being hit by one of these is, pardon the expression, astronomical. But if they're headed our way, the space-program wants to know, so they are tracked from the first day they're identified, usually by private hobby-astronomers around the world who do their star-gazing for fun, many of them quite advanced in their skills and science."

He paused again, taking a sip from a water glass and clearing his throat. "So, okay, you get the idea," he started again, speaking into the sound system. The audience was molting before his eyes, like a mist of hopeful faces and eyes, trying to understand. But many already knew, and the prospects caused them to lower their gaze or look away. "The research is not bull-shit. Many of you are very advanced in the sciences. Feel free to go over the documentation and proofs, which are available, and have just been released as complete reports. Bottom line, this asteroid is too large to ignore, and the projected paths of both the planet and the meteor, are indeed sobering."

Now an image flashed onto the second screen behind him at the front of the hall---an actual Hubble telescope-type image of the asteroid they were discussing. Dark, gray, potted with craggy peaks and valleys---a rock.

"It's one of those scary science-fiction type deals, I guess," Bowman told the audience. "U2753b is about the size of the state of Virginia, or larger. A hit anywhere on planet Earth would almost without a doubt have devastating, Extinction-Level results. But---it could miss, or it could be diverted, and really it's not a problem for the Mars program, in terms of avoiding a collision. That is not our mission, and not our arena, and we probably wouldn't be involved. Other international space-forces will try to stop the hit, and this could be done in different ways---blah-blah-blah, just like the movies, I guess. Also, don't start making plans for your own funeral. The asteroid won't even be near our world for at least another six years. I think Commander Rodgers-Smith said ten years, a moment ago. It's a long ways off, and it's slow, and somewhat unpredictable. For a local-space region meteor, this one is a monster. It's huge. So, what we're talking about today, is how the Mars Program, and the base on Mars, will be affected, and how our program will respond, and other difficulties and challenges."

The room was still, with only the combined breathing of 300 space-program staffers to be heard. Someone dropped a laptop PC on the floor with a clatter. Bowman paused. "Can I answer just a couple of quick questions at this point?" he said. "We've only got three hours for this meeting today, and then I guess another science-data conference tomorrow, so I want to move along quickly if I may."

A moment. "So we're all gonna die, right?" one of the space-pilots shouted into the room, somewhat out-of-turn, a bit rowdy.

Bowman smiled broadly. "That's about it, space cowboy," he said with at least some good humor. "But you knew that already, didn't you? You're a brave bunch and so am I, so take heart. We have six years to avoid that. But for the Mars Program, our base on Mars now starts to look like---dare I say it---our only hope for survival, or off-world human species survival, even in small numbers, if the Earth takes that kind of hit, and all those millions die in their pain and fears, I guess, and all the environmental stuff. Again, this would be like, at the level of sucking away Earth's breathable atmosphere for about 1,000 years. Stuff like that. With the Mars Base now fully functional and life-support working to maintain 200 or so people as a sustainable facility, even on its own---control and ownership of the base, is now a very hot item among international Earth military and space-program forces. So, that's our next topic. Let's break now for 15-minutes, please."

The conference-room lights flickered on again, and the group began to chatter. But now the melody of the voices, and the overall vibration, had turned to a different tune, at a lower key, as if a giant had struck a bell, indicating danger ahead.

Bowman nodded at Lynn Rodgers-Smith, and she looked down and away, somewhat grim, without a smile.

Cowan, Guy Reisling's Second, stood from his seat and stretched. "So, uh---not boring, then, right?" he said.

"I guess not," Guy answered.

## CHAPTER 5: The 2015 Spring Update

*"Mars is the subject of much speculation as to whether or not it is inhabited, because it's behavior is similar to that of the Earth. Mars is 141-million, 500-thousand miles from the Sun, and has a diameter of 4, 230 miles. The diameter of the Earth is 7,918 miles, so gravity on Mars is somewhat less."*
*--The Story of the Globe, Replogle Globes, Chicago, Illinois, 1933 ('Replogle Globes Are Better Globes')*

It was true what Cargo-Transport Commander Okman had said about disgraced pilot Guy Reisling's relationship with the woman who worked at the Molinari Deep-Space Dock. Her name was Lila Meetek. The Molinari Space Dock was created years in advance of the Snikta-Ridge Volcanic Basin Mars Outpost. Molinari was placed in permanent deep-space orbit as a mid-point rest stop for ships making regular voyages back-and-forth to Mars. It was like a small city in space, floating in orbit, in the vast gap between the two planets, a rather lonely form, shaped like a buoy one would find bobbing in the waters of an ocean-harbor back on some sunny beach or rocky coastline on Earth. Molinari was the size of a very large skyscraper building in New York, or even a small airport. And of course it was sustained and operated just like any space-ship or space-vessel, with life-support and breathable air and food, a large array of high-tech computer and communication stuff, and thruster-powered maneuverability. But it did not fly or travel, and remained at a constant distance from the Earth, in orbit forever, or until it died, or decayed, or was somehow destroyed, perhaps in 1,000-years or so.

Which was about how long Guy supposed he would stay in love with Lila. Who would not delight in a beautiful space-girl, the portrait of ideal health and vigor, as well as sexy intelligence, and a certain knack for grilling outstanding hamburgers? She was about age 38 years old, and it was Lila's job at Molinari to monitor and track activity in the abyss corridor on the Mars-Earth flight path. Lila had blonde, henna-reddish, blue-green, or brown-gray hair, long and feathery. Thin, athletic, and privately slutty, with an outstanding set of boobs and other body-parts that Guy often dreamed of, 1,000 years of her wouldn't have been enough for Guy.

*"Molinari! Ha! Your girlfriend!" Okman had said when Guy was decommissioned from his ship's command. "The data on the solar heat-flares was no better than your hot sex-chat and perverted pic-trading with Lila on official communications-links! No wonder you screwed up!"*

So, as the US Mars Command Mission Up-Date Conference for Spring, 2075 at California's Vandenberg Space-Port, continued into its third hour, Guy had to speculate about what would be going on at Molinari, where Lila was currently stationed, and how the news about Asteroid U2753b would affect her, and the others. The Mars Base was not the only off-world sustainable human habitat. At least one other was Molinari. But it hardly seemed to matter, with the Asteroid's near approach to planet Earth still six years away. Unless there was now to be some sort of international conflict for control of these same off-world resources. Which was the topic of the last part of the conference meeting, with Dr. Willy Atta-Bowman, Ph.D., as the Explainer-in-Chief, that chilly gray California day. Guy envisioned his beloved Lila taken prisoner by Russian or Chinese space-forces of a more military sort. *Meet the new boss, same as the old boss, he thought to himself.*

The conference meeting had melted like grilled cheese on a beef-patty into a long sequence of science-proofs for the claims about the asteroid. Photos, plotted orbits and intersecting paths, distance, speed-and-acceleration, trajectory, mass-density, impact results, timeline, and anticipated response or planned attempts to divert the meteor, were all quickly reviewed. The Mars program staff and various worker-bees were used to this kind of sharing. Dry and boring, technical and mathematic, all science and facts-and-figures---and yet at the same time, critically important for the lives and well being of millions. There were many yawns and aching backs as the experts went on-and-on.

"Let me introduce now our Mars program Security Specialist, Captain Branson Porter," Bowman told the audience, as the topic now shifted. Porter, the tough-looking Texan in charge of Program Security, came forward like an altar call at a Baptist revival, and took the speaker's podium. There was a pause, as everyone waited for his report. Many people in the program felt Porter was too harsh and military for what was essentially a long-term scientific research mission. But it was inevitable that the space-program would have its hawks. Porter even had a large, ugly-looking blue-black steel high-caliber handgun strapped to his belt in plain view of everyone, there at the front of the large conference room. There was no mistaking his job-description and grim intention to defend the program and its people and components---something many felt should never be necessary at all.

"Hi everyone," Porter started. He coughed and cleared his throat. His voice was deep and sandy, like grit. "As you know,

I'm Captain Porter, Security and Military Police Commander for the Mars Program. Right now I'm going to share with you about what we feel we know concerning the Mars Base, and why the news on the asteroid could cause us problems with our international Earth neighbors."

Now the image projector operator programmed a series of photos showing the actual Mars Base, to roll past on the large screens behind Porter. Everyone could see the base, much like a small city, with many buildings, structures, gates, towers, holding tanks, ports, etc., set against the stark, dusty Martian landscape. Sort of like a family memory-album for the program, and they had all seen the same images many times before.

"Okay," Porter said. "Well, we may be military-police, in my department, and we may be environmental-scientists, too, but we're not stupid, and neither are whatever enemies we really have, right here on earth. What I'm telling you now is classified, so try not to head out and do interviews with your local TV news-shows. The info we have here is from good old-fashioned spies and informants. That's right, the FBI never really died, it just rotted to a new shade of green. And obviously, it's the type of thing where you just don't know, and it's also incendiary, and by that I mean, a cause for conflict, battle, war, call it whatever you want. Hostile or war-like. Which is not for me to decide."

"Get on with it, Branson!" someone shouted from the assembly, probably one of the pilots, known for their antics. Others in the audience laughed. Branson stiffed, a little embarrassed, not used to public speeches.

"Right. You're grounded for that one, pilot," Branson said. More weak laughter. "Well, the report is simple enough. Bottom line---intelligence feels that an alliance of Mid-East Islamic and Russian-Ukrainian Space-Program forces are planning to take control of the Mars Base, sometime prior to the arrival of the asteroid, as a way to assure their survival and control of future programs, if any. The logic isn't hard to understand. If the meteor wipes us out, whoever controls the Mars Base would survive, even though in small numbers. The same is true for Molinari, and the ships, and various systems."

A long pause. Many in the crowd had not heard of this. "So, you might be asking yourself how we know this, or exactly what the Russians have planned, or how they feel they can get away with it, right?" Branson continued. "The intelligence community never really changes. No one knows anything. But we have various convincing indicators. The Eastern space-programs are just as advanced as ours here in the US, and in some ways more so. They have ships like ours, launch-and re-entry programs, highly trained crews and pilots, tracking and satellite control. But, Russia, and the Islamic space-programs, and also space flight out of India, have had too many internal conflicts, wars, and financial shortfalls, to really compete. The US Mars program was initiated as a global partnership, at one time, maybe 30 years ago. But that fell apart. There were agreements and treaties, however. The US went ahead, while the others fell away."

Now Branson paused again, clearing his throat. He took a sip of hot coffee he had with him at the speaker's podium, and idly rested his hand on the blue-steel handgun on his belt, as if not even thinking about it.

"In reality, the Russians and the others are putting out signals. That's how the game is played. There have been recent high-level meetings in Kazakhstan, in the Ukraine region, where Russian spaceports are based, as well as their nuclear bombs and rockets. Russian space scientists have gathered information on our Mars Base systems, flight-paths and orbits, our ships and really our entire program. None of this is actually secret, but much of the Mars Base technology is highly classified. Even more convincing----the smoking gun, if you will---was the recent acquisition of a secret document-file, stolen from Russian think-tank planners, in exchange for $50-million in gold held somewhere in the Netherlands by a private individual. Hey, it's spy-stuff, what can I say? This file, or document, however, represents a 200-page detailed proposal and specific plan, for Russian space-forces to attack and take control of the Mars Base, and Molinari as well. It's all there. This is only a proposal, only a paper, or electronic file. But they put a lot of work into it. It's all there. I've personally reviewed it, and made notes. Russian ships and crews would make the run to Mars, take control of the base by force, take hostages or kill anyone who resists them, and then squat out whatever else happens, at the Snikta-Ridge US Mars Base. And of course in typical Russian style, any explanation or apology to the world community would come later, if ever."

Atta-Bowman tapped the microphone at his seat at the long panel-discussion table. "Captain Branson, if I may?" he said.

"Sure, Doctor Bowman."

"How do we know this supposed attack-plan to take over the Mars Base is real, or authentic to the Russian space-command? Could it be a fake, or planted by someone else, or other enemies of theirs?"

Branson took a breath. "It's intelligence-community stuff, Doctor. So, it's true, we really can't know. From reviews and expert analysis, however, the file I looked at was very well researched and very well planned. It included details on the

Russian space-fleet and resources that would be hard to obtain outside their own staffers. The source of the document was connected directly to high-level insiders on the Russian side, so that's also a point. The science was also very accurate, something a novice or terrorist group probably couldn't master in a short time. The report was also attributed to known Russian or Islamic scientists and Ph.D. astro-physicists. Real people, we know their names. Additionally, other reports show Russian hardware, real equipment and gear, or actual ships, taking baby-steps towards this type of effort, like minor-level preparation. I agree, it's a sort of Cuban Missile Crisis deal, or a WMD-type report. Maybe no need to panic, that's for sure. So, to answer your question---how do we know their plans are real? Well, we don't. We don't know for sure, and we may never know for sure, until they go ahead, if they ever do."

"What about their timeline?" Bowman said. "From the plan you looked at, when would they be thinking of doing this?"

"The possible meteor-hit is at least six years out. They want control of the Mars Base well in advance. The stolen attack-plans were not specific. But any time in the next three years, or even one year, the entire might of the Russian-Islamic Space Program alliance could potentially launch a group of ships armed with various weapons and ground-level soldiers with oxygen suits and weapons, to take control of Snikta," Branson said.

"Would they just kill everyone? Could they actually destroy the base, maybe by accident during a battle? What would happen here on Earth? Would they try to excuse their actions at the United Nations, for instance? Or would they go to war with the West? Anything there?"

"It's all speculation. Anything could happen. As sneaky as the East can be, they might simply stonewall, and claim they have rights to the base, as participants 20 years ago, or as educational research. They could stall and drag it out. Or hold hostages. After all, if the meteor is headed our way, all they really care about is the survival of chosen leaders and persons on Mars---when we're all the rest of mankind dead and gone back here, or living in caves under a black cloud of ice-cold meteor dust, eating bugs for dinner."

A long stillness hovered in the air throughout the conference-hall. Now even the courageous pilots and space-jockeys were nervous. It all seemed unreal.

"All right," Bowman responded to Branson's remarks, into his microphone. "What about our plans for a defense, or to protect the base on Mars, or to fight back an attack?"

"That's another hour's worth, Doctor Bowman," Branson said. "For ten years, my department has only had to deal with protesters at the gates here at Vandenberg, drunk cafeteria workers, stolen toolboxes, and night-watch duties to protect expensive high-tech items stored outdoors. I'm not necessarily prepared to figure out a five-year space-war. However, if you give me another 15-minute break here so I can take a piss, I'll be glad to tell you all about it."

Weak laughter again from the audience. "Space-men piss in their flight-suits into catheter tubes, Branson. Everyone knows that," Bowman joked, More laughter. Now Bowman stood up and stretched. "Let's break again for 15-minute folks. This is all too much. Rest-easy, back in 15."

Now the room full of Mars-workers began to break up again, as the audience stood, or separated into groups, or grabbed coffee-and-snacks. Everyone seemed relieved for a moment. They had a lot to talk about.

## CHAPTER 6: A Certified Transmission

*" A philosopher said, once a body in motion, it tends to stay in motion. Once a body at rest, it tends to remain at rest. What the kid meant was, use it or lose it! "*

*--Comedian and burlesque performer George Burns ('Living It Up, or, They Still Love Me in Altoona', Berkeley Publishing, 1976)*

That weekend at Vandenberg would change the US Mars Space program forever. No one was to blame if the cosmos had finally pooted out a huge meteor from the Eternal Abyss that was likely to strike the Planet Earth, or even with inevitable certainty, would strike the Earth, with no real recourse to avoid disaster. An act of God? This was not the way they liked to think about anything much at all in the space-program, given the science-based nature of the work, the extreme dangers of space-travel, and a tendency for the deeply religious to have certain emotional problems associated with the work involved, in particular actual space-travel.

The rule they used was often spoken as, 'The Universe is actively hostile to intelligent life. Deal with it." This didn't mean the space-planners were heartless men devoid of any true feelings. In fact, as far back as the old Apollo program, when men first walked on the moon, there was a 'space-man's prayer' that was entered into the communication-record on flights, or prior to the many challenging launches and recoveries, such as the nearly-doomed Apollo-13 flight, in 1969, when the whole world watched a group of men very nearly die, struggling with a wounded ship and low-oxygen, to somehow navigate a safe return to Terra-Firma, from a voyage to the moon.

"Give us the knowledge, that we may pray with understanding hearts, to set forth the coming of the day of Universal peace. Amen," was the space-man's prayer in those days. And it hadn't changed much in 100 years since the Apollo program, and many felt it was a transit of souls, into Infinity, that was being answered every day, and not just for astronauts, but for all men. Or maybe the troubling specter of the heat-death of the Universe itself, called 'entropy', figured at many billion years into future-time. In contrast to all the high-tech science and lab-coat feelings, or space-suit stuff, this prayer once would resonate on the radio-link that reached the men headed to the moon, or while on the moon, or prior to launch. Maybe it was because every single one of the astronauts were risking their lives from the moment of lift-off---yet something mysterious in their hearts drove them onward with the greatest courage. Others merely tolerated this sort, and it was a bit of a tradition, in any case.

Guy Reisling finally heard about his denouncement and loss-of-privileges as a pilot, about two weeks following the Spring US Mars Program Update Conference. Enjoying some time off at home, a Certified Transmission arrived via Internet-computer, still in use for private citizen communication, and in 2075, now far more secure for major life-path business transactions, legal, government, banking, political-votes, and many other, it's promise having finally risen beyond the early abuses, porn, terrorism, fraud, crime, etc. By 2075, the Internet was a solid rock of modern lifestyles, as dependable as legal-paper and business-title, standard paper-mail or government-taxation, and even money-types. So, a Certified Transmission meant it was something important, and of course Guy knew right away what it was.

It was just within the twilight hours, there where Guy had his home, North of Santa Barbara, close enough to the Spaceport to keep him busy. His next-door neighbor, an 85 year-old Chinese woman, thin and tough and brown as a small tree, who enjoyed being outdoors with half an acre of organic asparagus, was working with a wheel-barrow to move a load of fertilizer into a compost area, just a few yards from the side of Guy's house, by a large window. The evening dusk-light touched the area between the two homes with shadows.

"Hoooooooooo-Yah! Wooo! Got it!!", she could then hear from within Guy's home. "I'm back! Wooo!"

The old woman paused, wiping the sweat from her forehead as she worked, casting a hard look toward Guy's home. "Stupid space-boy," she said to herself, again hoisting the wheelbarrow, huffing.

Guy was elated, inside his home, that he had been re-instated. The Certified Transmission was from Okman's office, and the Flight-Protocol Review Board. The Board now held that Guy had made errors on his last flight, but they were willing to mitigate their decision because he had been able to correct himself, and return with his crew safely to Earth, despite the difficulty. So, he was re-instated, and this was in the best interests of the program, given Guy's experience, training and loyalty. This meant that he was once again fit-for-duty, and he and his crew would be back into service with the next probable passage date for a transport. To lose his standing and pilot-qualifications would have been a disaster, and a disgrace.

Following the Conference, there was a great deal of 'scuttlebutt' about the findings, especially among those who would be involved in what came next. Guy and Rob, his Co-Pilot, got together later at the base to talk things over and also look at flight-logs.

"I guess we'll be running weapons-cargo or bombs over to the Martian Snikta-base for the next couple of runs," Guy said to his Second, Rob Cowan. Rob was an un-extraordinary type, a hard worker, well trained, mature and responsible. If Guy failed as Captain of the Penelope, Rob would take over, and for many tasks during the voyages they made, there were shared duties of all kinds. Rob was thin and tall, sometimes a bit pale, or seeming less-than-perfectly-healthy. But he was quite strong, and always ready at his work, which was a matter of personal pride to him, like them all.

"They shouldn't turn the Mars Base into a military facility, just because of the meteor," Rob answered him. "It's for research. If the Russians take control, it won't help if they're heavily armed. The staff at Snikta-base is not military. They'd hardly know how to launch a bomb or missile. All they do is soil tests and mapping."

"Maybe they'll have to learn" Guy answered. "Maybe we will, too."

"I'm not a soldier. All we do is haul the mail. Food, water and goods. If they send our crew on the Penelope up, and we have some kind of battle, I can only assume the Russians would be far better prepared. After all, they planned it that way. But we didn't," Rob added.

"Well, unless that changes," Guy replied.

Karen Tutturo, the Communications-Specialist assigned to travel to Mars, was now only two days away from her departure. The news from the Conference gave her chills. Not only did she now need to deal with space-travel, and all her fears and the hardship involved, in addition to repairs to the Mars Base communications-gear---now she also had to worry about some vague kind of Russian-Islamic intrigue, or even an attack. And even, eventually, to consider whether or not the Earth would survive a meteor hit, and her world and all she ever knew, would vanish.

Two days prior to a people-shuttle flight departing for Mars, Karen's life was all about preparation. There were medical exams, gear and life-support suits (which she had to learn to operate properly), and also her personal items, the plans and schematics she needed to work on the Mars Base radio-link for repairs. She would report to the base Launch-Control early in the middle of the week, to be ready for the flight: a flight-suit, waste or body-fluids elimination 'diapers' to be fitted (for launch-and-orbit sequence only), and then to get familiar with the ship, ship's crew, her berth, and also terms and conditions of the actual passage.

"Why don't they just give me a pill and knock me unconscious, and pack me into a bed or something, for the whole flight?" Karen said to her best friend, a biology-student, currently studying at Bakersfield State University.

"You're too much fun for the pilots and crew to chat up or flirt with," her friend said. "No good if you're unconscious."

"Not necessarily," Karen replied. "Not with this bunch."

They chuckled as only girlfriends can. "You can do it, Karen," her student-friend said. They hugged. "You're Number One."

"Well, you know. Save the world."

Another meeting, behind closed doors at the Vandenberg base, included Lynn Rodgers-Smith, Dr. Mehudi, the program specialist for sciences, and Winton Berle, the overall Fleet Commander. The Vandenberg Space-Port, now some 100 years-old itself, and having a new life since the year 2006, when funds were set in motion to create a high-end West-Coast US space-port, mostly for launches, but also some recovery or vehicle re-entry, and there were many other space-related functions, such as tracking, plotting flight-paths, prepping the astronauts, etc. By 2075, it was one of the world's major facilities, at that time only among about 40 or 50 such 'ports' on the Planet Earth, many of them far inferior to Vandenberg. The three 'Mars-Bars', as the lower-level type workers called them, gathered in secret, or at least with significant privacy, in a pleasant ante-room, at the back of a long hallway of cubicle-offices, where in the past, US Presidents and Dignitaries interested in the space program, would stop by for drinks, or to smoke their cigars, or to get away from the media, or hide-out a troubling blonde-bombshell affair or two. The room looked rather like an Old West bar room.

Lynn, ever the Texan at heart, had a coffee-and-brandy, and was walking back-and-forth at one end of a longish-green pool table with claw-feet. Dr. Mehudi had a plate with a pastry set on the green felt. No one was playing pool, but 'Kick', the only one of the three of them to have actually traveled to Mars, was working with a length of rope, tying knots he learned in the Maritime Academy, something he did when nervous for relaxation. His command-equals called him Winton.

"I'm sorry to point this out, Lynn. Maybe Dr. Mehudi can understand my point. It may not be clear, if you've never actually been to Mars, or the Mars Base, as I have," Winton said, seated by the pool table, in a recliner, his back straight. "I mean you both have a lot of knowledge, of course. But on the ground level, on Mars, conditions are way different. If there is any attack, of a military sort, conducting a defense, would be an opportunity for things to go from bad-to-worse very quickly. There's really no air that anyone could breathe without a suit, okay? So, beyond the walls of the base-facility, to stop the Russians---even a handful of men---it wouldn't even make much sense from a military point-of-view. The air-suits, or Mars-gear for movement on the surface, are not intended for any kind of conflict. They're fragile, really. Even a small tear in the fabric of the suit, which is an aluminum-mesh cloth-wire type---the slightest air-oxygen breach in the suit---your soldier dies for lack of air. Not from a gunshot wound. So a bunch of guys out there fighting---ha! They even fall down by accident, or a hard shove, or the other grabs him by the arm the wrong way---it's over. So it makes no sense. It's like sending a guy---a guy---a guy in a fireproof suit, to a swim meet competition, at the Olympics. It's not going to work that way---trust me."

"What about the Russians? Is there any way to fight them outside the facility itself, or stop them---just, hold them off? How will they reach the surface anyway? Aren't there suits like ours?" Lynn asked.

Winton now had successfully done square knots three or four times, while talking. "Their ships will have to enter orbit. Then they descend to surface-level in re-entry, you know the drill. Could be pods, parachutes, and gliders; something new. All the Earth-technology for suits is basically the same, but we have a lot more experience on Mars and have made some advances. But, you never really know what they might come up with. If we stop them on the way or while in orbit---much better. Outside the entry airlocks on Mars, into the base---without a suit---you're dead. It's like the Mojave Desert in the summertime---only double--or the dead winter of Iceland---minus-ten---but still in the Mojave Desert---but with no air---and a certain amount of ambient radiation. And no way home, except back into the base, through the airlock. Or a shuttle-launch up into orbit onto a ship."

A pause. Dr. Mehudi nibbled his pastry. "Those are nice knots you're making," he said to Winton.

Winton laughed again. "I do a good hangman's noose, too," he said. They all grinned slyly, understanding

## CHAPTER 7: Galaxy Baby

*"And when he had opened the Seventh-Seal, there was silence in heaven about the space of half an hour."*

*---Saint John's Divine Revelation, 8:1*

*"They used to call it Heaven, the skies and stars and planets, the infinite. I call it a pain-in-the-ass."*

*---Guy Reisling, Mars Base space transport pilot, 2075*

Lila Meetek felt herself something of a Galaxy Baby, one to whom working in space, near-Earth (but not near enough), was more-or-less a normal career, or a normal environment, something her generation took as granted, though of course dangerous. Guy, on the other hand, had simplified the term, and thought her in every way, a 'Galaxy Babe'. And true, she was. She was just 40 years-old, athletic and slim, real hard-body material. All the space-workers were in top-condition as an absolute conditional aspect of the rigorous work involved. It was by no means easy. Lila was among the best, although her job was somewhat sedentary, at least once she arrived at the Molinari Space Dock, via a transport-ship much like Guy's. Her official job-title was 'Deep-Space Traffic Corridor Environment Monitor', which meant that she sat at computer tracking and satellite-analysis data feeds, with various inputs back on Earth, and on Mars, and in-between, looking for trouble. There were more than ten people at Molinari who performed this function, without respect to gender. But it also meant that Lila was spared some of the more difficult tasks of space-travel, like space-walks, or re-entry, or suspended-animation sleep-periods, or planet-level oxygen-suit journeys and excursions. And this suited her just fine.

There was much to know about the work at the Molinari Space Dock. Lila had been there almost five years, making her a true program veteran. It made a lot of sense for extra-planetary exploration Earth sciences, as far as the establishment of the space dock. Despite the public view seen in film and TV, any planetary travel was laborious, and very slow. Depending on the relative position of the two worlds, it could take as long as a year for a ship to travel from Earth to Mars. So, one of the first choices the planners made, was to create a mid-point rest-stop, even before the first ships arrived on Mars, and began to build the base, now 15 years in-the-making. The same system would be used for Jupiter-missions (or, to the moons of Jupiter), almost like a ladder of platforms, or series of extended positions for the sustainability of life, always in danger in deep space. With Molinari in place, ships headed to Mars had the edge, for the unexpected. Pilots could dock, re-fuel, rest, board-and-offload, get information or corridor-conditions updates, and more. For emergency situations, it was a lifeboat. And this made Lila a very popular woman indeed with all the space-crews.

"God---there we are! I'm home!" Lila exclaimed. She was gazing out one of the view-ports on a people-mover transport, which was about ready for re-entry into Earth orbit, and then her shuttle down to Terra Firma. The Big Blue Marble, Earth, was like the Divine Mother---green with promise, fresh air, beaches and oceans, cities, people walking upright with regular gravity, trees and birds and animals---kids. The view-port windows were few on the transports, and much coveted for stargazing and dreaming. Lila was on leave from her regular work-shift at Molinari, and looking forward to seeing Guy again. But from where she now was, just entering orbit, he may as well have been an ant. Yet there was a connection between their two hearts, beating passion.

The transport ship seemed to glide above the planet like a sleek stone, or elegant knife, looking to be slow, but in reality moving quite fast, even thousands of miles-per-hour (which of course was not how the ship's speed was calculated). These ships were about 1,000-feet long or longer, perhaps the size of an old-fashioned deep-ocean cargo-ship, circa late 20th-century, like the Exxon Valdez, or a big oil tanker, in space but not so in appearance at all.

Mars-Labor Unions and also space program management, only permitted Molinari workers to spend six-months on duty at a time, for obvious reasons. Exhaustion, fatigue, and so on, took their toll, and efficiency suffered, which could cause mistakes. Six months on, six-months off was the rule, which was sometimes skirted just a bit, given the rarity of needed skilled labor. Prior to departure from Molinari, Lila had been in touch with Guy, via space-phone, a sort of video-audio-link, which could be set-up for one-on-one communications at certain kiosks.

"Geez, you look like crap, Lila," Guy said. "What the heck are they feeding you? I mean that in a good way, of course. You're beautiful to me, I mean."

Guy was at the Vandenberg base, where the same type of comm-link was available. No one had them in private at all. It was then months before, with Lila at a similar station on Molinari, floating somewhere in space.

"Kiss my grits, Guy," Lila responded. "I look great and you know it."

"All I see is this vid-screen in front of my eyes like a piece of plastic and glass and you on the other end of it, and you got your hair all messed up and your eyes look droopy. You all right?"

Lila brushed back her longish, thin hair, currently colored red-and-green. She sneered. Guy had a way with her, and he knew it. "Yes, Guy," she spit back at him. The radio waves traveled through space with a certain spin at that point. "I'm fine. I even had sex with two of the environmental men last week, just to piss you off, and it was great!"

"Two of them? Grow up, Lila," Guy responded. It was daylight at Vandenberg, but Molinari seemed always somewhat in darkness, even inside, where electric lights were always running. "You did not. That's a code-violation and you know it"

"We call it the Three-million Mile High Club," she said, and laughed. They both smiled and paused. There was something they shared, maybe knowing who they were, that was endearing to all their friends. Space-opera romance. Star-crossed lovers.

So, they shared the details of her voyage. Even though routine, it was still dangerous, as it always was. Her arrival time-and-date, shuttle-to-Earth landing, then her de-bugging and de-briefing, and finally her freedom. By the time the comm-link went dead, and their conversation ended, it was once again confirmed to them both, that 'love' could somehow survive, even in space.

Lila's transport performed flawlessly back into orbit, months later, and the shuttle back to planet-side was also seamless. Her de-briefing and medical review, and so on, took three days. Within another two days, she was staying at Guy's place North of Santa Barbara, back in his arms and in deep embrace within hours. *Absence makes the heart grow fonder, but there is truly no distance between hearts-of-fire in love, Guy thought. Even a million miles.*

A day and a night of lovemaking, food and drink, walks, hot-showers, current film releases, restaurants, gazing into each other eyes, sharing those moments that the written word cannot intrude. The clouds above the cliffs of Santa Barbara blushed red with embarrassment. You know, sex-like-athletes. The Right Stuff. Made it all worthwhile. Some things never change.

Rumors had of course reached both Molinari, and the base on Mars, regarding the Mission Program Spring Update Conference, and the 'news' about the Russians, and the approaching meteor. Lila also wanted to know all about Guy's re-commission to certified-pilot status, and what had gone wrong on his last flight back to Earth. They had a lot to talk about, there again in his backyard, where he seemed much-at-ease. Lila was making grilled burgers. The smoke winnowed into the air like souls.

"Local organic beef only," Lila said. "Less than a week off-the-hoof. Thick patties, but larger-around, flops over the buns."

"I like that," Guy said. "Flops over the buns. Got it. Let's try that later."

She smiled. *Right. What a lover-boy.* "You season prior to grilling, and I only use a special steak-blend from a steak-house I just adore up the coast. I have no idea what's in it. Pepper-and-onion, cloves, garlic-salt, chilli-powder, like that. So you season both sides. As you grill, the fire is not too hot, you go slow."

"Go slow," Guy mocked her. "Right. Not---uh— premature?"

"You're funny," Lila said. "I know about you and the gal from the base, Guy. Don't pretend."

"Which one?"

She huffed. "Anyway. So you grill until cooked well inside, all the way. Then on the bread, you use the sourdough from San Francisco, the big ones, but sliced thin, then grilled toasty-brown. I always want fresh-raw red onions, fresh iceberg lettuce, and decent sliced tomatoes. Pickles if you like, and mustard, or ketchup. Only organic. But you can make it up any way you like. I'm easy."

"Damn fine burger, girl," Guy said.

If a meteor was approaching Earth and the world's second or third remaining so-called Super-Power space program was planning to forcefully take over and control the Mars Base where they both were involved as workers, you wouldn't have known it. Their talk turned to those topics. Guy had been to the conference, but Lila knew most of the details, too. It was more a matter of opinions that would enable them to go on, or, how they would view such things, as worker-bees, the scuttlebutt, that seemed intense. The science was boring as hell. The real-life work and people---that was different.

"Maybe the meteor will be deflected, or maybe not do as much damage as they thought, if it hits," Lila mused. "I can't quite grasp it. It's like going to work, and you come home later, and your whole town is gone. Or your whole state."

"Every returning is a new beginning," Guy said. They had finished their meal, with beer, and also ice cream. It was for all purposes just another pleasant California day, or afternoon.

"What do you really think about it, Guy?"

"Uh---oh---end-of-the-world, I guess. You know. No more planet Earth. Or, a ruined Earth, like a dead-world, Ice-Age, thousand-year frozen dust-cloud, billions dead, ocean tidal waves washing away cities like children's toys, people floating away like ants. Or, vast regions of impact zone, ground zero, like a thousand nuclear bombs. Not good."

They paused in somber silence. Birds flew past, twittering.

"I hate when that happens," Lila said. A meek chuckle escaped between them.

"What can we do?" Guy offered. "We're not in command. We just play our parts. They'll find a way to deflect it, I bet. It can be done. The meteor is still five years off. What are they calling it now? Big Bertha, or something?"

"I heard other names for it," Lila said softly. "Bad names. People are likely to panic."

"Not my problem," Guy responded. "I don't care what they call it."

"It will be your problem if there's no place to come home to, five years out when you're on your run, if you still are. What do you think about the Russian-Islamic space program take-over on Mars? Is it real?"

Guy's mouth was stuffed with a big bite from his hamburger. "Mmmmppp---mmmm---just a sec," he said, chewing and swallowing. Earth food was way-better than the stuff they ate in space, for sure.

Lila reached tenderly towards him and wiped away a bit of ketchup from his bottom lip.

"Thanks," he said. "Well, look, I just don't know. About that. The Russian-Islamic space-program thing. They said it was real, but you just don't have any real way of knowing. They had documents and files and so-called evidence of a plan to attack. But so what? They always do. It could happen. Sure it could. It makes a certain kind of sense. They want their people to survive, and the Mars Base looks good. People like you and me will never know until something starts to happen, and we're needed to respond. All we can do until then is prepare. And don't worry, we'll be preparing, it's already in motion, as far as what will be needed. But I transport goods, and you monitor the planet-corridor for heat-flares and comets and tiny rocks. You won't have a gun in your hand, or be killing any Russians. Neither will I. And I don't want to. Some of my friends are Russian. They're good people."

"What about a ground war, here on Earth? Like a regular Earth-war?" Lila now was in political science-mode. Not very sexy.

"What about it?" Guy said. "If everyone panics, it's certainly possible. The base on Mars means survival, even if only a few hundred people. Who the heck knows? Regular war was outlawed by the Planet Authority-Federation, 30 years ago or more. Big deal. They break the rules when things look bad, they always have. If they want the Mars Base, and try to take it by force, even if the meteor is deflected, the US side will almost certainly respond at the Earth-level, or international. It can't be helped. More war, more death, more killing. I don't even care. It's bull-shit."

They paused again in their meal, relaxing a moment with the same heavy thoughts.

"I always saw the whole thing, my work, and the program, as just science-and-research," Lila said soberly.

Guy burped. "I saw it as an opportunity to have sex with you in a weightless-environment, personally," he joked.

"A multi-national, multi-trillions-of-dollars program based on thousands of years of advanced space science-and-evolution, so you personally could orgasm in a weightless environment. Great. You really are a philosopher, Guy. You really are."

He laughed. "Lighten up, Galaxy Baby," he said.

## CHAPTER 8: Think Tank Task

A Task Force was assembled by the various authorities concerned with the future of the US base on Mars, following the 'public' revelation about the meteor, Asteroid U2753b, now known by at least a few of the Mars-program regulars, as 'Big Baby Bertha' (for some reason). Again, the Mars program wasn't directly concerned with the approaching asteroid, now thought to be even larger than early estimates. It just wasn't their area. Instead, because the space-program circa 2075-76 included an Earth-Moon program, an Earth-orbit space-station program, deep-space docking platforms like Molinari, where Lila worked, a wide variety of ground-level bases, launch-sites, space-ports, and support industries, and also many deep-space and Solar-system probes and un-manned research vessels, and even early-stage attempts to mount a mission to Jupiter and its moons. With all this going on, the Mars Base program and the anticipated affect Big Baby Bertha would have on that facility, was just one part of an expanding whole, at a time when near-Earth space-exploration was finally satisfying its former era promise.

So the Task Force for this function had a very specific goal: if-and-when the Russian-Islamic space program masters decided to go ahead and 'steal' the Snikta-Ridge Volcanic Basin Mars Base, how would the US respond, and in particular, how would they defend the base, and prevent or turn back a take-over attempt? The rest was such a vast complexity of circumstances and situations, including the asteroid, that to be concerned elsewhere, or other than their own task, was a diversion of resources and manpower that would delay success, or cause possible failure. And failure meant the loss of control of the Mars Base to 'hostile' forces. This unthinkable idea might be compared to the loss of a major US property, like control of the Grand Canyon, or the Hoover Dam, or even a mid-sized US city, to a foreign power. This included the potential loss of US lives, and an incredible level of simple wealth, and the many years that had been devoted to creating the base on Mars. And not incidentally, the loss of the Mars Base to 'them' also would preclude the future population of the base from Western control, and Western ideas and staff, in the event that Asteroid U2753b actually collided with the Earth. Or, in simpler perhaps ethno-centric terms, Earth's only viable survivors following the meteor strike, would be Eastern-Islamic-Ukrainian astronauts. Not that there's anything wrong with that.

The group was known as the Mars Base Defense Planning Team. Winton Berle, and Branson Porter were on the list, and Lynn Rodgers-Smith was a behind-the-scenes source. Dr Mehudi, the science-lead, was part of the team as well. But of course, the Mars-program really had no military component. So the 'Mars regulars' found themselves seated with other types of 'big-shots'.

No one ever thought a military-division would be needed, and for as long as even 100 years, the concept of a space-war, or serious military applications of the space-program, was considered a very extreme error. There were many reasons for this position, mostly the presumption that any militarization of space, would defeat the 'real' purpose of space-exploration, and even make such exploration impossible. Space travel was hard enough anyway. With opposing sides trying to shoot down each other's ships, or placing huge bombs in orbit, serious new discoveries and new science, would be lost, perhaps forever. War-in-space was thought to be a total disaster, as far as future-planners were concerned. An absolute waste of time, energy, and high-priced resources.

Yet, here they were. Heavy-hitters from other US powers were brought into the Mars Base Defense Planning Team. US military, and Federal, also so-called intelligence community. Typically, the Mars Base battle-plan was now of interest to the global community as well, and at least one security representative from the World Council, was either at the planning sessions, or closely informed of details, with complete access. There were also Space-Technology and Computer-Science experts, and weapons experts, as well as people whom supposedly knew what the Russian-Islamic space-program planners were up to. Both male and female leaders were included. They all had a lot of experience, and for the most part were exhausted with the endless effort and data. It seemed to some of them a hopeless task, far too complicated to really predict, and far too dangerous for space-workers who were used to such levels of care that even their heart-beats and body-sweat were monitored for signs of stress while they worked. Safety first, in space, meant no one was trying to kill you, other than space itself. Or whatever was out there, but not anymore.

"They can't take Snikta, without entering orbit, and putting people on the surface of Mars," said US-Army General Price Fortuna, a large, even portly, Caucasian man, about age 60-years old, with a deeply lined face and tendency towards bombast. The General attended sessions 'in uniform', an impressive contrast with the science-types, who might wear short-sleeve shirts and khaki shorts. The meetings were held at this point at California's Hunter-Liggett military base, not at Vandenberg. "So that means we either stop them in space, or in orbit, or outside the base on the surface," he added. "Unless we stop them here on Earth."

Many members of the team, but not all, were present for this session, now a few weeks into their effort, after initial organization. There was a large table, or series of tables, in a stark, rather bleak-looking room. A secretary took notes. They had computers and other communications, and common items like food and drink. An air-conditioner bled cold into the room, with a sound.

"Of course, General," was the response from 'Kick' Berle, the Mars-fleet Commander. "I've thought about how I would handle it, if it happened on Mars. Let's say they reach Mars, with ships and men. Let's say they reach the surface near the base. What then?"

There was a brief pause.

"And now a word from our sponsor," joked a dark-haired woman named Melissa Envitra, one of the Computer-Specialists. "This extra-terrestrial gang-fight is brought to you by DuPont, makers of high-quality hydrogen rocket-fuel!"

Some laughed. Some didn't.

"Ha-ha, very funny," Berle answered her. "Repeat after me: not a movie. Not a TV-show. Real life. Say it often."

"So what's your idea about it, Berle?" General Fortuna asked, getting them back on track.

Berle rubbed his chin. "It will take time for the Eastern program to gear up their assault, launch, travel the corridor, and reach Mars, ready to do whatever nasty thing they have planned," Berle said. "If the people we have on Mars now can prepare defensive positions, outside the walls of the facility, in the non-air environment---if they start now, and create small kiosks or fox-holes, where men in oxygen suits and Mars surface survival suits can set up weapons, and survive on their own---we'd be able to hold our own, if they land on Mars, with soldiers to take the base. See what I mean?"

Dr. Mehudi now perked up, offering his take. "Winton, please, if I may?" He gathered his thoughts. "General---there are only eight main entry-portals to the Mars Base, from the outer surface. They have different functions. Some are just for people. Others are for cargo or large items. Other gates are only for things like waste-matter, or intake from the Martian air---useful airborne-chemicals. The air is thin, or even none. But there is some, packed with carbon dioxide. At the base, they use everything. Nothing is wasted. But anyway, eight main doorways. Some are even for excursion vehicles they use, more like hangars."

"Right, Mehudi," Berle replied. Everyone listened to their back-and-forth. "And each entry-port is sealed, just like an air lock in space, or at the space-stations. Somewhat different, for the gravity, but basically an air lock system."

"So they'd blow off the gates," the General said tersely. "Just blow them open. Enter in suits while everyone inside dies for lack of air."

"If they get that far, they might. It's very destructive, and they'd have to rebuild the air locks to survive themselves, later. Which is not easy. If they plan to destroy the base, that's one thing. Drop a bomb; it's done, over. But as we know, they want to live there. So yeah, they could enter by force, such as blowing off the air locks, and then enter in suits, with weapons. Pretty much the only way it could be done."

"But our people inside the base could put air-suits on, too, prior to the attack, and fight on equal terms, if they enter," offered Branson Porter, the Mars-Mission security chief. "Right?"

"Well, it has to be part of any defense plan, yes," said Berle. "That aspect."

"Agreed," said General Fortuna. "Secretary, please make a note. Okay, fine then. Plan ahead for an attack, and create external oxygen-sustainable foxhole positions to fight from. Good idea. And if the Russians try to make entry, our men inside get into their suits first. Fine. I want us to look at the Mars defense from three main fronts, people. One is the ground level, such as we're discussing now. The other is Mars-orbit and re-entry. The other is the planetary-corridor. And then I guess also the plans for an Earth-side defense, but that is much more a matter of diplomacy."

There was a pause again.

"More on the ground-level, on surface-level Mars," said Berle. "I thought about this, too. What if they make their takeover attempt into a long-term deal? I mean, instead of taking Snikta in a day, or a few days, or a few hours---what if they plan to take the Mars Base in the course of several months? Or a year? What if they're equipped to survive on the surface, maybe in temporary life-sustaining units, like oxygen-shelters of their own? And then shuttle back up to their ships to re-supply, or for materials? They'd use surface launch rockets, or like personnel pod-boosters, like the early Apollo moon-missions. Blast off from surface-to-orbit."

"That's certainly possible," said Mehudi. "That's what our people did when we built the Mars Base fifteen years ago. We had to. There was nothing on Mars. So we survived in temporary units, while they worked on building the facility. And just like you said, we had ships in orbit, and the men would go back up to re-supply, or rest, and so on."

"So you're thinking they could draw it out, like a stand-off, making demands, or taking hostages, or making assaults, is that it?" the General asked.

"Well, yeah, it's one scenario," answered Berle. "As far as what might happen on Mars. It's more efficient. A direct assault, a big, violent frontal conflict that would only last a few hours, or a day or so, would be very destructive to both sides. You have to remember how delicate the space suits are. If they take their time, or if they can figure out a way to go slower, and survive---taking control of Snikta wouldn't be that hard. They'd surround the base itself, set up their men and weapons. One side or the other would eventually prevail."

"Remember the Alamo," joked Envitra, the Tech-Specialist. "Uh, I mean, in Texas---not the old car-rental company."

Short laughter from the others.

"Are they still in business?" said Porter (a Texan). "Alamo is a nice town, if you never been there."

"Still in business. Just hydrogen fuel-cell cars, now, that's all," she said, yet the joker.

"Let's take a break," said General Fortuna. "Please, the meeting secretary will keep track of ideas and concepts for later review. Take an hour for lunch, folks. The base cafeteria has seafood today, I think. It's across the Flag-Plaza---that way." He pointed towards where he meant with a fat-pink bony finger as the group started to break up, rising from their chairs and seats, folding their laptops, or stowing notebooks.

The same sorts of meetings would continue for months. The Mars-Base Defense Planning Team needed to present the entire space-program hierarchy with a working plan---and one that would 'win' the cause. And they had to do it in short order. Needed was a way to defend the Mars Base, even though the people on the Mars Base now, were not soldiers, and had few if any weapons. Of course, the US would send her own space-soldiers, in ships, in equal or greater number than the Russian-Islamic space-soldiers. And of course, if there were to be any planetary flight-corridor space-ship 'dog-fights', or ship-to-ship battles, in an attempt to stop the enemy ships while still on their way, those would be planned for as well. But few if any of the space ships used for these purposes were intended for shooting at things, or firing missiles, bombs, or laser-beams. They weren't fighter-craft. They were research vessels. The Mars-orbit and re-entry 'battle lines' were also drawn. They also had to defend the Molinari space dock. Like any military campaign, they planned for the worst-case.

Somewhere out in the abyss of space, moving towards planet Earth, a rock the size of Texas---perhaps in the shape of every modern, college-educated person's worst nightmare---tumbled through the emptiness, like a granite Buddha, silent, eternal, and dead on course. Like a rolling stone. An Ozymandias of space, from Percy Shelley's poem.

*"Look on my works, ye mighty, and despair!!"*

## CHAPTER 9: KK-F/Region-Six

*"Earth-crossers, or Apollo objects, orbit in a path around the Sun and towards the Earth, then back again, in a journey of about five years. There are about 40 of these known Apollo objects. Some, such as Hermes, have come to within twice the distance to the Earth's moon, about 770,000-kilometers, of our planet. A direct hit on the Earth may happen only once in 250,000-years, and some experts feel such a collision might happen only once in a million years. Such an impact, however, would produce an explosion as great as 20,000-megaton hydrogen bombs. Scientists feel it was this kind of meteor strike that led to the extinction of the dinosaurs, 65-million years ago."*

*--- 'Our Universe' by Roy A. Gallant, National Geographic, 1980*

Two weeks before the US Mars Defense Plan Task Force met at Fort Hunter-Liggett, in California, where it was warm all year, dry, and oak trees everywhere, along the sides of gently sloping brown hills. Away beyond the cusp of awareness, also hidden, in another land, so different and far-off as to almost be viewed as 'another planet', or 'another world', which was the Southern Ukraine region, formerly a part of Mother Russia, and the vast USSR before that; here it was that the Russian-Islamic space-program, for at least a small part of their various efforts off-world, held court, made their plans, and dreamed their dreams. It was cold here, often with snow on the ground, and high, Rocky Mountains, and woods very different and deeper or darker than those of California. To those who knew where to find it, a Space Port known as KK-F/Region Six, had been built years past, and now rested, there among the woods, mountains and snow.

Somewhere among them, inside the facility, their own team had assembled, this for the eighth or ninth time in a period of several months, with the same set of science-facts and research, as the US Mars Program had. Among them were four of their regular space-flight pilots, all husky, large men, pink-skinned with dark whiskers, or shaven, dressed for warmth, in casual uniforms. The Commander was Rudolph Terchenko, an older, mature man, and veteran of Russian space flights for many years. In addition, the on-going Russian-Islamic 'think-tank' participants included so-called Islamic Renaissance scientists, and Resource Managers from Saudi, Iranian and the Northern Indian sub-continent.

This alliance for space-research was much different than the American program, and formed itself from a wide pattern of states, nations, science-Universities, military bases and spaceports. In an odd way, the Russian-Islamic space-program was far more resourceful and 'tougher' than the US program. They managed the same accomplishments and feats in space as the US-side, but working with less. Longer space-walks, greater distances beyond the moon, faster launches, and rougher landings, were the rule, and a matter of great pride among the men working in that team.

Commander Terchenko laughed and rolled back his chair from a long wooden desk. It was a chamber for his rule over the space-program arena he was in charge of. He had himself very well equipped with comforts many of his countrymen did not have: food and drink, plenty of vodka, warm heaters, computers and communications, servants, and a real wood fireplace. His assistant, a slender young woman with a stiff laugh and dark hair, gathered his papers and books, and laptop computer. She knew where he was headed.

"No, Milana, it's not true, what he told you. It never was that way. They tested the bombs, yes, and then the areas were sealed for contamination. But anyone who lived in the region was evacuated," he said.

"But he said there were deformities, and still-births, and cancer, and diseases, from the nuclear tests, Commander," Milana replied. "I'm sorry to repeat it again, he was very insistent."

"Do not believe lies," he answered. "Unless they are mine."

They left his chamber, and proceeded down a hallway. This was a simple complex of offices and administrative centers, and also research-and-science, associated with the Russian leadership portion of the Eastern space-program alliance. Terchenko talked as Milana walked with him, toting his stuff.

"I was only curious, sir," Milana said.

"Never mind," Terchenko said. "This meeting ahead will decide our final choice about the US Mars Base. That is, if we plan to take it, or not. You understand. So please, keep yourself quiet about anything, and just take notes, or get my meal. It will be a long meeting."

"Yes, sir," Milana said.

"The Iranian military space-program leadership will be present. They have great power, and very specific equipment, and also trained men, and clearance. Also, my entire staff. We have trained our teams for months, but there is no command to launch. If we launch, it is war. A space-war. You must not discuss this sort of thing with anyone, dear Milana," Terchenko said.

"Never," was her terse reply. They continued down the hallway. Ahead were double-doors to a large meeting-room. Soldiers nearby in uniforms, and armed, kept watch; a needless guard, given the obscurity and hidden location of the KK/F-Region Six spaceport facility. You would have had more luck passing bodily through the Wailing Wall in Old Jerusalem, then entering here, un-welcomed.

The double-doors opened, and they passed inside. The room was a busy place, with an entire complex of long worktables, perhaps ten or twelve long areas, with seats, nameplates, computers, and covered in long sheets of dense white-blue cloth. Each man at his seat had papers and books, and beyond the back of the room was a large projector screen-image, where they all could view data, statistics, and graphs-and-charts. Commander Terchenko and his assistant, young Milana, took their place at one of the tables. A plate of dried apples and cheese, with hot black coffee and brandy, was at his left hand.

At once as Terchenko settled, a small Eastern-looking man with dark skin and a gray beard walked nearby toward him, much like a scientist but perhaps 'some sort of egg-head', as Terchenko mused within himself. "Commander," the man said. "Please, just a moment. Before we start."

"Yes," said Terchenko. " You are---??"

"Doctor Martin-Sarcasian, with Central Planning. You don't believe me? Here." He produced a small leather-bound packet with his immediate ID inside, on a nylon cord around his neck. Terchenko viewed it briefly.

"Yes, I know you," Terchenko said. "An egg-head."

"Just a word, sir, before the meeting. I'm troubled by the direction we are going, on the council team. There is an aspect of reality here with the planners, it has been discussed, but I think there was no fair hearing about the matter. I want to review it again. But it is very sensitive. I don't even know if there is time for a full review. I'd like yourself as local program Commander to---maybe---just bring it up, with the group---at the right time."

Now Terchenko had seated himself and was having his coffee. "Well, fine. Tell me first, and I will decide."

"You already know," Doctor Martin-Sarcasian replied.

"The meteor? Yes, we know," Terchenko said.

"No, no," said Sarcasian. "I'm talking about the Edinburgh Society contact we've had. The Scottish group. What the panelists don't recognize is the long-term motivation for taking control of the base on Mars."

"To survive the meteor strike," Terchenko said. "Is it not?"

"Well, yes, on the face of things. But we can't survive on Mars forever."

"With our men on Mars, after the meteor hits, if it ever does, we can send survivors back to rebuild, or work recovery, and so on," Terchenko said.

"You are not familiar with the Edinburgh group. Our people have been considering off-world information---off world, I mean, from other planets. Not Mars. Worlds far out into our galaxy. Inhabited places. That is, you would say---aliens."

Terchenko paused. He refreshed his coffee. "Go ahead, Doctor Martin. Please be brief. I don't believe in aliens. They don't exist."

"The planning team has not recognized that once we take the Mars Base, and if Earth is smitten of the meteor, with heavy damage, that the contacts through Scotland, would be re-established in the future, on Mars, with our people who survive there. In other words, the human race could survive. We'd go on. I know it's far-fetched. But you see---I have studied this aspect. I know a lot about it. It has a high level of probability. The information is secure." Sarcasian continued.

Terchenko laughed again. He had a big jovial laugh, spreading his hands widely on the table. "Maybe we just want to survive, anyway!" he said. "Maybe we just want to survive the damn meteor and the hell with your aliens!"

Doctor Martin-Sarcasian seemed angry. "That's not the point," he said. "Of course we want to survive. If the meteor hits, we want people on Mars. That's not the point. What I'm saying is, we need to plan ahead for this aspect, so that when-and-if we arrive on Mars, or take the base from the Americans, that we will be prepared to deal with the Edinburgh Society findings and radio-telescope deep-space communications---for the survival of all mankind! We need to plan ahead so that we can accommodate this---it's important!"

"Even if it's all horse-shit?"

"Damn you, Terchenko! I'll bring it forward myself! Good day to you!" Sarcasian now walked away. Terchenko smiled. He had been a part of the Russian space-program a long time. Men like Sarcasian had big ideas, big dreams, and radio telescopes to listen to for years on end. But it almost always meant nothing, so as a practical person he never trusted them at all, or their information. He made a mental note, and felt the man's idea would probably come up again. It had already been discussed. Sarcasian seemed unsatisfied that the planning team was not thinking 'his way' about it. He probably wanted to be assured that any future Russian-Islamic stake on Mars would include the gear, technology, man-power, and resources, needed to re-establish whatever his so-called Edinburgh Society had accomplished. As if Terchenko could plan ahead to build him a radio-telescope on Mars, at the same time they were over-powering the US Mars forces, and taking control of the base, with as little loss of life and damage to the Mars-facility as possible. And also, Terchenko himself had no real faith in the Edinburgh Society, and certainly no faith in talk of any aliens. For real space-men in 2075, it was a joke.

The rest of the meeting proceeded as planned. They wrangled over the issues and topics for hours, shouting each other down in native Russian, or sometimes other languages. The ships and men were trained for the mission. The plan-of-attack was prepared. Some members felt an attack on the Mars Base was premature---the meteor was still years away, and might never even hit the Earth at all, or be deflected. Others saw it as an opportunity, but the political-wing fully understood the ramifications for Russia and her alliance-in-space, when they had to explain to the global community what they had done.

Later it became clear that the fear-based military side was winning the argument. The game was played such as to launch or not-to-launch, and when. By giving themselves a year's advance at the base on Mars, or longer, well ahead of any meteor---which well they knew about, and were tracking as closely as the rest of the world---the notion was that they could secure their goals, made to sound lofty and noble, or in terms of saving humanity--- when it was true enough they also wanted to save their own skins, and avoid a future-life on a dead-Earth, smitten by the meteor, with untold damages, a new Ice-Age, an unlivable world, the environment uninhabitable. Even if only a few hundred people, at least it would be 'their side'.

A secret vote passed from table-to-table, hand-written on scraps of paper. It was late, they were all exhausted. They were collected and tallied. The results were brought to Terchenko, who as Commander of the local space-program was placed in the role of meeting co-coordinator, and announcer of colossal mistakes.

"Thank you," Terchenko said to the Aide, after the tally was gathered, and the vote was done---yes-or-no to launch, and also yes-or-no on a spectrum of launch-dates, which had to be arranged in harmony with the position of both planets, within the coming year. The Commander's smile evaporated.

"All right then," he said. "The vote is done. The answer is to launch our forces, to the US Mars Base location, in three months, which is in May, to accommodate the position of the planets. So that is the vote. We will launch. Done is done. Thank you."

The room descended from anxious silence into hushed chatter in every corner. With all their brainpower and eggheads, all their information and data, and the space ships, they had chosen to attack. For the good of all mankind, of course.

"Get me another plate with the apples and cheese," Terchenko told his assistant Milana. "And more coffee."

She looked down as she scurried off quickly towards a facility kitchen.

## CHAPTER 10: Mars Virgin

*"There's no goodness in me equal to all the badness of the world. But deep space---what could go wrong?"*
*-US Mars-Program spaceship transport pilot Guy Reisling, 2076*

Like a dream of globes, or gigantic stones, or spinning tops in the hands of a child-deity, formed from infinity, yet round and lovely, spheres, the first solar planetary object, the second, and then third through ninth, and beyond, had been dancing delightedly for so long, few could truly remember their origins. In fact, no one could. But it was long ago, for sure, and the furnaces of creation, the formation of matter and energy, and Guy Reisling's ancestors, somewhere in his blood-stream and DNA, silent yet eternal; those burn-bins yet lingered in the rear-view mirror of himself and mankind, ever-curious, super-chimps, as one philosopher said, like ants on our glorious Earth, home forever.

And there away on Mars, after long years of hard work and learning, the US had, at one point now past, established the Mars Base, under discussion in their meetings. Why do we do these things, mankind might have mused, taken whole? What's the use? Who cares for Mars? It's dry and boring and empty and barren, only rocks, no good air, cold and hot both, in extremes, without a single tree. For every single person on Earth, there was no question. 'Thanks, but no thanks, I like it here'. Even the Earth's bitterly poor had a handful, otherwise not long to live, and adventures of their own, humble. Yet the few, the proud, the US Mars program space-flight workers, and the other Earth space-explorers, really numbering only a few thousand people, but a tiny fraction of humanity; where few had gone, few would ever go, and among them who did, most learned not to ask themselves why, for their own self-respect.

Dinner on Mars, at the so-called Snikta-Ridge Volcanic Basin US Mars Base, was a family-type affair, and tended with the perpetual concept of keeping spirits up and overcoming exhausted workers and depression or other off-world emotional troubles; same as at home on Earth, just without the view. Much like any demanding service that trained and skilled people would undertake, such as military, the food was prepared with the very best quality available, for this reason. On one evening, Earth date then at January 23, 2076, in the food-court within the safety of the base itself, the meal for about 20 or 30 people, was once again completely vegetarian. The main reason was that for the base to be self-sustaining, they needed to raise their own food, difficult on Mars to be sure. But it was essential, they would all die if the shipments of goods that Guy Reisling and others were responsible for, failed for some reason, for a significant period. Rice was the most successful. It grew easily in shallow water, was acclimated to heat-changes, high in carbohydrates, and also preserved well. They had many other types of crops, carefully tended in long, very large hot-houses---beans, corn, organic melons, onions, carrots, and so on. The plants also produced free ambient oxygen, exchanging $CO_2$ for $O_2$, for photosynthesis. But the chefs were stocked with wonderful spices and ingredients, and the meals were truly quite good, with many variations. Once or twice a year the Earth-transports brought in a supply of familiar meat-products, too.

The crews ate in shifts, around the clock. About 230 people occupied the base at this time. The number at the base changed, but not by much, as Earth-bound passengers departed, no more than five or six at a time, and others arrived. Thus, there at the dinner-table, on that day, the Communications Tech-Support Karen Tutturro, now, at last, found herself enjoying her first meal on the new planet, and she was filled with wonder, and joy, though cautious, at everything she was learning about this new world.

"Do we always eat here, or is there another dining area?" Karen asked her guide, whose name was Juno, a Belgian-German base security man, whose real job, since they were never under any actual threat, or hostility, other than that of Mother Nature, was mostly to maintain the ebb-and-flow of civic and family life there at the base, which was complex enough that minor disputes could be disruptive and needed to be dealt with. But that was rare.

"There is another one, three corridors down, and then a third as well," Juno said. Each one will feed 150 people at a time. But there is no need. They work in shifts. The second food hall is smaller, and the other one about this size. How was the transport voyage? How is Earth these days? I haven't been home in six months."

Karen laughed. "Well, it's still there," she said. She continued eating. The dish was rice-risotto, with pickles and nuts, and veggies, and a soy-based fruit-type nutrition drink, with other items as treats. "The transport was fine, but demanding. You know. I liked everything, except the bedding was too rough, or, just not very comfortable."

"That was your first time, then?" Juno asked her.

"Yes. I'm a Mars virgin."

"You'll get used to it. Take nothing for granted. It's still dangerous, even today. A few miles beyond these walls, certain death." He just smiled. "Even for Mars virgins, it's cold outside."

The dinner-hall was an echo of pleasant voices. People laughed or chatted. Mars-TV, as they jokingly called the in-house communication system, played popular entertainments on a screen-surface. The most popular were nature-documentaries from Earth, but also many others, agreed on by committee.

"My job is working on the Inter-planet communications," Karen said. "I was sent because there were troubles in the system that made failure possible, which might cause troubles or mix-ups in essential services."

"Yes, I know," Juno said. "It was more than a year ago. They needed a communication series regarding a research inquiry program, and complex data. But they messed up somehow. It was a geological survey, with samples, too. But it failed badly. They panicked. And now here you are."

"In the words of Ringo Starr, I'm glad to be here or anywhere," Karen replied. "I'll get to work tomorrow, today I rest from the voyage. It took the ship more than 120 days from Earth. I'll have my hands full. It's a complex system."

After a while, they left the cafeteria, and entered the rest of the Mars Base complex, as Juno would guide her around for the next couple of hours, to view the 'tour' that newcomers enjoyed. The various halls of the facility were arched, large enough for small-motorized carts, and went on for even miles, taken together, though only a total of about four miles in all. Karen seemed to take charisma with Juno, and they laughed a bit, and became friends, as everyone at the base was encouraged to be.

The Mars Base planners wanted to create and fabricate a long-term facility that would serve several main purposes: to host the science-and-research team, the pilots and crews, and equipment-computers-machines; and also for work involved in the hot-houses and with growing plants, processing chemicals and oxygen-water, recycling, process-generation; and also of course for purposes such as housing, offices, astronomy, and operations. Beyond the main walls, were modest launch-pads, and a series of ramps and short drives leading to the eight main gates; there were also observation equipment-stations, raw-materials containers, and ladders or steps that could access the upper decks, windows, pads, observations stations, and then the roof.

The entire layout looked very patchwork and military, like a puzzle of squares and shapes, and various components, small roads and containers. The actual Snikta-Ridge Volcanic Basin site had been chosen by the base designers because of the geology structure of the rocky region. The ridge was a solid 100 miles long, North to West on the planet surface, in the upper-equatorial area, and included the unique aspect of containing an underground ice-flow, or H20-pack of frozen moisture, perhaps millions of years old. There was some frozen 'dry-ice' on the surface as well. This incredible find meant the base could be self-sustaining within only a few years of completed construction, which were about 2064.

Karen and Juno turned a corner deep inside the base, about two corridors over from where they had eaten their rice-risotto. They moved into an area with a room full of computers and tech-gear, mostly for communications with Earth. An Asian man, named Boji-Than, met her by appointment. This was the Mars-Base Commander, far more a man of science than a soldier. He was taught-looking and slender, keen-eyed, fast-talking and very wise about all things related to the base.

Juno, the escort, dismissed himself. Bojji-Than took Karen's hand, and she smiled gratefully. "We have much to discuss, I'm so glad you're here," Boji said. "Come with me."

"Pleased to meet you," she said.

Karen knew a good deal about the type of gear and tech they used to communicate back-and-forth with Earth. The system was essential for their survival. Her unique area was in microwave and basic radio, directional and power-supply, as well as antenna-array, and data-compression, and similar. Of course she had studied many years, with special education and knowledge. Likewise, Bojji-Than was completely appraised to these tasks.

They moved through the equipment and he showed her various monitoring-screens and simple ways to appraise the system's functionality. "It was no one's fault, the way it broke down," he said. "We were in the middle of a research project, mostly a geo-survey. None of the information was very important, which is typical here on Mars. It was a six-month project just to send the findings back to Earth, which were on-going."

Karen gazed at one of the monitors, a computer that measured the flow of signals from one point to another, their receipt and content-stability. "You see?" Bojji-Than said. "The origins and transport-flow are fine, apparently. But when you reach this point—"

He showed her one of the sections on the screen. This icon indicated a certain collection of computers and processors that routed pre-compressed signal data to their final output via an antenna-array. "It's blocked, and we don't know why," he said. "I've been working on it personally for weeks. By the time you were called up, it was a mess. I can tell you more, but that's the outline."

So, they talked more, mostly tech-stuff about the system she would be working on and various specifications. After half-an-hour, they settled down in an office in the same area.

"What are your feelings about the news from last year's Spring Up-Date at Vandenberg?" Karen asked Bojji-Than. "About the meteor and the Russians?"

Bojji-Than relaxed a bit, and folded his arms. "Of course we heard all about it. As base-commander, it is very significant. We'll have to adjust to whatever the Earth planners decide. My feeling is the situation could become a disaster, if the Russian-Ukrainian-Islamic forces in their space-program do indeed arrive. This base has no defenses for that sort of thing. No one ever dreamed we would need them. I am not a military person. So, we'll see. Hopefully things will resolve without a problem."

"The end of the world was never a problem before," Karen joked.

"Only a few times," Bojii said. "If something has a beginning, it has an end."

"Let's hope not," Karen said. "It is home, after all."

"For me, too," Bojji-Than said. "I have not been back in almost a year."

They continued to chat, and Karen was able to tell Bojji-Than about some more mundane aspects of life-on-Earth---new films, sporting events, new car models, and things in the news, celebrities. This cemented their friendship somewhat. The work before them was demanding, and would take a lot of time. The mid-point processor that had broken down was complicated and highly technical. Karen had brought a good deal of the back-up tools and analytical equipment she needed. After a while she was guided to her quarters and other new friends she would meet, while getting to know her new world.

"Just like Earth," she said to herself, as she was finally alone. "Nice planet."

## CHAPTER 11: The Snikta Ridge

*"Three enormous volcanic mountains line up northeast to southwest, the Tharsis Montes. Each one is about twice the size and height of the volcanic island of Hawaii. Almost hidden in the shadow of Olympus Mons, an even larger volcano. Above it, wisps of water ice-clouds hover. Farther north, the pole displays a shrinking cap of carbon dioxide snow typical of early spring. East of the Tharsis Montes is a system of giant canyons that stretch some 5,000-kilometers, east to west."*

*-Roy A. Gallant, Our Universe, National Geographic, 1980*

Such was the Snikta-Ridge Volcanic Basin region, where the US Mars Base was built. The Snikta Ridge was a minor formation, and not particularly impressive, and of course the builders needed a hospitable, level, 'flat' or non-mountainous foundation, where the construction, approach of manpower or staff, and goods, would function. The view from within the base, or anywhere nearby, was quite beautiful, though barren and formidable, even deadly. Mars was hard to love, or hard to enjoy, or find lovely. But the staff and workers who lived at the US Mars Base, almost to a man, eventually realized that they were among the very few ever of the Earth, to live and work on another planet, and this alone endeared each person in dreams and reverie.

The red planet clearly once had a very active geological past. The volcanoes were dead, but not the deadly dust storms, which could be vast, even planetary-scale events, as the ancient reddish sand and dust was swept aloft by 'winds' that created huge clouds of terrible power, and could be seen from Earth by telescope.

By comparison, Deimos, the red planet's outer of two moons, is only 15-kilometers wide, at its broadest point, which was irregular in shape and not perfectly globular. So Deimos was only one-fortieth as broad as the Olympus Mons volcano on the planet surface.

Mars was the Roman 'god' of war, stereotypically represented by the 'circle-and-arrow' sign, and often used to portray the male. And macho it was, for a planet where anyone would dwell or live, certainly compared to Earth, far more nurturing and feminine, with it's near-Paradise of plants, life, water, people, food, cities, oceans, creatures, and so on. Only the strong could survive on Mars, it might have been thought. Yet, with the help of science-technology and good-old know-how, many of the residents at the US-base, were women, and even a few children (who had visited temporarily in the past). The planners of the US space-program knew well, that with women joining together with men, as they voyaged into space, that morale improved overall, and depression and anxiety decreased. Yes, strong and healthy, athletic---but all-woman, feminine, nurturing, the second half of mankind's Adam-Eve dichotomy, which was eternal for humankind.

A 'Martian Year' would last for 687 Earth-days---almost two Earth-years. The planet has 'seasons', which are also irregular in length of days---in other words, the Northern Fall season can vary almost 60 days longer than the spring season. The planet is significantly smaller than Earth, about one-third as large, and one-tenth the mass (weight) of Earth, and less dense internally. Thus, gravity on Mars is just slightly more than one-third the pull of gravity on Earth. A 200-pound man, on Mars would weigh only about 70-pounds. Like astronauts who first walked on Earth's moon, this was a delight, or, at least, ease and convenience when carrying heavy loads, or thick, heavy space suits, or gear, etc. In a strange way, as their hearts yearned to play or glide and jump or leap, or fly about, or do incredible athletics, in the lesser gravity, the staff and workers on Mars at the base knew they never would be able to do so, for fear of the deadly atmospheric conditions. Naked, or in running-shorts on the surface of Mars was not an option, except in dreams.

At the closest point between the two planets, Earth and Mars, as they move in the dance of orbits, are about 56 million kilometers apart. This solar-system intimacy, or closeness, could take as long as two years to happen, and was purely a natural event, caused by the orbits and positions of the planets. A Martin 'day' is just slightly more than 24-hours, oddly enough, providing the Mars Base staff with a sense of Earth-like normalcy.

For what water there is on Mars, nearly all of it is found frozen beneath the ground. Both planetary poles had ice caps. The polar caps are actually 'dry ice', or frozen carbon dioxide, with only a little actual 'water-ice'. And, of course, much of the planet-surface is pockmarked with craters, large or smaller. The soil is somewhat like that of the soil on Earth, with silicon, iron and magnesium. The famous reddish color is apparently from iron oxide.

The Tharsis Montes region is 1,000-times the distance from New York to Los Angeles, in its breadth. The four giant volcanoes are three times higher than Earth's Mount Everest. To the east is Valles Marinarus, with canyons deeper than America's Grand Canyon, but longer than the US itself from coast-to-coast. Petrified lava-flows, sedimentary canals long dried and turned to dust, endless vistas of barren rocks, sand and rises, and many other formations, are everywhere. Yet not a single tree, bird, lake, grassy field, cow, horse, natural waterway, fish, beach, or indigenous life-form, anywhere on Mars at all, that had ever been found since exploration began.

Maps of Mars show formations called such as Chrysae Planitia, Sinai Planum, Arsia Mons, Hellas Planitia, Elysium Mons, Du Martheray. There's no liquid water anywhere on the surface of this entire world. The atmosphere is very dry, and any water vapor that does exist, will not turn to liquid (like rain does on Earth). Even if all the water vapor in the atmosphere of the entire planet of Mars were reduced from air-born vapor to liquid, the entire volume of it would only fill a small lake. By comparison, of course, Earth is host to vast oceans over most of its surface, which though salty, condense and lifts into the moist atmosphere of the Earth, eventually turning into rain, or other water-forms, providing the basis of life. Mapping, measurements, geology and innumerable observations and records continued without end, as one of the US Mars Base's principle objectives--- to learn and record all there was to know about this 'new world'.

The base itself was, from the exterior, rather a fortress of technology and survival-means construction. From the ground level, it seemed somewhat like a common military installation of some sort. There were numerous buildings and gates or entryways, large tanks and vats or towers, numerous antenna-arrays with gaudy high-tech spider-webs of dishes and spindly formations that could project radio and other signals all the way to Earth, and many other aspects. There were also glassy, or clear-view formations attached to the base, more or less like plant houses, or patio-like gardens, for restful viewing---but of course the entire inner-world of the base was protected, airtight, from the harsh outer-world. And in this sense, although 'home', the base was also, and always would be, a prison, from which escape into Nature, meant only death for the natural human creature.

But life at Snikta Ridge was by far more comfortable than life on-board the transport ships that made the voyage from Earth, such as the one Guy Reisling piloted. The residents enjoyed fine meals, fresh water and air, good plumbing and bathing, regular personal quarters and housing that allowed for sexual relations, and off-hours of even a week or more at a time to relax, or 'vacations'. There was plenty of entertainment in every form, also a library, and small performance theater. After ten years, the crews at the base started a vocal choir, which then petered-out, to be replaced by a jazz-band and a small classical string quartet, and other forms of arts, as staff found time and inspiration.

As ships entered orbit around Mars, a standard re-entry, or shuttle-to-surface descent, was initiated---not by the transport pilot, but by a specially trained 'space harbor-masters', to whom this process was not a mystery. To the pilots, or any person arriving on Mars, the view from orbit of the Snikta Ridge Volcanic Basin US Mars Base was indeed spectacular. Telescopes and magnifiers provided digital-screen views, and the naked eye was really not much use, even through the thick transparent-aluminum port-views 'windows'. Maybe it was because after months in deep-space transport, travelers knew that when they arrived at the base, they would again enjoy 'normal' gravity, walk around in open-air interiors, have private rooms all for themselves, and so on. Or maybe it was just the wonder of it all.

In any case, what they saw from above, when entering Martian orbit, was a spread of about five or ten square miles, laid out like a patch-work of squares, circles, and other shapes---the same air-tight fortress which from the planet-surface rose up beneath the dark cliffs like a strange specter of the power of modern science and the survival spirit of humanity. There were launch pads, too, and landing-areas, and roads between the useful platforms or storage for fluids, liquid-oxygen, or H20, and then areas for surface-to-orbit rockets or 'lifters'. Complex hardly described what was needed to survive in this way on Mars, in the year 2076. As an achievement of human consciousness, Snikta Ridge was equal to the Great Pyramids of Egypt, or any of the Earth's great cities, or other wonders 'back home'. Yet it all seemed as lonely as the silent and dead planet upon which it rested, a tiny spark of human life, against the face of the Universe, a fortress of the living, as strong as any ever conceived---an 'outpost', in the true sense, as may have been established in the early exploration of the American West, or the early European exploration of the 'New World' of the Americas, or Marco Polo's first voyages to China and the East.

Sunrise on Mars was oddly unique and just as wonderful as the rising Sun on Earth. More distant from the Sun by millions of miles, Martian sunrise was more distant as well from the source of life and heat and warmth---the Sun. So it seemed ethereal, somehow hollow, or lacking a certain familiar mighty blaze, almost like tin compared to brass, or silver compared to gold. The reddish dust and dark-reddish mountain cliffs, the huge volcanoes, higher than any man would probably ever climb, set against the rising Sun, distant to the east in the early hours of each morning---there was no doubt this was 'another world', however much one wished to go home. Base staff could watch the sunrise event from glass-domed patios, and at various view-ports, and also on digital camera screens.

As a regular task that needed to be done daily, a base perimeter air lock seal and atmospheric facility breach check was performed, in three crews of two men each. Comfortable in their Mars-Suits, the men rode on small electric carts suitable out to the surface soil and inclines. Communicating by internal radio-links, they were equipped with gauges and detectors that would reveal any oxygen leaks, mostly at various points that were most likely to decay, break down, or release internal air-pressure due to wear-and-tear. As they stopped the carts, at some twenty points along the foot of The Castle (as they sometimes called the base), they looked like rock-hunters, with magnetic metal-detectors in their hands, which in fact were for reading chemicals associated with any leaks.

There were many daily dangers and environmental threats on Mars, but an internal air-pressure breach, or leak, was among the most feared. So the men worked carefully, every single day, to find even a very small leak. A small leak could become a large one, and a sudden loss of internal sir would result in many deaths, before it could be controlled or contained, if they caught it in time.

"Nothing here," said one of the workers, through his inter-suit radio, to his partner. "It's clean. Not even a molecule. Let's try the footing seals on Number Two."

"Got it," his partner responded. They moved back to the little electric cart, like figures in a dream, against the distant Sun rising behind the giants, tin, not even silver.

"Are you ready?" Bojji-Than, the base-commander, inquired of Karen Tutturro, who he found that morning at the cafeteria. Karen had maintained her schedule meticulously since her arrival, more cautious than perhaps she needed to be. She wanted the other crews at the base to respect her, of course, which was no simple matter. "I want you to meet some people you'll be working with," Bojji-Than added.

"Certainly, commander," Karen said, getting herself up from the table where she was eating breakfast, along with 20 or 30 other shift-workers, at one of the cafeterias.

"Please just call me Bojii," he said. "Everyone else does."

"Okay," she smiled back. "Bojii."

They walked out together from the dining-area, into one of the halls. The base was alive and vibrant with life, work, and a pulsing truth that sustained them all.

## CHAPTER 12: Roll Call: Mars Base

Commander Bojji-Than led Karen away from her breakfast, again now into the maze of cavernous recesses, rooms, hallways, offices, tech-rooms, and work-areas, that were unanimously regarded as his 'kingdom', as leader of the US Mars Base on-site program, in 2076. He was a benevolent type, not given to tyranny of delusions of power and pleasures, far more interested in the science, research, progress and knowledge. Karen followed him like an obedient duckling, and he the drake, she somewhat of an Alice in this new wonderland on Mars.

"This is the Command Center, where I keep regular offices," Bojii said. "This way."

Bojji-Than had inherited his job at the Mars Base much as any of the space-program staff and leadership had been commissioned to their various posts---the pilots, launch-specialists, navigators, communications, technical, planners and policy, science-and-computers, rocketry, and those at work at the Molinari mid-point space dock, such as Guy Reisling's lover, Lila Meetek. Years of training, education, and rising through the ranks, and each had proven themselves, and earned their jobs, which were highly prized. For Bojii, it had been through years of work with the program, mostly in launches, inter-planet navigation, and physics-science. Another egg-head, another adventurer, and a man beloved of his 'employees' for level-headed decision-making, human-compassion, and good-judgment, backed by solid knowledge and experience.

ROLL-CALL: these are the men and women who live and work at the US Mars-Base at the time of the discovery of the approach of Asteroid U2753b, now circa 2076.

Commander Bojji-Than: male, about age 62-years, Asian. Responsible to oversee all operations at the base (on Mars).

Installed as base Commander in 2070, at work on-the-job now six years. Thin, dark-skinned, and muscular for his age, the Commander enjoys playing classical violin as a hobby, and his collection of fine wine (a rarity on Mars).

Juno Amorrossi: male, age 45 years. Juno is the muscular, masculine and athletic base Security Officer, French-Belgian in heritage. He is affable and friendly, trained in marital arts and especially judo, at the Master's level. Work for Juno at the base on Mars is rather boring, due to the nature of the crew-and-staff who work there. He is only rarely needed for personnel disputes, sometimes misbehavior or intoxicated residents who over-do things, and minor disciplinary actions. Most often he would act as an event crowd-control manager, or public host, and then of course in a safety-and-security capacity, concerning matters such as proper air lock function, or passage of people through air locks safely. It must be said, Juno was a big hit with the ladies at the Mars Base (ciao!)

Vinces Grant: male, age 49 years, in the role of Mars Base Science-and-Research Lead. Work at the Mars Base had always been intended as a platform for discoveries about Mars, including anything and everything there was to know or learn thus, a vast arena, even an entire new world. With a staff of about 20 science-specialists, which also changed as needs arose, Vinces organized each long-term or short-term exploration--- mapping, geology, life-and-water search, planetary physics, atmospheric and radiology-solar, soil values for potential agriculture or other uses, planet history and archeology, and so on. These were on going, and all collected data was analyzed and recorded, or sent back to other researchers on Earth. Vinces Grant was a husky-looking Latin man, variously multi-ethnic in his DNA-origins and ancestry. Hobbies include stargazing and astronomy, and he also seemed to have an endless personal memory for sappy one-liners and jokes.

Chassidy Katola: female, age 28 years. Chassidy's job might

have seemed at first less important than others, but Mars Base residents knew her as their primary source of Health-and-Wellness guidance in this strange world. Though young, she was a successful and advanced Wellness Therapist, and her work included nutrition, exercise, medical-holistic, emotional, and other health areas. Chassidy was a Black woman, very beautiful and sexy, and with quite dark skin. She also had an MD in a broad spectrum of 'wellness' knowledge. The base staff included many other MD's, various types of doctors, and dentists, others. Maybe it was because her discipline included all areas of general health, that Chassidy was a touch-stone of help and guidance for all 230 base residents, in any area of physical-emotional distress, pain, fatigue, or illness, that she was so popular, or maybe simply because she was a fun and beautiful person, effervescent and joyful to be with. Any illness, or viral infection, even a simple flu-bug, could spell disaster for them all. Fear, ideas and rumors, and overall staff-and-crew fitness, was critically important. Chassidy encouraged regular and vigorous sexual release for all the Mars Base residents.

Matt Curisonn Van Templar: male, age 42-years. Matt, or 'the Templar Knight', as he was sometimes called, was the Lead-Person on Rocketry and Launches, for the Mars Base. This included any re-entry, or shuttle landings, ships in orbit, transports to and from Earth, and 'lifts' from the Mars-surface to ships waiting in the orbit-space above. Matt was a 'white man', not very physical, though fit enough, and trended towards 'nerd' in appearance and disposition. His job was critical; without the skill and knowledge needed to launch or land ships and people or goods on Mars, and from Mars, they would all soon perish. Or, if they didn't die, life would be very difficult until launches and rocketry were restored. Like anyone, Matt 'the Templar' took pride in his work, and there had never been a disaster or crash, under his command, that being some 15 years (including work back on Earth in the same arena). Crashes did happen. Ships would fail to re-enter atmosphere and gravity, or even explode, with all hands lost, for whatever reason, mechanical, human error, environmental. Matt was like a watchful hawk over each and every launch, keen for any hint of error or failure and for this much he had earned the respect of all. He was also a homosexual, with a male-lover commonly known to many of his friends, who was a Safety Worker at the Mars Base.

Charley Barron: age 53-years, male, Mars Base Environmental Safety and Atmosphere-Integrity Officer. Charley's job was to maintain the Mars Base facility internal environment such as to remain humanly habitable and sustainable for life-support, here on this hostile planet. His staff was one of the largest from among the 230 base inhabitants. The breathable air, the C02-scrubbing and oxygen recycling, temperature-control, drinkable water and waste-processing, crops and hot-houses, imports of supplies, raw-materials, chemicals, and also internal energy-systems, and much more, were all under his authority. And he knew what he was doing; never forgetting that all their lives depended on the inner environmental-system integrity and functionality. An air-leak to the outside, a recycling failure, a water-supply loss or contamination, an energy-power system failure for air-circulation, or even a tiny meteorite from space, that somehow penetrated the external shell of the base-structure; all of these dangers and more, would wipe them out in days, if not hours. The entire base had back-up's as far as most of these systems, but each required constant attention, monitoring and adjustments. The US Mars Base was remarkable precisely because it was 'self-sustaining', and could basically exist all on its own, pretty much as long as the residents could keep it all going, even without Earth transports. So Charley was in charge of making that a reality, Twenty-four/seven, flawlessly and without any surprises. He was a short-statured man, rustic looking, and given to parties and drink in off-hours. Everyone who knew him encouraged his happiness and parties or girl friends, given that he held their lives in his hands.

Of course the Mars Base include many, many others. The Safety-Workers, the Reserve-Pilots, the Surface-Workers and Excursion Commanders, the Suit-Suppliers and Suit-Maintenance, Food-Workers, Communications, Satellite-Traffic, and on and on---all fascinating, healthy, colorful people, male and female, with much to offer. The oldest man at the base was nearly 70-years young, was involved in water-research. The youngest person who was a Mars-regular was only 23 years old, a Safety Worker. In the past, Earth children had sometimes visited the base, in groups of about ten at a time, as young as only about ten years old.

Karen Tutturro followed behind Bojji-Than, into the Command Center where he was expecting another boring day. Boring is usually good, in space-travel. The oval-shaped room was much like an Air-Traffic Control Tower at a large Earth jet-airport. Various staffers, working long-shifts, manned numerous monitoring computers---they kept track of everything from the external base-perimeter, to Mar's twin orbiting moons. If a dust storm was kicking up in the Southern hemisphere, 3,000-kilometers away, they needed to know. Communications from Earth were constant, but only a few types of messages had any real importance---the communications equipment and actual operating systems that Karen would be working on were elsewhere. Earth-communications were still functioning, and had been all along, but not at the level they needed for transmission of research data in large enough batches to make the effort successful. She had come a long, long way to get the job done, and it was important enough work that she was temporarily a minor celebrity at the base (as all new arrivals were).

"Everyone, please," Bojii addressed the room full of people, about 20 in all at various stations, in a loud voice. He stood at the head of the room on a small observation platform, and of course got their attention. "Please welcome our latest visitor from back home. This is Karen Tutturro, a Sci-Tech in Communications from Vandenberg. Hopefully Karen is going to repair our communications system."

Karen blinked and smiled. "Hi---everyone."

The room called back, some laughing, with 'hello's', 'howdy's' and welcomes. It was a rowdy bunch, mostly veterans who recognized a Mars-Virgin when they saw one. One man honked a small flatulent-horn he had at his desk. Others tapped their coffee-cups.     "What's wrong?" Karen asked Bojii. "They don't seem to like me."

"Of course they do," he replied. "With most of these, if you had brought a 12-pack of beer from Earth, and barbecued ribs, it would seal your fate with them forever. Have no fear. They'll be your friends in no time. Especially if you can fix our antenna problem."

They moved off the platform at the head of the room, strolling slowly around some of the posts and tech-stations. "I don't think it's an antenna issue," Karen confided to Bojji-Than. "I was looking over some of the data-schematics and system-analysis. What I need to do is shut the entire Earth-link down for about two or three days. Then I'll be able to isolate various components and their functions, and find the gremlin. Once I find the gremlin, I make a repair on just the part that isn't working well, then hook it all back together and start it up again. Then you start-up a new data-stream to Earth of the type you were working with. With any luck, you'll have no problem."

"Sounds like a plan," Bojii replied.

"I'll also need an assistant or two, or co-workers, under my direction, from your Telecomm staff."

"Certainly," Bojii replied. They paused. After a moment, Vinces Grant, the Research Lead for the base, approached them both and introduced himself. Grant worked in a jump suit that would have been more appropriate on a fishing boat off the coast of Catalina, back in California. Not tall, but rather wide, he had a slinky-masculine appeal that was irresistible and charming. He took Karen's hand. "Vinces Grant," he said. "Pleased to meet you."

"Vinces does control for all the base research teams, exploration programs, science-and-data. In fact, it's his programs and the data they have for the teams on Earth, that comprise the content for the Telecomm-systems to transmit, that failed---prior to your trip here to Mars," the Commander said to Karen. Bojji-Than had known Vinces for many years, and could anticipate one of his one-liner jokes, before Vinces even opened his mouth.

"Why are there no uncooked hot-dogs on Mars?" he tossed out a line, in Karen's direction.

"Huh?" she said. "Oh---uh---I don't know. Why?"

"Because Mars is the god of war, not the dog of raw."

Karen took a moment to try to understand the humor.

"Not one of your better jokes, Vinces," Bojii said.

"Come on, you get it---'raw' is 'war' spelled backwards, that's all. See? Laugh, would you! I worked on that for hours."

Karen complied and chortled a bit, if only for his attitude and emotions. He was a pleasant man in any case. "No, really---I get it," she said. "No raw hot-dogs."

"The base cafeteria does not usually feature meat-products, due to storage and preservation issues with importation from Earth, and also the inability of the base here to produce any meat products of our own, like chicken or fish," said Bojii.

"That's the other reason," Vinces added.

"I don't really like hot-dogs anyway," Karen said. "You know, the nitrates they use. Unless they're organic."

They seemed to pause in the small talk. "Well, nice to meet you," Vinces said. "Let's just hope those dick-heads in Russia and the Islamic-Hindu Space-Program Alliance don't figure they'll be gobbling us up here at the base like a handy little inter-planetary snack, in the near future, in anticipation of any meteors out there. Know what I mean??"

They both nodded. Everyone pretty much already knew about the approaching asteroid, and fears about the Russian-Islamic Space-Program. Even on Mars, it was a cold topic.

"Nice to meet you, too," Karen said.

## CHAPTER 13: Karen's Mars-Walk

Within only another day, Karen found herself in a large fitting room, where Mars Base staffs were suited up for any external Mars-surface work or travel. The Alice-in-Wonderland effect was wearing off, and she felt confident and assured about her work. The voyage across the abyss, or planetary corridor, was itself demanding and unpleasant---just something about it, as any space-traveler of 2075 would confess. Now on Mars, her physical body functions returned to normal, and her mind was sharp. The Earth-Mars communications system, from the Mars-side, was of course complex and high-tech, which was why she had come to Mars, given her training and education. After clearance and de-bugging, and orientation at the base, Karen enjoyed a short time at the exercise-area, which was a gym---to her delight, the one-third gravity on Mars seemed to bestow amazing physical prowess, and her usual yoga-and-dance routine was somewhat like that of a 'tiger on the moon'. She had never leapt so high in her life or felt so powerful. But it was deceptive and she knew it. Otherwise, throughout the Mars Base, the lower gravity was not really a 'problem', and folks could walk about normally, as a 170-pound man now felt himself a mass-weight of only about 60 to 70 pounds, but still functional. Cargo and heavy items were easy to move, and sometimes people would play gravity-games or pranks, like one of the janitorial workers who had learned to run and quickly climb up a tall wall and even run quickly across the ceiling, like Donald O'Connor in the film classic, 'Singing In the Rain'. But it never rained on Mars.

Each surface-excursion was carefully planned and prepped. Every footstep beyond the safety of the airtight base was dangerous, if only for lack of 'normal' air. The base workers enjoyed speculating about how long a healthy person could last, without a suit, on the surface, such as in regular outdoor gear, maybe a thick jacket and boots or a thick hat. Even a strong person would only last a few minutes, it was unanimously agreed. Attempts to breathe the very thin, oxygen poor Mars atmosphere would seem like a sprint-runner sucking his air-supply through a soda-straw, and within only a few minutes, you'd collapse. Additionally, the Mars-surface temperature in the region of the Snikta-Ridge Volcanic Basin, at the feet of the Tharsis Montes volcanoes, towering higher than Mount Everest on Earth, was very cold. A warm day might see 20-to-40 degrees Fahrenheit, and cold days could quickly drop to far below freezing. The Sun, so blazing with it's heat and glory on Earth, here on Mars, was hollow and distant, fainter, and not nearly so intense or providing as much ambient surface heat. Complicating survival matters was the fact that without a dense atmosphere, certain types of radiation easily penetrated from above, and surface-walkers on Mars would find themselves 'burning up'---or 'sun-burnt', without proper protection, from UV rays, and other exposure.

The Suit-Room and Surface Excursion Crew were ready when Karen arrived. Her 'surface-work' today was simple. Herself and two others, would hike North along the South-facing base of the facility, only a matter of 300-yards or so, to inspect and review the antenna-array and satellite-communications radio-transmission tech-stations, that were necessarily built exterior to much of the operational base. In this way, Karen would have a first-hand look at what she was dealing with, rather than from mere schematics and books. No one expected her to go much farther than just those 300-yards and back, she wasn't an astronaut or explorer, and was not very experienced with the suits. Other Mars teams would commonly go on excursions of even a few hundred miles (in the motorized carts). So Karen's outer-work today was 'easy'. Nevertheless, every detail was planned-and-prepped carefully, scheduled, timed, and monitored, for complete safety. Back-up rescue teams were always ready if anything went wrong.

The Suit-Room was large enough to prep and outfit as many as 12 men at a time. The 'suits' were stored 'ready-for-use' in closets like row-after-row of coffin-like bins and chutes, and each bin was equipped with electrical charging, oxygen-tank re-charge, safety-testing, and other. They were not 'moon-suits', like the Apollo astronauts wore. Mars was a significantly different environment, and so were the suits and the needs. But they were similar, and also more advanced in terms of the materials and techniques applied for the comfort and mobility of the walkers. They looked rather like floppy, aluminum or metallic, cloth-like head-to-toe jump-suits, thick with padding in rings, and with joints and elbows craft-worked carefully for movement and strength. The boots were more common and not very unusual, except that they connected to the legs by an airtight seal, also rings. Similar with gloves, and then the headpiece, a full-cover helmet, with large plexiglass type rounded hood for 280-degrees viewing, and an internal radio-link to both inside-base monitors, and other walker-team members. The walker could control various features through a small panel of buttons on the right forearm sleeve. Once suited-up, a back-pack connected to the suit, carrying more than 24-hours of fresh oxygen, and also water, and a heating-system, and internal body-monitors. Amazingly, they were fairly light and easy to use. But running a great deal, or any kind of vigorous repeated movement such as an athlete, was somewhat unlikely, due to bulk. Some people at the base liked them, and some didn't.

Two Suit-Helpers worked with Karen, as well as her teammates. John Balker was an experienced surface-walker veteran, more-or-less her guide, male, about age 45, with three years on Mars (not counting Earth-vacations). The other walker going with them was Bob Johnson, a former transport pilot who was recruited to work on Mars with the ships and launches, as a 'space-harbor' pilot, or what they called a re-entry pilot. Bob was only 38 years old. Both men were of the same husky astronaut 'right-stuff' mode that they all aspired to, jocular and up beat.

"We exit at Gate-Three, the large air lock," said Balker. "That's off the main-entrance you came through when you arrived, just to the South about five hallways. Once we're outside, we'll just walk. We could take a cart, but I thought you might enjoy the experience. So, we just go South along the facility perimeter, 250-yards or so. Nothing out there, just rocks and dirt, and cold. Then we move up by the antenna-array, and just beyond are the other communications stations."

Balker and Johnson were already suited-up, and were testing their gear. Each person always took full responsibility for their own suits, and knew them well enough so that any emergency would not be a mystery. Karen, however, was maybe not-so familiar with the set-ups.

"Slip your hand under, then over, then through the hole," said one of the women Suit-Helpers.

Karen complied, and then figured it out. "It's a lot like mountain-climbing gear, isn't it?" she said.

"Very similar, yes," the woman answered. As Karen finally got herself completely into the suit, she laughed. It was a locker-room atmosphere, given that the walkers would put on their suits, with only 'long-underwear' underneath. "How's that?" Karen said.

"Looking good, lady," Bob answered.

"Let me just check you out on the safety-systems one more time," said the woman.

"We're still on schedule. Base-Control doesn't have us leaving the large air lock for another hour. A hallway mobile-cart will move us to the air lock so we don't have to walk through the building wearing these things," Bob said.

"I'm also going to need my testing-equipment and hand-computer," said Karen.

"Your bag," said John.

"What?"

"You'll have your carry-bag. If it works inside, it works outside. Just make sure any electronics are not going to freeze, and have charge, and that you can push the buttons with your gloves on. And nothing with sharp edges, like knives or blades. Other tools may be okay. I'll look at what you have before we exit."

Now all three of them closed down their rounded plexiglass helmet hoods. The helpers booted the life-support systems, and the suits went 'live', a spritz of new-air filling the insides like elegantly comfortable balloons.. For the next five hours, each of them would exist in a tiny, closed-system inner-world, sometimes sweaty-moist or uncomfortable. If they had an itch, it wouldn't scratch, and pee-and-poop went into an often less-than-perfect diaper-and-catheter system, with large fitted-plastic male or female 'rump kits', that routed substances as appropriately as possible. John tapped his helmet as their radio-link went live.

"Read me? Karen? Bob? Testing---"

"How do I look?" Karen said, posing.

"Ready for Hollywood," Bob answered.

"Not a movie. Not a TV-show. Say it often," Karen joked.

After more adjustments and waiting, a hallway mobile-cart came and took them from the Suit-room to Gate-Three. Other base-workers were used to the sight, just another external gig, function, or labor of some kind. By the Martian clock, it was early afternoon, a good time, because the distant Sun had by now heated the surface to about the daily high-temperature.

Gate-Three was an air lock gate, operated by three workers. It looked much like an Earthside jet-air travel gate at a large airport, perhaps wider, and more industrial, or even military looking, not for tourists, to be sure. Walkers passed through the first inside door, onto a mid-point platform with railings. Then the first door closed behind them, and was sealed for air. Once sealed, the operators opened the second, outer-door, now by machine-power, and they could pass on foot to the outside, the actual surface of Mars. This door then closed and sealed again behind them. It was quite safe. However, sans the machine-power and air-pressure seals and pumps that filled the mid-point area with fresh air, there was no way a person could use the gates, or pass through. In other words, you could not go through the air lock gates 'manually', or operate them by-hand. There were other 'hatches' and portals into the base from outside, that could serve that purpose, should the need arise. Inner air-pressure leaks and safety were obviously a critical safety issue at all times.

As instructed, Karen, John and Bob, processed through the gate, which took about five minutes. When they stepped outside, Karen could not help but slip again into Alice-mode. Her inner-child thrilled. Here she was, walking on the surface of another planet, Mars.

"Good heavens! It's beautiful! Oh my god!" she said.

The other walkers chuckled, which she could hear on the radio-link.

"Yeah," John said. "Definitely. Not Earth, that's for sure. You'll get used to it."

Once-in-a-lifetime moments for Karen had included her first sexual encounter, college graduation, and the first time she smoked marijuana with friends. They paused. What she could see was the region that stretched out before the face of the Mars Base, but the view extended literally hundreds of miles, as they were rather high up. Tharsis Montes extended like the gods themselves, to the East and far out of view, jutting skyward as bands of rock and colors, a geological marvel, yet so large as to be almost dizzying. The planners wanted permanence for the base they were building, so the facility seemed almost to be fitted into the rock itself, yet at a much lower level, flatter. The mountains did not have 'snow', but seemed to glow with a crystalline sort of shimmering, especially up higher, as if the ages and ages of 'dry-ice' moisture had coated the stone, like a far off silver. Clouds and mist were also not really present in amounts one would think of as 'weather', or much like Earth-clouds at all. But as tall and high as the mountains were, one could see that the Martian atmosphere had a substance of some sort to it, as if carried by the wind, which was hardly even the slightest of breaths.

In the other direction, downward off the flatland where the base was built, Karen could turn and see that the surface of Mars was indeed barren. Rocks and sand, or dirt, with only a little variation, spread away outward from her feet, towards the distant Martian horizon, perhaps hundreds of miles. Even this vista was stunning, much perhaps like California's Mojave, or Death Valley. Slopes, and ridges, small rocks, patches of colors, various rising areas. Far beyond, more mountains, also of great size. The sky-color was an odd patina of blue-green-red, for daylight hours. The Sun was visible, too, distant, smaller than on Earth, somehow cold-looking, seeming to strain with long slender rays of life-energy extending towards them.

"Let's get to work," Karen said, after her moment.

The two men chuckled again. The inner-helmet radio-link had a quality tone, and operated automatically with each spoken word. Each team-member could hear any conversation. "This way," John Balker said. They started the hike in the direction he was leading them.

He touched a button on the forearm panel on his suit. "Hello, base. This is Balker, with Karen Tutturro and Bob Johnson. Please recognize."

A moment. They could hear the Inter-base Communications Monitor and Excursion Team Link, a woman named Sally.

"Hi John. Base Inter-link here, I'm Sally. Got you covered. How's the weather?"

"Paradise never changes much," John said. "Beautiful day!"

"Of course," Sally answered. "You're on my screen, looks good. Hi Karen, Bob."

Karen and Bob both answered with greetings. They were walking, over the sand and rocks, with the walls of the facility to their right, and the buildings and structures looming near, comforting. Their boots crunched on the sand: swish-crunch-swish-crunch-swish, like the sound of walking on a rocky beach in roller-skates. With the lighter gravity, the walking wasn't too physically demanding, yet was hard enough that Karen had to breathe deeply and focus. She had her carry-bag with testing equipment, and she had a good idea what she wanted to accomplish. Bob and John also knew why she needed to look at the Communications outer stations and antenna-array, but neither of them were trained about those systems (though they knew them well-enough).

It took them an hour to reach the outer stations and small buildings and structures that housed the communications gear and antenna-arrays and more, which she wanted to look at. They stopped to rest several times along the way. Karen was eventually stricken with one painful truth about Mars, as she again enjoyed the view. If there was anything truly alien about the place, it was the total absence of any form of life, or natural plants, or any kind of foliage, or running water, or trees. Beyond the base, peering through her helmet visor, viewing, she blessed her own private memories of Earth with its abundant life. Nowhere could she even a single twig. No life. Except themselves.

"This is the lower of the two Off-World Radio Dish systems," Balker told her, when they reached their goal. "You can see, it has the three large radio-telescope type dish projectors, and then a series of smaller ones. The large ones go to Earth, controlled by a targeting computer, so the positions change slowly. The small ones can signal to orbit, or also to surface. Some of the small ones can reach Earth, too, but not as efficiently. Our helmet-link radios are actually running off smaller antennas up there, on the second platform building, I think."

For Karen, it was like coming home, her turf, technically. She recognized all the types they were using, and their functions, power-sources, routing, and computer-controls. There were large platform pads, with outbuildings suitable to the Martian environment, and then the dishes above, and arranged all around. At the highest point the set-up reached upwards about two stories, not really all that impressive for size at all.

She tapped her helmet. "Can we get any music on these things? You know---some Beatles tunes or anything? I like to listen to music while I work."

"I'll see what I can do," Balker replied.

"The old Beatles, not the late stuff. I never liked their darker works."

Both the men with her laughed, then found repose on rocks or against the buildings, as Karen began what only she knew needed to be learned and discovered, to begin work on repairs to the base's vital communications-link to Earth. The Alice-effect loosened its grip as her mind focused on familiar, highly involving tasks. Within another few minutes, their inter-link excursion helmet radio system treated the three of them to gentle, low-volume versions of 'When I'm 64', and 'They're Gonna' Put Me in the Movies', and other early Beatles tunes, courtesy of Sally's ingenious monitoring.

## CHAPTER 14: The Molinari

*"Come on then, follow me. We'll track the data on the transporter log, and go after them," Captain Kirk said.*

*"But to where? That dead moon is just a hunk of solid rock," Doctor McCoy replied nervously, now walking quickly through the space-station hallways.*

*"To---wherever they went," Kirk said.*

*"The signal was from somewhere inside the planet. We could end up re-materializing in solid stone."*

*"Then you'll finally get to take that long vacation like you've always wanted to," replied Kirk.*

*Star Trek: The Wrath of Khan, Paramount Pictures, 1982*

The Earth's Solar System of planets was at long last being fully explored by 2076, at the time of the conflict over the US Mars Base. Science-and-technology had progressed enough, and Earth's own ecology and peoples were in a state peaceful and sustainable enough, that progress could be made. In any other era, the opening doorway into the Solar System, with its endless mysteries and resources, though obviously Earth's nearest and most attainable neighbors---in any other era, the door would never have opened at all. And, as some science-exploration fans pointed out, wars and disasters on Earth, could easily create a circumstance in which the progress and learning of previous generations, and the means and will to go beyond planet Earth even only as far as Mars, would be lost, for whatever reason. In other words, if things went corrupt on Earth to the degree of open nuclear warfare, a serious meteor-strike such as the anticipated U2753b asteroid, large scale Earth-environmental changes, total collapse of civilized life and energy or communications--- anarchy---if these things took place, all would be lost as far as space-exploration, and possibly never recovered at all, in another million years of Earth history. The door had opened, and people like Lila Meetek and Guy Reisling, passed through as naturally as they would have boarded a city bus.

After Lila's 'R-and-R' back in California, and considerable enjoyments with Guy, there along the coast in his cozy love-shack South of Vandenberg, it was 'back to work', which meant, 'back to space', for her. All the regular space-workers took regular breaks back at home, to maintain health and well being, and it was not too difficult to run the ships across the yearlong passage, and move people around. The ship-transport system was no luxury-liner, and conditions were still quasi-military, or just plain demanding---not for everyone. But standardization, and up-dated techniques, made it all much more reliable. So when it came time to re-enter Earth orbit from a launch out of Florida (not Vandenberg), Lila knew exactly what to expect, and how to handle herself. From orbit they docked to a transport deep-space cruiser-transport, and transferred people and goods from ship-to-ship. The deep-space vessel was configured for navigation away from the Earth's gravity-well, and then preceded into departure and attainment of speeds equal to the journey, as was normal. Not normal at all, from a passenger's point-of-view, such as Lila, but also never really much more fearful than any other mode of machine-travel, even on Earth, where a road-crash in a hydrogen fuel-cell speed-car, was just as deadly, as anything that might happen in space. Once in motion, at rest, the journey was spectacular, to say the least.

But that was long ago, on Lila's personal biological life-clock, having moved into the vast emptiness between the two orbiting planets, known as the Earth-Mars Corridor. Long ago for Lila, her life rather young and fresh, meant five months' travel in space. More than five months, more accurately five months and ten days. She was not traveling to Mars, but making her 'commuter run' to work, at the Molinari Deep-Space Station.

Alberto Gonzales Molinari was a US Military General-Commander, who at about the age of 70-years, retired, was recruited by the space-program, to conceive, design, plan to build, and anticipate the use, sustain-and-support, and future history, of this important space docking station. It was his baby, and he was variously qualified, even an outstanding space-researcher and planner for the US. Molinari died before the space station, the largest and most advanced ever, was completed, and set into orbit forever, halfway between two worlds. In so doing, progress into deep space was assured or even possible, and a new approach to near-Earth exploration was established, by which deep-space journeys to Mars, Jupiter, Saturn and so on, could be achieved. Planners felt that by creating a series of similar deep-space rest-stations---even ten, twenty or more, like a string of pearls between Earth and her neighbors---in this way, the very long voyages were now possible for frail human astronauts. So Molinari was remembered as a much-beloved figure, who more-or-less invented the ways-and-means to do this.

On the deep-space voyages, the 'view' was quite boring. No blue-green Earth below shimmering with puffy white clouds and green-brown lands, and dark blue waters. No planet Mars to gaze at, with its dusty red terrains and polar caps. No spinning comet-trails, like pixies or angels, fairy-mothers with wands of fire to dream about. No spinning giant asteroids, like the movies. No other ships, no aliens-in-saucers, no UFO's. One could hardly even sense the forward-motion of the space ship, though there was a vibration-effect. And true enough, passengers such as Lila really had no lazy patio they would sit on and gaze at the stars while sipping vodkas. There were external video-sensors that could create inner-ship images on electronic screens, and actual transparent-aluminum 'windows' or 'view ports'. And then the pilot's room, or navigation deck, which had it's own view-system. And of course they didn't fly by line-of-sight anyway, which would have been madness and certain disaster. But when docking, or approaching a world like Mars, and for other reasons (like any necessary space-walks), ordinary eyeball views were useful.

But for five months to Molinari, and at least 11-months to Mars, if not much longer, depending on the season, there was nothing to see. The deep-blue indigo Eternal Mother Space, and yes, stars, much like seen from Earth, far off twinkles---and that was all. Either Mars or Earth were so far off as to appear to be only stars. Molinari could not be seen at all within less than 100,000 miles. Blue velvet, like a deep sleep, peaceful beyond notion. Working in space was like working in a coal mine of infinitude, silent and eternal, awesome. By the time Lila was finally back in her own rooms and work-area on Molinari, she could not have been more ready to share with her co-workers, and get back into the busy life of the space-station---never a dull moment. The transport ship connected successfully, other work and transferred-goods and items or data went back-and-forth. The ship's crew could rest and enjoy life at the space station too, and other travelers could connect to their destinations, and so on. Molinari humanized the whole affair, and was greatly appreciated by all.

Externally, Molinari was impressive, though maybe not beautiful. Space-station design over many years included all sorts of approaches---one would choose a wheel-hub design, another would be like a series of long connected tubes and machines and solar-energy collectors, with dock-ports, and telescopes or antenna. But Molinari was big---it had to be. The mid-point space dock needed to house its own crew and staff of about 60 regular workers, and also be able to safely dock with numerous ships, sometimes as many as 20 or 30 a year, one-after-another. It also functioned to measure, monitor and analyze, as well as transmit back to Earth, or to Mars, information and data about anything and everything going on in that region of space, that would effect safe working research and transport. So the floating space-buoy was out-fitted with endless sensors, telescopes, antenna-dishes, radio-transmission, and so on. And deep inside, people like Lila Meetek, worked to keep all that going and up-to-date, watching computer-screens, tracking comets or asteroids, heat-flares and solar conditions, Mars-weather, planetary position and orbit, ships and people, goods and materials, or cargo, and also any related details of various ship movements or research programs, that might contribute to success with anything they were doing---or alert to failure and disaster.

"If I was African-American, you'd call me Uhura," Lila once told Guy, in a private moment. "Earth-Mother, Eve, the voice of the Word that keeps us alive forever. Is that arrogant?"

"Huh? I never got that stuff. But, yeah---you do good work, babe. Let's find a cute little planet somewhere and start a new--- uh, a new—you know---Genesis."

"Most people just call them babies, or infants," she answered. "Yeah, I know."

"That's it---that's what I meant. A new human being, like a kid—and then another, and another---I'm pretty good at getting that going. You know. Start-up. Quick-start---species. A new world. Somewhere they don't got wars and---death."

"Right here ought to do, for that," she answered, and he knew what she meant, as they rolled again in the hay-that-pays, there back near Vandenberg, in Guy's love-shack, in each other's arms, to that orgasm we all adults endlessly enjoy to share or discover, the wound that never heals. Within only a few weeks of their love-making and muse, Lila would be shot into space, as Guy began to re-invent his career, and his next trip to Mars, having won the challenge to his pilot-worthiness and judgment, by virtue of a mid-course flight-path correction, now a year behind him.

Molinari hung like a Christmas-ball ornament, in the nothing. It had five pods for docking and other work, around a central hub that dropped below, with a tower-like structure above. It was quite large, about the size of an ordinary skyscraper in New York, or a jet-airport in some large city. The materials were 'new metals', and obscure, ideal for deep space, and secure to sustain life. Power-sources, solar-harvesting energy-collectors, communications---it might have been a deep-ocean underwater 'base', or an odd kind of bell, or a unique automobile engine part, from an old-style car (in shape and form). Lila and the others there called it 'home'. It had taken 40 years of planning, and then work on building and fabrication, begun in secret, and then revealed to the world as a major achievement, about year 2050, with new hopes and promises. And of course the Earth-world community yawned---"What's in it for me??"

Lila moved through one of the inner hallways. Yes, the space station lacked for normal gravity—it was a zero gravity environment through the entire structure, except for areas where there were centrifugal doughnut-shaped spinning rooms for certain purposes. But mostly, workers used slippers, with magnetic 'Teflon' coated bottoms or soles, and movement in hallways, and work-areas, had magnetic floors and also walls. The system worked fairly well, with a few complaints---it could even be fun. There were also handholds and rails. You could pull yourself along the halls by hand, floating at length, or use the slippers. Seats, toilets, dining-areas, any 'people stuff', all with magnetic strips, and then also gloves, clothing, and so on. It worked, and they learned to live with it.

As she entered the Earth-Mars Traffic and Environmental Monitoring Work Room, her co-workers, who had not seen her for months, turned and cheered. "Medusa had returned from her mirror gazing!" said one man, a chummy type. "Here she comes---Miss America!!" another man started to sing. Others greeted her with applause or hoots. Five people worked the room pretty much around the clock. There were numerous computer-monitors, devices to detect various space-corridor conditions, communications, sensors and telescope feeds, kiosks for one-time video-communication to either Earth or Mars (somewhat rare), and many other services they provided.

"Yes, it's me," she said. The dress-code was 'space dock casual', which meant most of these sorts of workers wore ordinary street-clothes, but were supplemented with additional features, like a harness that clung tightly to the body, with the magnetic strips or other useful items for movement about the station---like comm.-links (inter-base cell-phones), timers, body-function monitors, small food bars, and so on. Molinari was no elegant shopping mall, no tourist destination or space-hotel. It was all strictly business, and safety came first both within the dock, and for the beneficiaries of the work they did---the men and crews and passengers attempting the yet-dangerous and truly mammoth voyage, from Earth to Mars.

"How's Earth?" said her best friend at the station, a woman about her age, with a similar job. This was Eve Morton, from

Illinois State University's space-program, part of the US internship-development partnership. Eve was an often giddy, yet a very intelligent and wise woman, given to flights of fancy, and who loved poetry, like Percy Shelley, or Anne Sexton, Robert Frost---a big hit with the guys, too, sexy. She had only been at Molinari two years, compared with Lila's almost five-year run.

"Big, round, blue, and stupid," Lila joked.

"Some things never change," Eve replied. "Missed you. We had a meteor sprinkle a few weeks ago you would have enjoyed. No flights in the corridor, so it meant nothing. A lot of teeny ones like the Leonid."

"I always try to keep in mind that we're in the corridor ourselves, here," Lila said, now settling down and adjusting to her workstation. Before her was an array that Star Trek's Uhura would have envied. She sighed a familiar breath. Home. "I mean, teeny meteors we track to protect the guys in the ships can hurt us too."

"Yeah, but our dock is much heavier with the titanium-abstracted plastics. It's tougher than anything. But you're right. We're in the corridor too. Good to have you back, girl. I hope Guy was making life work for you down there, yeah??"

"Uh---big, round, pink and stupid," Lila joked. They both laughed at her slightly erotic humor. "Mind you own business, Eve. Geez."

## CHAPTER 15: Star-Crossed Lovers

Work is work, whether in space, back on Earth, or whereverpeople must survive, Lila Meetek was thinking. Space-travel, in films and TV shows, or books from past writers, has been portrayed for whatever reason as very fancy and up-scale, even decadent.  Artists have dreamed that Mars was occupied by six-legged horses, ten-foot tall green men with tusks and four arms, or beautiful princesses who lived in elegant and extravagant palaces complete with servants, silk pillows, wine and food, or massive art-collections and amazing adventures ('John Carter on Mars', by Edgar Rice Burroughs). Or that star-ships were so limitlessly provided-for, as to resemble Paradise Cruises, with kushy malls and bars, gardens and astronomy view-patios, gymnasiums or holo-decks for amusement, wide hallways and automatic doors, and of course beautiful stewardesses, mixed drinks, and endless exotic visitors. Not to mention perfect, if not absolute safety, unless the 'bad guys' attacked for some reason, usually with laser-beams. Perhaps, thought Lila Meetek to herself, later, as her work routine began again, now back for another six months at the Molinari Space Dock---perhaps it was because of the perception that space-travel, if it were ever to be a reality, would be very costly, or expensive, given the reality of rocket-launches and air locks or high-tech space-suits and computer navigation systems. And of course, it wasn't cheap. But fancy? A Las Vegas-style romp? Lila had to laugh. As-if.

Yet, had the designers of the Molinari station, the base on Mars, or other ships and bases (like the one on the Earth's Moon, known as PlanetView-2), created environments that were very harsh, dark-and-dank, military, metallic or inhospitable for heat, steam and vapors, and so on, no real residency would be possible, or not for any length of time. Lila's workplace was somewhere in-between. The Russian-Islamic Eastern-block space-program was known for much harsher environments inside their ships and cabins on their orbiting stations, etc. They could be like cramped, iron diving bells, intended only for survival in extreme conditions, without so much as view-ports or windows or bed-cushions. But much to the pleasure of Lila and her co-workers, Molinari was not really uncomfortable at all, and had many creature comforts, and accommodations, that made long stays tolerable, or even quite nice.

And the great thing about the work they did at Molinari was that it was inevitably slow-paced, with a bird's-eye view of both Earth and Mars, and beyond, so that much of what they did was to monitor various instruments and telescopes or scanners, for indications of changing conditions that would affect inter-planet travel. Earth's solar-system is a fairly active place, but also of course vast, and even giant meteors like U2753b (Big Baby Bertha), moved very slowly, given the distances involved. So it was all very stately and graceful, and also un-changing, or like a vast machine, set on it's clockwork path from an original Source of Virtue and Miraculous Provision and Creativity. Comets, meteor-showers, solar-flares and heat-flares from the Sun, planetary orbits, moon-orbits, planetary conditions, even gamma-rays and neutrinos, or other cosmic phenomena---truly a celestial dance and endless wonder, vast beyond the scope of the mind of Man (or Woman).

So, Lila settled into her 'shift'. Six months and possibly longer, monitoring her slate of scanners and telescopes, radios, and communications, and other chores. Lila's specific tasks included: 1) planet-corridor ship or vessel transport logs and monitoring or tracking. This was done much as any space-launch would be tracked from Earth, such as the old 'Houston Mission Control' would keep track of the Apollo voyages. It was essential, and it was also how Lila came to be intimate with Guy Reisling, whose transport ship command caused them to be in regular contact. 2) Lila was required to organize and handle various day-to-day communications relays. And yes, this resembled the job of the fictional character 'Uhura' (with respect to actress Black American Nichelle Nichols, who made the mid-1900's fictional role famous), from the now ancient and laughably inaccurate Star-Trek TV shows and films. Lila handled 'calls' from Earth to Mars, from ship-to-Earth or Mars, and in routed person-to-person contact along numerous various paths, linking-up much-needed information sources by radio-signal. She was not the only person who did this job, but it was among her assigned tasks. 3) Lila's other job was to track and log 'weather conditions' in space, specifically those originating outward from the Sun (rather than any coming 'inward' from beyond Mars, or the outer planets). So this meant she watched for all kinds of changing conditions, the heat-flares or solar alterations and variations in fields of meteor-flecks the size of pebbles, also in orbit, like marbles gliding soundless on a skate-rink, round-and-round---and many other things the scientists felt needed to be watched, which also changed as new things were learned about the astonishing space-environment.

Lila and all the other Molinari workers also had many other much simpler and easier duties associated with life on the space dock. Health-matters for any crew were essential, so many details were logged daily. Any virus or infections, even a common cold, could wipe them out. Crew maintained their own housing-quarters, articles of clothing and equipment (like the magnetic hallway slippers and gloves or butt-pads), and also grooming, diet, entertainment or studies. Anyone working at the base was considered a science-researcher by default, and indeed they were. They all knew their jobs were of that nature, and that their lives were not their own, in a sense---so every hour at the base took some form of service to the greater purpose, and nothing was wasted, as we might have thought in the imaginary world of space-travel that more resembled resorts or casinos, or luxury-hotels, with wine-and-song, hot-spa treatments, etc. The outer-space environment was indeed suitable to such functions, very serene and beautiful, exotic, full of wonder. And no doubt within a few centuries, enterprising corporations or business-interests on Earth would develop exactly those kinds of uses. A space-bordello? An orbiting space-station religious cult? A deep-space health-club for anti-gravity athletes? An artist colony?

Molinari had telescopes that were the envy of Earth-based astronomers, and the view was unique. The famous orbiting Hubble-telescope from the early 2000's and late 1990's, had expanded the astronomer's view of the heavens in ways that Galileo or Copernicus, or Isaac Newton and other science-pioneers, would have much-enjoyed. Molinari had several similar space-telescopes, and a staff of highly qualified attendants and astronomers. In this way, the Universe Creation, galaxies and stars in truly limitless variety and species, worlds beyond worlds, heavenly formations to boggle the mind, could now be viewed and photographed, cataloged. Mankind was slowly and tediously becoming a citizen of the galaxy, at least as far as awareness was concerned. Whereas the writers of early Earth's Abrahamic Bible, Saint Paul, or even Christ, or Moses, may have enjoyed believing in a very simple misconception, such as that the Earth's atmosphere extended endlessly beyond the mountains and lakes of Gaia (Terra), by the time Lila was again working at Molinari in 2076, those comforting assumptions could no longer be appreciated. Any high school or college student on Earth was now forever a heretic, by virtue of the scheme of things observable to anyone 'with eyes to see and ears to hear'.

Yet, spirituality was encouraged at the Molinari Space Dock. Traditional forms of worship, and also more esoteric sorts, such as yoga, Zen-meditation, Buddhism, and a variety of credos and expressions---each was permitted and allowed, with meetings as well, music, even 'priests', though in reality formal services did not occur. But it wasn't banned or outlawed, and in general, work in space was thought by most of them to be a deeply moving and transcendent experience. Both Molinari and Mars were visited by spiritual leaders or visionaries, who enjoyed the view as well, in their natural appreciation of a newly available platform to celebrate belief, or to expand consciousness for art-works, inspiration, etc. Most of the astronauts and space-workers, being dedicated science-researchers, kept a distance from these types, feeling they may become deluded under the heady influence of space-itself, weightlessness, the endless dark and stars---a sort of Darth-Vader 'warning' or doubt, such as megalomania, delusions of grandeur, and similar 'G-D'-consciousness effects of the human creature. The planners knew it was a danger, psychologically, sad but true. But for the sake of health-and-wellness overall, and morale, there was no real strict prohibition. So most of the space-jockeys preferred the athletic model---big strong guys, husky and buff, pink with vigor, sexual, modest in opinion, laughing at it all, and likewise the women.

Eve Morton, Lila's University of Illinois-originating gal-pal, had a word for her, about a week later, harvested from the daily logs. "Lila, listen to this," Eve said, moving from her own workstation to where she found Lila three hours into her shift, there in the pod-like Earth-Mars Traffic and Environmental Monitoring Work Room.

Lila's station had no 'window', other than electronic screens. It was as big as a household bedroom or living area, even only Lila's area, and each of the other work-stations were similar, connected by short ramps and catwalks, and not closed or locked-in by doors. Inside, it was crammed full of detection gear, computers, communications-links and so on---desks, chairs, hand-rails, and Lila's personal decorations (in her case, a very nice collection of stuffed toy-penguins and penguin-art). Even in her work-outfit, Lila was a nice dish of womanhood, having recently died her hair green, and outlining her eyes with green-and-purple cosmetics.

"What-cha' got, Eve?" Lila said, turning in her chair. Eve literally floated into the room, pulling herself towards Lila by hand-rails in the exquisite anti-gravity, like Ariel or Uriel, a somewhat absurd-looking image---a large, healthy female human creature floating towards her like a very large and very odd fish. Eve had a plastic binder with some files inside. As she reached Lila she deftly pulled herself into a vertical uprightness, then held herself down with the magnetic slippers with a 'click', metal-to-metal.

"Vandenberg launch schedule for the next nine months," Eve said. "And guess what? Guy's transport is slated for launch four months from now."

Lila smiled. "Wow, really? That means he would be here in about ten months."

"Don't worry. You'll be back by then, I mean, from your next furlough," Eve said. "What are you going to tell him about Tommy. You know?"

Lila winced. Tommy was a Molinari regular, one of the external-hull repair guys. A true astronaut, Tommy's job was indeed perilous. Any work outside Molinari---minor repairs to the hull or exterior of the base, adjustments to the fixtures or joints, things like solar-panel up-keep, and even sometimes emergency work---anything like this, and Tom Bordino, along with ten other fully-equipped and trained men, were called into action. They were basically the Molinari Space-Walk Team, and it was every bit as demanding as any other space-work, if not more so. So of course they had the suits, lines, rigs, oxygen, mission assignments, and so on.

Tommy and Lila were somewhat regular sex-partners. It was inevitable, to an extent, and sex among the workers at Molinari was not prohibited, within reason. But it was 'prohibited' between life-partner lovers like Guy and Lila, for obvious reasons common at all ages of Man. Lila and Guy had even talked about marriage and children, and she was much more serious about Guy, for reasons only they understood. They thought themselves mature enough to allow 'other partners' along the way, if only for freedom and health. But the jealous beast within never really rests, and Lila loathe telling Guy about Tommy. The fact that Eve and others at Molinari knew about Lila and Tommy didn't help matters.

"That's not the question," Lila answered Eve, who now had re-positioned herself and was floating upside down near her, or slightly sideways, still holding the files. "What I'm really worried about is what you or anyone else here who may know about me and Tommy, will tell him. You know---gossip. What I may or may not decide to tell Guy is my business, girl."

"What's to tell? Sex is sex. Big deal. Guy's not really like that, is he? Jealous."

"No," Lila said. "He has girl-friends, too. There's this food-worker at Vandenberg he sleeps with, and she has big boobs and a big butt, too. Margie, the little slut. But, he does get his feelings hurt, I guess."

"Love hurts," Eve said. "Anyway, this came over on the launch-schedule from Earth yesterday, then I realized Guy's ship was listed. I'm glad he got his ship back. I never felt he did anything wrong, when they brought him up on charges like that. He could have lost everything, as far as his job."

"It will be his first flight after the board-review," Lila said.

"What choice did he have? The data on the solar flare was incomplete, and he had to protect the ship and crew. So he changed his flight path and fuel-type. Anyone would have done the same thing."

"It just screwed things up back on Earth for the navigators and re-entry crews," Lila said. "But I agree. Of course, I'm not a pilot."

Eve handed the file-binder to Lila. "Back to my cave. I'm

watching the progress of a comet that won't be within a million miles of us for ten years. But—uuuuh---it's a living, right?"

"Thanks, Eve," Lila said. "And please, don't talk about me and Tommy. Especially not with Guy."

"Didn't you know Tommy put your last romp on video and sent it over the satellite?" she laughed. "Not really."

Lila frowned. "Just float yourself out of here, bitch," she said.

They both laughed. Lila opened the file-binder with the California launch-schedule, as Eve pushed herself the other way down the room, shoving off from a wall. The computers and machines and monitors buzzed and bleeped and winked all around.

CHAPTER 16: The Barefoot Pilot

Within four weeks of Lila's conversation with her friend at
the space dock, Guy's transport launch was bumped ahead to a
much sooner departure, mostly due to the work communications-
specialist Karen Tutturro was doing at the Mars Base. So his next
flight would ferry hardware and gear she needed, and other
common shipments for the Mars Base. By the fifth week, Guy
and his crew were prepped for launch. So within 45 days of
Lila's expressed consternation that Eve and others knew about
her affair with Tommy and how she would handle it, her one true
love Guy Reisling was catapulted back into the abyss aboard the
'Penelope', on his way to her side again, into her loving arms.
The Western Earth-US space-launches were actually rather rare,
mostly Earth-orbit satellites. The inter-planet launches happened
only once every few months. Everything depended on timing the
launches with the position of the orbiting and moving planets.
The Molinari space dock was also in orbit, and thus in motion.
So a launch window was carefully calculated for each transport.
A trained pilot like Guy was a valued resource, accounting for
his re-instated pilot status as forgiven for the flight-path error,
now more than a year behind him. The launch went well, and he
left Earth orbit on track with his navigation plan. The journey to
Molinari would take about five months, by this scenario, given
the planetary-system positions when he left Earth, which was
very favorable, somewhat faster than many trips.

Six weeks after he left Earth, the US Mars Program planners
and decision-makers received startling and troubling news.
Earth-scan tracking and monitoring of planetary-orbit pathways
clearly indicated that the Russian-Islamic/Ukrainian-Hindu, or,
Eastern-block Space Program, had launched a series of ships.
Five inter-planet capable ships left Gaia-Earth from two launch-

sites, one in the Ukraine, and one from China-Mongolia. All five were launched over a period of about ten days, then stabilized in orbit, and then escaped Earth-gravity, headed for Mars, by all appearances, much like Guy's ship. The tracking-monitoring records and data were checked and checked again, and analyzed endlessly for days, until there was no doubt remaining. From the US Mars Program bases at Vandenberg, and also in Florida, Texas, and Puerto Rico, their worst fears had now happened. Like a space-rocket on a chess-board (or, five of them), the Eastern-block Earthside space-explorers had made a bold move.

Ibrahim Mehudi, the Science-Lead at Vandenberg, spread his hands across a large, table-sized view screen that showed the local solar system and planets. "This will lead us into conflict," he said. "The world's first war in space."

"Move the image to show me the estimated position of Big Baby Bertha," said Lynn Rodgers-Smith, the program director. The two of them were studying the situation, from a secondary navigation-and-pathway workroom, at one of Vandenberg Launch-Commands. The tool they used was a flat computer-screen linked to a model of the solar system, that approximated the daily or weekly position of the planets and other cosmic objects---an astrologer's dream. Mehudi could fairly easily manipulate the view from various points in the circle of solar objects, the Mars-Jupiter asteroid field, planets, moons, the Earth, and also any ships, large satellites, etc. Within a few minutes, they could roughly estimate the relative locale of Asteroid U2753b, viewed as a small spec or dot on the screen, also with a computer-text data-line, indicating its 'cosmic ID'.

"It's here, apparently, from this," said Mehudi. "U2753b is moving much like a comet would, in an oblique orbit, not parallel to any planets, or not circular, you might say. So it's cross-wise to other solar-pathways, cutting across the planet orbits from the side. So, from this view, it's on the other side of the Sun, probably 10 or 12-million miles away from Earth. But, of course, as Fate would have it, our science-prophets have plotted the two paths of both Earth and Bertha, to a collision point---now I guess roughly five years from now. Supposedly. I mean, a collision is a one-in-a-million chance. But this is the science about it."

Rodgers-Smith, the so-called Commander of Angels, glanced over the table-screen with a scowl. "Five years," she said.

"Give or take," said Mehudi. "Five and a quarter years."

"And we have only the one transport-ship in space-track corridor to Mars now, and Molinari, and then to the Mars Base with their ships there, in terms of any immediate response or defense to the Russian ships that we feel were launched to Mars-track pathways last week," she asked. "Is that about right?"

"Well, if you include all of our space-able ships out there right now, that would be about right," said Mehudi. "Keep in mind, the Mars Base has three ships of their own, but they are not used for regular transport. They can enter Mars-orbit, and they can reach Earth with preparations, mostly for emergency evacuation, if ever it was needed. And then Molinari has similar lifeboat type evacuation ships, two, or three. And then the currently in-transit transport, from Vandenberg. We launched that about two months ago. But that's it, as far as ships-in-space right now."

"PlanetView-2 has ships," Rodgers-Smith said.

"True, but those would never reach Mars. Local only, Earth to moon," said Mehudi. "You know, we have ships that can launch to Mars. Realistically, our side can launch maybe---oh, I guess, as many as eight---eight space ships that we could get on path to Mars. It would take time to set it up, for launch-window, navigation, preparations, crews."

"Eight? I thought we had more than that, almost a total of fifteen? The ones that can reach Mars."

"I don't know, Lynn. Just a guess. I mean ships that are in good enough condition and ready. We have repairs and lost function on a few of the others. Some are out-of-service, too."

They paused. The two of them withdrew from the high-tech Planet-Plotter, and then settled across the large rectangular workroom, by a window where they could look out on the base from above, more than four stories over the view. It was very early in the day. They had been summoned-to-task past midnight, still in the dark of night, when the reports reached the base-system data-review hierarchy. The morning light from the West beyond the Pacific as the Sun rose in the East created an orange-red-pink-gray spectacular against the distant and hazy-blue sky. Beneath, the towers and buildings, high-cranes and machinery, communications-dishes and antenna, power-transformers and huge fuel tanks, and ship hangars---it all lay before them, a shared kingdom, powerful, dormant, resting. Low-laying green-brown hills, small roads, fences, mossy oak trees and pines, and the occasional beef-cow munching on his grassy breakfast in far-off solitude, at the edges of the vista.

"Is it really a war, Mehudi? Must it be? So what if they launched to Mars? They have before. It's a free world---or, a free solar system. Surely we could work together, both sides helping each other. We're scientists, researchers, explorers---not military," said Lynn Rodgers-Smith.

"Ask the Russians, that's all I can say," Mehudi replied. "I totally agree with you. I'm a dove. But the hawks always take the lead. It's inevitable, especially from the previous intelligence on the Russian plans for Mars. They didn't even announce or report the launches of five ships, which is required by law. No flight-plan. No pre-launch air-space precautions. No global aerospace community safety review. Nothing. Essentially, the launches from Ukraine and China were done in secret. What does that tell you?"

They paused. "All right, " said Rodgers-Smith. "What a shit. I want a high-level meeting with all department heads by this afternoon. We have to do---something. Not sure what, I'll admit."

"The Mars Base has to be informed, and Molinari," said Mehudi.

"And the transport pilot already on his way. He certainly needs to know."

"I know that pilot. Reisling," said Mehudi. "Good man, fairly young, a bit wild. He was the one that Okman brought to the review-board for a flight-error, last year. But he was cleared. He should be about halfway to Molinari by now."

"Six weeks ahead of the Russians," said Lynn. "A single transport ship. I think I know him, too. Reisling---they had a shipment of communications tech-gear for the Mars Base."

"Yes. The Snikta-base has had a significant communications problem for quite some time. We sent a specialist, Karen Tutturro. She's been on Mars for six months or more. Her first trip. It's a mess. Power-supply failure obliterated data-transfer, not sure what is causing it."

"I know." She sighed heavily, holding her hand against the window. She was a large, busty woman, but very healthy and even quite attractive for her age. A Texas-gal is sexy all her life, in her ways. The two friends just waited, side-by-side, a knowingness between them, a fateful dread, and yet a call to honor and critical duty.

Somewhere beyond---and beyond-beyond, there in the depths of the infinitely peaceful abyss, a deep-indigo blanket of nothingness enveloped the space-transport Penelope, Guy's ship. One never looked back, after a launch, or, one never kept his thoughts on Earth, where home-and-hearth waited, comfy-cozy, with food and drink and friends. Of course the ship was in contact with Earth. But in his heart, Guy, and his crew, and any space-worker, were always aware that any such voyage could be their last, ending perhaps by accident in sudden death. So the view was ever outward, into eternity, as if Earth never-was, and only the unknown ahead. *It's just a job*, Guy was thinking. *Same as always. And I'm damn good at it.*

The Penelope hummed like a top. During his board-review furlough period, she had been re-fitted and fully up-graded for the usual maintenance and repairs. Like any flying machine, only absolute mechanical perfection was adequate for launch and the rigors of space-travel. The so-called Condrum 21 Monsanto-DuPont Inter-Planet Space-Cruiser was a mighty piece of work. They weren't short-term vessels for only one or two voyages, or in any way disposable. Many previous former-era spacecraft were designed with numerous discarded parts, even life-support pods and engine-parts that would be jettisoned into space, for efficiency during the trip, such as re-entry. By the time the astronauts arrived back on Earth, there really was no 'ship' remaining, at least not as they had started out. It had all been dismantled on the way, with each part falling away into the nothingness, until all that was left for re-entry was a single life-support pod that could somehow land the men back on Earth. But those types of ships were no longer is use. The Penelope was one solid block of materials, made-to-last, intended to function for many years, rather like a much more old-fashioned idea of any vessel, be it a car, a jet-airliner, or a large ship at sea. *She is One,* Guy thought. *She is mother.*

He was in good spirits, and so was his crew. Rob Cowan, his co-pilot, relieved Guy from his watch every nine hours. The other crew were all the same men, except for one 'new' navigator, who was of course experienced and trained, and had been on several successful voyages on other transports. Everyone was glad to be back on-duty, back in space, back on-the-job.

"All is well, Captain," said Cowan, entering the pilot-deck, a large control-room at the front of the ship, set high above the front or prow. Almost any control anywhere on the ship could be accessed from here (although many functions were never manipulated from the pilot-deck at all).

Much like Molinari, the environment was weightless, or null-gravity. Thus, Guy was seated and held down by magnetic-strips and a small strap, and Cowan pulled himself into the chamber by handrails, then could settle onto the magnetic floor-strips with a tiny 'click' from his metallic shoe-soles.

"Thanks, Rob," Guy responded. "Entering pilot-deck command transfer to you at 1800-hours, Day-52. She's all yours."

"All systems are at ideal status," Rob said. "Engine three will rest-and-restore in ten hours, then cycle through engine four and back to engines one and two, in sequence. Vandenberg-command has some kind of pow-wow going on. Earth-link telecomm ticker says you need to be set-up for a voice-to-voice at 2200-hours. Your second comm-systems-man is working on the hook-up, should be fine."

"Thanks," Guy said. As Rob was providing his brief report, Guy was busy re-fixing the magnetic shoes on his feet.

"For god's sake, Guy," Rob said. "You're the only pilot in the entire Mars fleet who prefers to fly barefoot. For as long as I've known you, you jockey your ship with nothing on your damn feet! Is that even sanitary, much less protocol?"

Guy finished with the magnetic bootstraps. The boots were somewhat heavier than the slippers they wore on Molinari, like thick, mechanical-looking tennis shoes. On the bottoms they were metallic or magnet-sensitive. The men could move this way through the ship fairly easily---click-clack-click-clack. More commonly, they pulled themselves through the passageways or tube-tunnels by handrails, floating-weightless. After many voyages, for experienced crewmembers, the novelty soon was past.

"It's my ship, and my command, dude," said Guy. "My feet get hot and sweaty. The life-support workers on below deck work in their fucking underpants. Don't worry about it. Was there anything else from Earth? What do they want? There's nothing new on this pass. Steady-as-she-goes. No news is good news."

Rob was settling into the command-chair (there was a second chair or station for the co-pilot, not Guy's). He checked several of the gauges and monitors and standard-tracking views. Typically on a command-deck shift, he wouldn't touch a single control or change a thing. But they had to be watched by an experienced pilot ready for anything, which he was, just as well as Guy.

"Not sure. The Earth-link ticker doesn't tell you much. They want to talk to you. That's all she wrote," said Rob.

Guy chuckled. "Yeah," he said. "So does my mom."

They both smiled. "You're relieved, Captain," said Rob. "No problem. Get some sleep. You got four hours."

## CHAPTER 17: World Council of Nations

*"It is a joy sometimes, when a needful matter is called a sin. Christ the Lord was familiar with wicked ways and means, the brutality and savagery of life, and the unexpected qualities of life's long journey. If our sins are as holy as our good deeds and religion, then without an open door to errors and mistakes, the door also slams shut to visualize or realize a better tomorrow, for the absence of change and dynamic circumstance. I don't believe in mistakes. To disallow for illness, broken legs, or broken minds, perhaps from a higher loving point-of-view, it does not follow we shall no longer have hospitals or doctors. The future will have many mistakes, and the Mars Space Program includes every variety of joys. Our astronauts are not robots. We are men, not machines. There are no sinless astronauts, Your Excellency. It is not a requirement for space-travel."*
*---US Mars Space Program Director Lynn Rodgers-Smith, to His Excellency Imam Mohammed Petrarch-Jinn, Interpreter of Global Sharia-Law to the World Council of Nations, 2077*

This man, Imam Mohammed Petrarch-Jinn, was part of the World Council of Nations, and was sent from Iran. In this era of 2075-77, and beyond, when the impending disaster of the approach of Big Baby Bertha, Asteroid U2753b, was now fairly common knowledge among world-leaders on Earth, and educated types, Iran was a much more powerful world power, by reason of religious philosophy common to billions of people. Global Sharia-Law, though not beholden to much more than one-third of Mankind, was a considerable impendence to the space program in the East, the so-called Russian-Islamic Space Alliance. By 2077, Iran had expanded its borders to swallow up many other regions, including Mesopotamia, the Tigris and

Euphraties, Babylon, the Cradle of Civilization. They were part of the space-travel community. By the time Lynn Rodgers-Smith was addressing a closed-session of the Council, the Easterners had by then launched their five ship to Mars, now confirmed by most sources. It was a violation of International Law, as usual for times of crisis, and Big Baby Bertha certainly qualified as a crisis in everyone's thoughts. His Excellency the Imam was busily accusing the US side, during the discussions, as a reply to defend the Easterner's actions with the space-launches to Mars. As absurd as it was, under Global Sharia, the interpretation was that the sins of the West had produced the Wrath of God in the form of the meteor. Lynn, the busty Texas woman, Commander of Angels, could only shrug him off as irrelevant. But the World-Council had to hear all sides.

"They're not quite sure if weightlessness is a sin or not, under Sharia," she confided privately to Ibrahim Mehudi, the Science-and-Technology guru for the Western program.
"The theologians are still studying that one. Nothing they do can ever be wrong or cruel," said Ibrahim. "Global-Sharia means their astronauts are also Saints to Islam. We are all inferior in their minds, so taking the Mars Base is their right and duty, by any means."

"I could give a fig," Mehudi replied. "We have far more serious problems than pork products on the Mars Base dinner menu, for God's sake."

The five deep-space cruisers were now two months out, at least, trailing Guy Reisling's ship by eight or nine weeks, on the passage to the Red Planet. It was a bit of a race, but ship's pilots could vary their engine speeds using the common hydrogen-thrusters, and also by changing course for shorter lines of travel, as the planets were moving. Because the Russian-Islamic launches were done in secret, and lacked the usual clearances and permissions, the World-Council was shortly to be consulted. It was a drab affair, also conducted out of the public-eye, with many meetings in musty-stale rooms full of angry people alienated by culture and goals. And whatever their conclusions, nothing would change anyway. It was a done-deal, the ships were on their way. No amount of argument and legality was likely to produce a radio-transmission to the Russian ship-pilots with orders to turn around and head home, which was not even a navigational maneuver that was possible with any safety, usually.

So it was a stand-off. Following the Russian launches, the West gathered their intelligence and data-facts, more-or-less proving the matter to world space-travel authorities, as far as what the Russian-Islamic program had done, which was to send five deep-space vessels aloft and on their way to Mars. Inter-planet diplomacy was now a reality back on Earth. There was nowhere else for those ships to go, and much like the laws of the high-seas, Earth-powers had rules and treaty-agreements about that sort of thing. The main reason was not philosophical, but instead was related to the success of orbital satellites and Earth-Moon commerce, a big business interest by this time in 2077. So they had broken laws, but everyone knew it was bullshit, a turd-in-transit, irreversible.

After they were called out on the deed, a predictably vague response from the East declared that the Russian-Islamic powers were ignorant of the launches, or the accuracy of the Western intelligence. Official denials pretended they would "look into it", and tossed out the bone that there may have been some rocketry in the hinterlands, that was somehow less than authorized. By the time the World-Council took sessions on the matter, it was too late, and of course the Easterners knew this. At the same time, the World-Council continued to discuss the approaching meteor.

"The asteroid can destroy half the world. If it hits, the impact might wipe out entire continents, billions of people. An ice-age winter of darkness will descend for 1,000 years. Any survival at all would be almost unbearably bleak," Ibrahim Mehudi testified to the Council, as one who supposedly knew. "From this day-

and-date today, here-and-now, we may be no less than four to six years away from such an event. That is the science and the truth as I know it."

And so did everyone else among them. Back at Vandenberg, as international diplomacy continued its useless and heated course (the meetings were held in Switzerland), there was a quick and inevitable decision to put into action the plans and preparations of the US Mars Base Defense Task Force. And like dropping a hat, this meant the US side would also launch ships, as a defensive response---also in secret. Communications, decisions and choices, like waves of gloriously dubious destiny, were sent back and forth, continent-to-continent, coast to coast, and planet to planet. Older generation leaders recognized the signs---it was starting to look a lot like war. Indeed, as things were now in motion to secure the Mars Base, leaders as high up as the US Presidential Seated Council, were nervous as hell that war would also break out here on Earth over the whole thing.

"A rock," commented US Presidential Council-Seat Mark Renolds, a wiry and spry white-male of about age 65-years, known for his freckled face and background as a farmer. "This entire ape-shit world-fucking crisis has been caused by a rock."

By the year 2077, the American White House and the President's Office, had been divided into a Presidential Seated Council of four individually-elected persons. So, in other words, America had finally made significant structural leadership style changes and Constitutional adaptations, thought to be for the good of all, and having occupied previous leadership for many years to accomplish. Elected in a rotating four-year cycle, such that at no time was any member of the Presidential Seat without at least two years of on-the-job experience, the group now included Renolds, a Southerner named Boline Bouvier (a Black man and educator from the University of Texas faculty), a Caucasian woman named Martha Hazlett, who at age 45 was basically a widely popular athlete (swimmer) with a law-degree,

and a youngish Hispanic man from a large farming family, also
with a law-degree and business background, named Martinez
Jeses-Garruero (age 38-years). Renolds and Bouvier were the
more senior members, and more mature.

"A very, very big rock, Mark," replied Reynold's frequent
foil, Bouvier. They held a meeting one day later, in the
traditional White House (still in use), during the heated fallout of
the information about the Russian space-launches. "But just a
rock, none-the-less." All four members of the Council-Seat were
present.

"It's not just Mars," added Martinez Jeses-Garruero, who had
a fertile and active mind, very keen on the practical, ever the best
of the utilitarian American character. "If they won't back down,
there may be retaliation or military response here on earth we'd
have to respond to as well."

"I don't see why," said Hazlett. It's only a base, just---an
outpost. If the East went into war-mode, what would be the
motivation? They'd still have to deal with Mars. It wouldn't
solve a thing."

"It never does, does it?" responded Bouvier.

"With the meteor---decisions are likely to be irrational. A
meaningless panic," said Renolds.

"Great," said Hazlett.

"Just tell Military to authorize Vandenberg's response, and
also I guess Texas, Florida and Puerto Rico, and any of the
others. We already have plans in place. Obviously, we have to
protect Mars, and avoid an East-West global war or crisis at the
same time. They sent five ships, we'll send ten," said Renolds,
who tended to take the lead. "Agreed?"

"Ten, or eight, or nine," said Bouvier. "Yes, agreed."

"Agreed," said Hazlett and Jeses-Garruero, almost in unison.

It was by then another morning, and the World Council
meetings about the Russian-Islamic space launches, and efforts
to deflect Big Baby Bertha, had now been underway for three or

four days in Switzerland. The Presidential Seated Council at the US White House met together in a large rotunda, complete with all the traditional US-White House décor and flourish. Washington, D.C., was typically muggy and warm, but the lilt and clouds that met the new day and hot sunlit dawn, reminded even these four world-leaders, as close and dear as any family, in their office, that life goes on, no matter what happens.

"Make it official, send down a directive. Let's get back to work on what's being done about this fucking rock out there in space that's going to kill us all in a few years," said Renolds.

Bouvier, the Black man, who was somewhat younger than Renolds at age 57-years, and had a very round, cherubic face, as dark-skinned as any Afrikaaner, leaned back in one of the deeply cushioned chairs they used. "Speak for yourself," he said. "I choose to survive."

And so it went on, in the halls of power and empire-Earth, that year in 2077, once again with an impending war and disaster. The lighted paths of hope for the future were dimming, closing like the iris of a camera-lens, or as the iris of the eye shrinks in light, or widens in the dark. None could predict the outcome. But if the meteor was a certainty to strike the Earth, be it the Wrath of God or no, the base on Mars, and other outposts on the Earth's moon, and Molinari, and the ships-in-space themselves, and Earth-orbiting life-support stations, could be vastly important, perhaps even the only surviving advanced helps for whatever remained. If it took 100 years, following a devastating meteor strike like BBB, some sort of sustained extrinsic presence, with the technology and knowledge retained in their computer libraries, could potentially begin to rebuild a civilized Earth, the only home mankind had ever had. Thus, a somewhat passionate dilemma.

Likewise, and not easily dismissed, away in the Southern Ukraine, in the chilly crags and rocky woods of Spaceport KK-F/Region Six, Rudolph Terchenko, the Russian Commander of the launch-site and secret facilities there, well knew and understood, that the fallout of their decision to launch the five Mars-bound ships, would pester and boil for a long while, as everything else was happening. The Region-Six 'think tank' was on full-power now, and Terchenko found himself oddly strengthened and even delighted at the chaos. His personal assistant, Milana, the gorgeously youthful and also quite educated young woman from Saint Petersberg, could not help but admire the older man's vigorous thrill and rush of power, as he handled all the various difficulties. She was secretly in love with him, as a less-wealthy and lesser-stationed common Russian girl who had found her way into a cool job. Rudolph Terchenko was big and bold, but also at least somewhat kind-hearted, and wise.

"Are we really all going to die, Commander, if the meteor hits us?" she asked him.

"There is no death, girl," Rudolph replied. They were in his private office. Milana had somehow gathered many reams of paper data-files showing every known radio, telephone, TV, Internet, and print communication or research, or email transmissions, radio-logs from satellites, and even collected recordings from 'bugs' placed in various embassies and certain hotel rooms in Switzerland, with reference to the crisis. The two of them had now been hours at work going over each of them, to learn what was known, or supposed, as to how the 'real' actions of the Eastern space-program had filtered out into the un-real world of mass-communications, and governmental sources.

"Please, Commander Rudolph," Milana said. "People die almost every day."

"That is their choice. They are---or, we all are---30 billion suicides."

"30-billion suicides? 30-billion---people? 30-billion people who committed suicide? You are joking, Commander. The population of the whole world is only---"

"There is no death, girl," Terchenko re-affirmed. "Only

change. Don't be a pest. These radio logs are useless, I can't even read them. What are you good for then, you tart? Just put those down on the table, and run out for my dinner if you would, please. I am not a philosopher who can comfort your fears of death. You'll be fine. There is no death."

She did as she was told.

CHAPTER 18: Mars Cargo Transport 'Penelope'

*"Yes, now I can remember. It was---great pain. Yes. Please don't remind me, you know?"*

*--Charles Lindbergh, pilot of the world's first trans-Atlantic prop-aircraft flight, commenting on his fame, and the later kidnapping and slaying of his child*

The US Mars Space Program launched eight deep-space cruiser vessels, similar to the 'Penelope' (Guy Reisling's transport), within three weeks of the closure of the World-Council sessions to decide on a course of action concerning the Russian-Islamic launches. Like the Easterners, the US launched their ships in 'secret', but the reality was that there was no real way to keep such things secret for long. The reason was that even at the amateur level, orbiting satellites and space-objects like meteors, were observed daily by a hundred different astronomers, meteorological societies, universities and students, news-media, governments, and so on. If large ships were hoisted into orbit, and then programmed for navigation and departure to the far-off red star, glowing brightly enough that even children could spot her (or, 'him', being thought as Aries, the Ram, or god of War, in mythology)---it was hardly something that Earth-dwellers could conceal in 2077.

Also, like the Russians, all eight ships from the US program were launched in a sequence, carefully planned and executed by then-familiar routine, over about two weeks, from several different launch-sites, some of them unknown to the public. The Mars Base Defense Task Force had done their jobs well, and following the Russian action, it was a no-brainer the

US and West would respond. But it wasn't popular, and it was a huge cost, and an extreme effort. In 20 years of active Earth-born travel to Mars, the program had only launched this many ships all at once, during construction periods, when large numbers of men, and large amounts of materials and gear, were needed over a relatively short time-period.

Each of the US space-ships was outfitted for the mission: to protect the Mars Base from the supposed Russian takeover attempt, still not confirmed or announced as an 'act of war'. But it wasn't doubted that this was what was ahead. So, the US ships were full of soldiers, of course, equipped for exactly what the planners felt would be needed to defend the Mars Base. The same was true for the Russians, but perhaps not exactly the case for either team, that these were 'warriors', something Earth-based space-travel had never seen in all of Earth-history, and thus 'new' or 'different'. The planners didn't quite know how to deal with the concept, as Winton Berle had tried to explain to the Task Force. Conditions in space or on Mars were so hostile, that the suits and mobility were simply not accommodating for men who might wish to make war, or bash each other's heads in. It seemed more that things would happen like a slow-motion dance of astronauts or moon-walkers, with very determined intentions, who on either side might have to hurt each other to reach the goals of the powers involved. Certainly true enough on the ground level, or Mars-surface. It was true, and would probably never really be much otherwise, that the space-men were such an elite and highly trained group, that even as 'enemies', any of them were be loathe to kill one another. But, true-to-form, each side had real weapons that would really kill, and were trained to use them---and would use them.

It wasn't hard to calculate the number of men on the ships, since the types of ships could only hold so many bodies. So,

Winton Berle and other planners could realistically assume the Russian-Islamic forces to number about 200 or 220 'space-soldiers', not all of them fighters, but many with very specific high-tech roles, because the job was so complex. With eight ships, and also the men at the Mars Base already, the US side could count on as many as 300 men or more. So, this advantage was fortunate. But on the other hand, the Russians might send more ships or more men, if the struggle dragged on for many months, and they might have some sort of 'secret weapon', or method---bombs, for instance, targeted at the Mars Base from orbit. Things could go either way in such a thing as a military assault on a Martian research outpost, in space ships from Earth filled with very determined Russians, fearful of the End of the World, (Earth-world, that is).

There in the Abyss, Guy Reisling had often taken more than one critical communication from Vandenberg, his link to home. Strategically, his ship's position ahead of the others could be an important edge. The journey to Mars would take a total of eight months, for most of the ships, varying somewhat as the planets moved in their orbits. It was now, as of mid-year 2077, almost an armada: eight ships from the US-side, five from Russia, and Guy's ship, for a total of 14. And the Russians could launch more at any time they were able. So the Vandenberg mission commanders wanted Guy to be ready to handle whatever came up.

There on-board the Penelope, it was business-as-usual as far as the flight. She was only six weeks from Molinari, something that would mean a rest and refreshment, and time for Guy with Lila. They really did care for each other, and it was a joyous thing for them both, sort of a romp, but also wistfully romantic, 'love'. Now, with the hostilities, and the meteor, it all began to fray and heave, for them all. Guy thought of it as a sort of deep-space version of an old World-War-2 film, like Humphrey Bogart in 'Casablanca'.

"If that deep-space transport leaves space dock, and you're not on it, you'll regret it. Maybe not now, maybe not tomorrow. But soon, and for the rest of your life," he mused privately. "Here's looking at you, kid."

For most radio-transmissions to Earth, or elsewhere, ship-to-shore as it were, Guy could use the basic radio set-up from the pilot's deck on the Penelope. Radio-transmissions, or EMP (electro-magnetic pulse) deep-space near-Earth solar-system communication, were actually quite speedy. Under most circumstances, the radio-waves were delayed significantly enough to create a means of back-and-forth chatter, where one side would talk, and the other would wait as long as 15 minutes or so, to answer. It was reliable, and engineers had learned how to compress the signals and transmissions, so the time-lapse was smaller, and chatter was easier, more normal. Guy's ship, like the others, had an array of antennas, and other gear, that were essential to the function of the ship. So for three weeks, transmissions from Earth to Guy's ship, the Molinari space dock, and the Mars Base, were flying the language fast-and-furious.

*It's just a load of tech-gear for their damn Mars-radio system*, Guy thought. *I don't have soldiers and I don't have weapons. These guys will figure I'm General Patton from now on with this.*

So once again, the Captain of the Penelope found himself sending back-and-forth messages of shattering importance, to Vandenberg planners. His immediate superior, Commander Okman, the head of the transport division, handled many of these links, as they were his men and crews, and he knew them best. As the signal reached his ship, scooped from the emptiness by the Penelope's antennas, an alert-sound notified himself (Guy), or Rob Cowan, his second, or other crew. With the time-delay for the transfer, they could often take their time to get ready to 'chat' with home base. From his seat at the helm, the radio-desk looked like a fancy video game. The sound-monitor was sometimes thick with static-noises, and at other times crystal-clear.

"Yes, Commander, I understand," Guy was saying. "No contact whatever with the Russian ships. Radio-silence until further notice. That's not hard, but of course you realize it's easy for them to monitor our communications anyway."

"That's not the point, Guy," said Okman. "I have very specific instructions from Berle, all the way up to Rodgers-Smith. They do NOT want you to chat things up with the Russians, or handle things on your own. It could create even worse problems. So anything we tell them must be cleared, or we say nothing at all, and if they pick up our chatter, big deal, they will anyway."

"Some of our systems will encode messages. What about that?"

"I'll find out. The lead ship on the Russian side is months behind you, some half-million miles or so. The military stuff will doubtless be encoded for radio-security, like any battle," Okman said.

This series of comments had been compressed so their language seemed to flow naturally. Now there was a pause, with static-noises, like a background cosmic hum-and-glow from space itself. After ten minutes or so, the alert-sound 'beeped' again, with a glaring red-light LCD on his control-dash.

"In-coming from Earth," said Cowan, also on the flight-helm

at that hour. He was curious as to the details where the Russian ships were concerned, too, and needed to know. "Second series. Vandenberg. Okman."

Guy activated their side of the system, glancing at his co-pilot. Rob seemed edgy, even un-well, but he supposed it was only normal stress. They were now 78-days in space. Long haul.

"Vandenberg, US Mars Radio Comm, Okman speaking. For US Mars transport Penelope, Captain Reisling. Please respond."

"We have your signal, Okman. Go ahead."

A long pause. "Three other directives for you to handle, Guy. This is from Command, so please document for later review, comply and implement. Number one: the Penelope will continue to Molinari, and onto Mars Base, as per previous flight-plan, with no variations because of hostilities. The reason is, your transport carries essential communications gear. This is your first priority. Number two: please make preparations for any negative ship-to-ship interactions, should they arise. This basically means you need your ship navigators and life-support, and tech, to figure out how to lock you down for any disruptive encounters. They probably won't try to shoot you down. But they might fuck with you, somehow, as the gap closes. So you need to make preparations to protect your ship. Number three: as of now, and upon arrival at Mars, the Penelope is under the command of Winton Berle, the lead for the eight US ships, also on their way. So, there will be codes and data to connect to his command. But by the time you reach Mars, and maybe before, you'll be drafted into whatever happens, along with your crew. We need all hands. That's a ways off, and transport protocol will handle your cargo mission as always. So, that's the main commands from Vandenberg for the Penelope with this transmission."

A long pause again. The information and recorded voice-signal was downloaded to storage. "Information received and logged, Commander Okman. Uh, Commander, did you say Fleet Commander Winton Berle was actually on-board the lead US ships, sir?"

"Acknowledged," came back Okman's voice, over half a million miles distant, in the comfort of the Vandenberg communications center.

"Commander Okman, this is co-pilot Rob Cowan. Questions, please?"

"Go ahead."

"What does US Mars Earth think will happen on Mars? How long are we expected to hold out? Our original mission would have us in orbit on Mars only for a month or so, at most."

A long pause-delay. Some of the voice-signal from Okman sounded like a fish in a jar full of royal jelly. "We don't know, Cowan. You have the basic info. Berle's armada is behind the Russians by two months, or more. The Penelope will be in orbit way ahead. You guys get to be on the welcoming committee, along with Snikta-Ridge."

"That's not funny, Okman. We don't even have a missile or a gun on this thing to shoot back with," said Guy.

"You may be asked to delay, or negotiate, from what I hear. Same with Snikta-Ridge. Who knows? Maybe things will change. But, to your question, Rob---you won't be coming home on-schedule. You should definitely get any supplies or fuel-resources, or re-charge, that you need from Molinari to hold out as long as possible, not to mention your ride back to Earth. Mars will also be able to re-supply the Penelope to an extent. It's just a gamble. Try to be prepared for anything, is all I can say. Obviously we're not going to let you waste away out there. We'll get you home."

Another long pause. The radio comm system seemed to boil a bit with a strong static-energy for a moment, fluttering, an electronic wind. Both Guy and Rob tried to adjust for the error. After a few minutes, the time-compressed conversation continued.

"Mars-transport Penelope, signal to US-Command Vandenberg, Please respond. Reisling here, over to Okman, re-connecting to you now. Okman? Hello?"

More static and a pause.

"Okman here. I'm logging off, Reisling. The guys are telling me the damn planet moved and our signal is screwed. Sorry. Next window for your signal is---wait a minute---next communications-window in three hours, I guess. It's three in the morning here, Guy. Just hang tight. Okman out."

Now the system shut down the link, with various alerts on their control-dash.

"Welcome to the world's first space-war," said Cowan. Guy took a deep breath, leaning back. They both quickly reviewed their flight-controls and ship's systems. All was well, on-course to Molinari space dock. Without extensive and time-consuming scanning, they really had no evidence the Russians were even out there, or the position of their ships. This was common---deep-space flight was a lot like a very long trip in an airtight elevator-lift, or a windowless high-rise office complex. There wasn't a lot to see.

"All right," said Guy. "Log all that bullshit and we'll inform the crew."

"Sure," said Cowan.

Somehow, the joy of space flight had taken a hit in their hearts they might never recover from. It just wasn't fun anymore.

## CHAPTER 19: Rob Cowan

On a small ranch in Montana, about year 2030, a banker-accountant and his wife, who worked the property with small crops and farm animals, had a boy-child they named Robert, after his grandfather, Robert Laverne Cowan. So Guy Reisling's co-pilot (Rob), was actually named Robert Laverne Cowan, II. The eternal Earth never really changes much, and Montana was yet by the time Rob was birthed, a mountainous wilderness-type area, fit for cowboys and folks with horses or who enjoyed fishing, or tractors, wood-splitting and 'critters'. Rob had three other siblings, two girls and another boy (younger than himself). They were quite a bunch there, secluded and safe from the ways of the world, just as their parents wanted, for life on Earth in 2030 was ever-the-same as now; a wild and half-mad scramble of humanity for wealth-power, survival, treasures of lust and pleasure, and yes, love-compassion. The privileges of natural-living and simple work, and the bliss of beatific-vision childhood, wandering around having fun, or facing coyotes and raccoons, gave Rob a strength and intelligence that stayed with him all his life, his strongest-best self. And this while many co-passengers on Spaceship Earth were grown under far inferior circumstances, even grinding poverty, starvation, war and terrorism, Rob learned all he could at the best schools, and with athletics, and at university. A background like this gave him and all the space-program workers and astronauts an excellence that was much needed, and even a long-term hope for mankind, that space-research could eventually improve things for all. Yet, no one was ever really that strong, and 'the moon is a harsh mistress'.

As a grown man, Rob Cowan was tallish, lanky, and a bit hairy, with a sort of sunken-chest full of hair, bony, but strong.

He had an under-chin, or an 'under-bite' that made him seem humorous or somehow chummy and funny, and he loved to laugh, and joke, at almost anything, with his big hearty chortle. The arts and music, poetry, literature, and so on, were not lost on him, but after his youthful years he abandoned the intellectual-side, in favor of the call to service in military, air-force, and then the space-program. But deep inside, Rob had a Bohemian animal-nature, that made him both tough, smart, and a dreamer. He had a thatch of dark hair, and sometimes a fast-growing beard. His arms were like tight wires of fleshy-muscle, and he could swing a large hammer to knock down a wall, for many hours at a time. As he entered service to military and air-force, this side of his personality was 'cleaned up', and Rob was re-invented as the 'ideal astronaut', because there were no 'fucked-up' astronauts in the program, with personal issues serious enough to endanger others by virtue of the stress and responsibilities of space-travel.

Within the US Mars space-program, Rob was enough of a veteran to co-pilot for Guy Reisling ('Oh, Captain! My Captain!'), and he did a good job, very detailed, cautious, and skilled. They were friends, and enjoyed a lot of rowdy time together off-duty, such as at bars near Santa Barbara, south of Vandenberg, and then at times for barbecues at Guy's home nearby. Rob kept his home on the Montana ranch, so it was not such as a daily closeness. They kept private jokes and views, vaguely rebellious as all good astronauts are, more-or-less at odds with the government, and figuring philosophical about the world and life, into the small hours of the night over a brandy or bud of quality marijuana, which by 2070 was legal for personal use in California, with limitations. But the program frowned on any drug use, for the enlightenments of recreation, and they were mostly very limited adventures for any workers in the program, and if not, they were quickly found out and dismissed or rebuked, for the good of all concerned. Laddish ways, as the British say, the astronauts needed those joys and romps, to keep their souls from withering and dry-death in the monotony of their jobs.

By the time the Penelope had launched with her cargo for transport to Mars, in 2076, it was a given that Rob would co-pilot, along with the rest of Guy's regular crew, and a few changes. Guy's Condrum 21 Deep-Space Local Planetary Cruiser from Monsanto-DuPont was like a temporary home for all of them and they knew her well. Similar to small aircraft, the ship needed to be more or less mechanically perfect---and not less. Failures in deep space would kill them all quickly, sad-but-true, it was no joy ride or cruise-ship, no walk-in-the-park, not a picnic. And they knew it and accepted the risk. The pre-launch also included a work-up on each crewmember, by now routine with each ship, but still a requirement. So for a week or more prior to launch, aside from flight-plans and cargo, the men were examined for health. This took place at the Vandenberg base, where Penelope would launch from (or crew and navigations, with the ship herself already in orbit).

Rob's turn came, like the others. He was feeling fine, in general, and anxious to get 'back-to-work'. But there was a problem, which he had been working through with various doctors. It seemed minor: Rob was suffering from the loss of a testicle due to athletic-stress. It wasn't cancerous, but urologists told him there was a blood-flow problem, leading to swelling, leading to the loss of the organ, which was easily removed by simple surgery, and then healed for cosmetic and sexual function, with on-going therapy and prosthetics. It had been two years since these procedures, and Rob was quite fit overall, even in the strength of his groin-muscles. (Rob was married at this time to his second wife and they enjoyed normal, vigorous sex, with two children).

The physician attending the examination looked over Rob's file, as the pre-launch prep-period went ahead. The examination room was typically cold and somewhat sterile, with white walls, green curtains, a few monitors and tools, an examination-bed. There were much more sophisticated medical diagnostic-gear nearby, and Vandenberg had a very complete hospital. Rob was in dress-down half-robed, having been probed a bit. The nurses found him delightful.

On the topic of the testicle loss, the doctor wanted to be clear. "There seems to be no real problem at this time, Rob," he said. "I know you feel good, and strong, too---and you are. With your right testicle, it's basically healed from the surgery to remove it. But you are taking the on-going pills for anti-septic, or anti-biotic, is that right?"

"Yes," Rob said. "There was some pain, and then a minor infection. Not the testicle---I mean, there is no testicle, but the sack and remaining vesicles. It had moved up into the lower bowel somehow, and was sensitive and soft, like a hernia. But it was very small, and then reduced. So the urologist is using the anti-septic bacterial pills, to avoid any further problem. For right now, it's fine."

"How long ago did that appear?"

"Uh---this was, now, I guess---nine months. Back---last year. There was the swelling, pain, not that bad. The doctor refrained from more surgery, said it was normal. The antibiotics since---four months ago."

"Yeah, that's fairly normal. The same anti-biotic series we use now also prevent a wide variety of other problems---flu, cold, diabetes, inflammation, and angina. They are very advanced. But I'm not sure the antibiotics will function in total harmlessness in deep space, you know? Do you have any other symptoms?"

"No," he lied. "Not really."

The physician waited a moment, pondering. "You have a few days before the launch. I am going to spend some time and look at the medications, and your blood-work, and other tests, and compare with previous space-flight records of other astronauts on antibiotics. It will take a day or so. Okay?"

"Sure," Rob said. "I think it will be fine. I mean---what could happen?"

"Just let me look into it," the doctor said. "I know you love to fly. It could---it might mess you up. The stress and null-gravity, the food, the radiation, and the other chemicals we use to help you on the space flight---it's a mess, if the antibiotics conflict. It might not be wise."

"Sure, all right," Rob answered. "When will I know?"

"Within a day. Before launch, with a window to schedule a replacement co-pilot, if needed. I'll also inform the launch-command, as per protocol."

And that was that. By this time in the year 2076-77, antibiotics had advanced significantly beyond the old days of penicillin or other types. For some, antibiotics were in-use for years at a time as a daily health-matter, and they prevented more than just colds and flu. Other doctors used these more sparingly, feeling they altered the body's natural-immune systems. AIDS (Acquired Immune-Deficiency Syndrome), the scourge of the late 20th-Century, and the many millions of deaths attributed to that disease, had been conquered, and was quite treatable and survivable, even in Africa. The completion of the Human Genome-Project, and developments and advances in medical applications based on the complete detailed mapping of the human DNA-map, meant that a healthy person like Rob, could anticipate a very high-quality of life as far as longevity and overall health. In Rob's case, with a potential infection and inflamed or painful testicular removal, and the swelling or minor surgery, the new antibiotics were par-for-the-course. But, in the deep-space environment, his doctor simply wanted to review any known or foreseeable effects that might come up. These could include dizziness, disorientation, and even psychosis, which sometimes happened to some workers in space.

For Rob Cowan, it was a cautionary note that he knew well he needed to be attentive to personally. That haunting dread a person sometimes feels, when the edge of sanity or the depths of organic failure start to loom large, planted its seed---but he also knew how to quickly up-root his worst fears. In space-travel, the mental forms were peculiar, to say the least, and a dreamer, or religious person, could begin to experience euphoria, or fugue-states. Rob's doctor also was aware, and behind the mask, Rob simply didn't want to lose his job, or be denied the position he had worked so hard to attain, as an astronaut, a pioneer and even a 'hero'.

By the time they launched for Mars, the Vandenberg-base doctors cleared Rob's health and his medical use of the genetic antibiotics. Other astronauts were using the same chemicals in space without any trouble, and the various parameters were reviewed, along with any other concerns about his testicle-loss. "Just forget about it, "Guy told him in private. "If you have a problem, don't keep it a secret. I'll watch your back. You'll be fine."

"Thanks, boss," Rob told him. They really were close friends, but work was work, and space is no place to screw around with those kind of problems for a person in a position of responsibility and stress.

In a dream, Rob was back in Montana, in a heavy downpour of rain, there on the family ranch. He was a child, in the dream, and the rain was so heavy his parents and family were scared. The horses had to be led to their stalls and barns, and other creatures, and the windows of the house rattled with wet, and the sound on the roof above them was like a freight train. Then somehow, walking down a hall, it grew dark, and 11-year old Rob, just a boy, entered the realm of his worst fears, like a miasma of vortex-powers, swirling around, no longer simply another room in his parent's large ranch-house, no longer simply another storm in the mountains, no longer the simple fears of mother-nature and broken tree-limbs that crushed cars, or animals swept away in flash-flooded gullies, or sheds that fell apart, or muddy drives that had to be cleaned again for regular use in the sunshine. Those fears would have been comfort, natural fears, for they were strong folks and well prepared. In their place---a personal nightmare of un-reality, for which he also needed to be strong. Only a dream, one night, but with a message from his sub-conscious. "Beware of thoughts that linger," the dream told him.

## CHAPTER 20: Up or Down

On their flight in 2077, that year with the conflict over the Mars Base, Rob Cowan and Guy, in charge of the Penelope and her cargo, anticipated upon departure, no conflict, war, or hostile Russian astronauts. Both of them were experienced enough to realize that yet another passage to the Mars-planet, would hold various demanding duties, challenging, and dynamic challenges, and dynamic circumstances. It was always that way, and there was nothing new they felt might happen, although the politics back home regarding Asteroid U2753b, were present to their thoughts like a looming dread. And that was really nothing new by the year 2077, either. So when they launched, circumstances were mostly 'normal', like an infant-child setting forth into life itself, into the womb of the cosmic Mother. In this case a space ship, just like in the movies. But the child is un-troubled, peaceful, not crying.

Life on board the ship, during the passage, was boring, and this was far preferred. Each crewmember had a set of duties allotted in regular 24-hour periods. Each knew his function, and was highly motivated, trained and ready, if only for the survival of them all, in the most hostile of all possible environments, that of outer space. The ship's interior was Spartan and efficient. But they kept themselves emotionally happy and positive with games, jokes, decorations, exercise and minor recreations. The all male crew mostly refrained from sex, and the passage was almost a year in time, roughly nine or eleven months, again, variable according to the position of the planets, and their relative speed. None of the men on the Penelope engaged in homosexual activity, and they were all as fit as any athletes.

So the big attractions as far as enjoyment or pleasures were meals, music and 'magazines', or other reading and recordings.

Like an Earth-sea cargo ship, the crew was provided with the best of everything, in each category.

The Penelope was much like a large Earth ocean cargo ship, in both size and character. It was quite large; easily five or eight times the size of the 1990's NASA space shuttle, and never intended for atmospheric re-entry at all. It was almost like a super-size space shuttle. So, it was easy to build these ships, or easier, being constructed in weightlessness. The Monsanto-Dupont Local Planetary Cruiser (the Condrum 21) was very massive and strong, and operated for its entire lifetime in null-gravity. Although space was airless and void of life, with extremes of cold and heat, it was a low-impact environment as far as weight-stress on the high-stress 'new' metals and alloys, and material surfaces. As the engineers had learned, it was just as easy to move 20-million tons, as to move half-a-thousand, in null-gravity. Throw-weights were essentially equally mass heavy, though overall density, or total mass-weight, would affect the rocketry. With all the science and technology, under control of routine systems and established practices, life on-board the Penelope was simpler and more comfortable for the crew, than a non-astronaut may have imagined. But it was no picnic.

Rob Cowan's main function was that of back-up commander. Standard for any such flying machine since the days of the jet-aircraft, it only meant that Rob would take over if Guy were disabled for any reason. They all could actually pilot the ship, any one of the crew, even the Life-Sustain Specialists, who worked only on life-support systems during the voyages (including the toilets). This only made sense. But for Robert Lavern Cowan, II, he was required to second for Guy Reisling during each 24-hour period, in-shifts, and was basically second-in-command, for the entire flight.

What were those duties? In planetary orbit, entering orbit, docking to passenger-shuttles and moving people from shuttle-to-ship, or at Molinari, the mid-way space dock, and also when leaving orbit and tracking early navigation for long-haul passage, the tasks were very demanding and came in fast-paced sequence. Errors or flawed moments of decision, were critical, so these maneuvers could be intense, much like a large rocket 'lift-off' into orbit from the planet-surface, but not quite.

The two men, and other crew, operated like a precision-team, pushing buttons, monitoring engines, communications, handling momentary decisions based on current flight-data, adjustments and off-ship controls or instructions from Vandenberg. They were in charge of an astonishingly powerful machine, the first of its kind for human travel to another planet, local to the Earth's solar system. Each step for each procedure was scheduled and well mapped, so to accomplish orbital re-entry, for example, or to leave orbit on the proper track to Mars, with all the delicate navigations involved. It was like a sublime dance---man, mind and machine, and 5,000-years of science advances. Rob and Guy only did their jobs as well as possible, and would sigh with relief when each action was successful. Failure was not an option, but sudden death and great loss, were an ever-present factor. Such was space-travel in 2077, and probably would ever-be.

Rob's first shift was from 0300 to 1000 hours, or seven hours. Guy typically retired, after a brief conference. His first role was to check and re-check all the ship's most essential functions, from the pilot's deck, or helm, which had lines to all the data-flow monitors that kept track of each vital part of the entire machine. This took at least an hour. The ship had four main engines, using the hydrogen fuel for much of the flight. It had been established early in the space-program, that by tuning the engines to a consistent thrust-energy emission, the vessel would arrive on Mars at a certain date and hour. To keep this going, because the energy levels were intense, the engines had to be carefully maintained and cycled through cooling and cleaning, and then re-started. Working with the ship's engineer, who coordinated his hours for that routine job, Rob, or Guy, would work through the needed 'off's and on's', the shut-down sequence, the re-start, and also the 'rest' period, when the engine up-keep was tended. This work could take almost the entire seven hours of Rob's shift, and was somewhat tedious.

There were other mundane duties. Once in the inter-planet corridor, the flight-path was simple and rarely changed at all. So, there was no 'steering' involved, or loop-de-loops. On a good night, Rob, or Guy, on their respective shifts, could do the essential tasks, and kickback while the machines did the rest of the work. Stargazing was a favorite past time. On a bad shift, well, you never knew what would happen, or how demanding it would needs-be to save their lives.

Rob could recall some of the flights he had made with Guy, and how things had happened. If there was any structural-integrity violation of the ship's hull, such as they might encounter with a shower of tiny meteor stones the size of marbles, or smaller, it could mean many hours or even days of in-flight repairs or damage-control. Obviously they could lose life-sustain very quickly with an outer-breach. Or, with the data-antenna and communications, and also solar-energy collection panels, and other external apparatus, if there were a problem they recognized, it would call for a space-walk, and all the tools and know-how needed, to get the systems up again.

On one flight, three years past, such a thing happened that Rob would never forget. It was an antenna-array part that had somehow vibrated loose, but it was too important for them to ignore.

Communications were compromised with Earth, Molinari and Mars, for a short time, meaning they could easily get in trouble, especially with tracking and Molinari's up-dates on Earth-Mars corridor conditions, such as solar-flares, meteor-showers, other ships, planetary movements or the unknown and unforeseen. So a space-walk was needed to repair the part.

Given the relative speed of the Penelope, it was just as easy to conduct a space-walk with the ship in motion, as it would ever have been to drop-speed to a dead stop. But it was somewhat unnerving. The ship was moving faster than a bullet from a gun, perhaps 1,000-miles an hour, approximately, for reference.

Rob was the most experienced at that time, as far as space-walks. He and another man, the communications-expert on that flight, named Peter Frigalle, would do the work. After reviewing the procedures, they suited in their outer-void suits, the same type of heavy, cumbersome and uncomfortable life-suits the program had used for generations; 'old reliable', never-failed, and also improved somewhat over many years.

"A pain in the ass," Rob commented, working to prepare in the 'ready-room' for external space-walks, and other purposes. They also had to get the tools they needed into the kits, with tether-lines for everything. Peter grinned beneath is mask, a light-reflective orb that hid his features. He was a fine-looking astronaut, now like a faceless droid intending some fantastic effort that would involve shielding both their faces from the Sun in its raw-glory.

"Mine's fine," said Peter, linked by an inner-radio to both men and the helm of the ship, where Guy would monitor the excursion, along with another crewmember. At this point, on that particular journey, which Rob recalled was about 2073, they were between Molinari and the planet Mars, closer to Mars, but still two months out from Mars-orbit. The situation was not life threatening. But the void is the void. For a space walker, it was mind-blowing.

Into the air lock, the door to the inner-ship closes. Hisses and whispers of air-pressure as the atmosphere was equalized, or created to vacuum. Within moments, Peter and Rob were in a totally airless space. Seals and locks were checked, and then the outer-door opened.

"Watch your step," said Peter. "It's a doozy."

As Buddhist philosophers have observed concerning the reality of being-ness in the space-time continuum, there is no real sense in looking for 'up-or-down' markers, in the depth of the void. Infinity below, infinity above, infinity in every direction, endless, meaningless. Like on the Martian surface, the Sun was now much more distant, at that position, but there is no ultra-ray atmosphere filtering you would have on Earth, or even on Mars, so it was alternately very hot, about 130-degrees, on the sunny side of the street, and rather cold, about 20-degrees below, on the shadow side. Indigo, the Mother Night, held its wonders as well, the stars, and among them the familiar planets, both home (Earth), and Mars, distant twinkles, Mars now closer, reddish. But the men could not enjoy the view at all. They had a job to do. Dreaming was their enemy, or the heady rush of emotions.

"You got it, you got it," said Guy, from the helm. He could follow their actions with a simple external viewer-camera, rigged easily for the situation. "You're drifting aft, towards the engines. Use the suit-thrusters, and tether to the fourth-tier array binding cells. Each cell has holder-hooks."

The Penelope's main-engines were turned off, but it didn't mean they weren't in motion. If the engines had been left burning, a fried astronaut or two could certainly happen by some accident.

For just a moment, maybe as in the life of any astronaut, Rob had a giddy sort of self-awareness, wondering, inwardly, what the heck am I doing here? Local boy makes good, he thought, but it's a long way from Montana.

So, it was this sort of event that made each voyage unique. They repaired the antenna, did the needed work, and then found their way back into the ship's safety, back inside. The systems were checked and re-checked, and the communications link was re-booted. The ship's engines were successfully re-started, and they reached Mars with no other problems. Their cargo was mostly supplies. But the space-walks, and other sudden terrors, were not for sissies, there in space.

As his shift for that period ended, Rob was feeling good about himself, and the flight, and even the news about the Russians. Whatever happened with that would happen on its own. Their part was small. It was only by a fluke that the Penelope was now the lead-ship, on approach to Mars, during the apparent Russian-Islamic-Eastern Ukrainian incursion. The meteor was a 'what-if' as far as he was concerned. They'd figure it out. They had a lot going on for that sort, in 2077. Big Baby Bertha was being tracked constantly, and was still maybe four years out; he hadn't reviewed the current day-date-hour-minute of the End of the World. It could be diverted from a collision with Earth, maybe even fairly easily, or maybe it would miss. They had time.

Guy soon appeared to take over, and they exchanged formalities.

"Anything on the bad guys?" Guy inquired.

"The program is keeping track. There was an up-date at 0830-hours. Nothing new. We're a month ahead of their lead ship. There are five ships. Eight US ships are two months behind them, roughly, maybe more. Winton Berle is commander of the lead US-ship and the US-group. We're under orders to do nothing. Our mission hasn't changed. I guess the only new item is that there was a big conference in Europe about the meteor and the space-program in general, three days ago, or about that," reported Rob.

Guy took over the pilot's helm, and Rob retired until his next duty.

## CHAPTER 21: The Krenika Draws Near

*"Damn these fuckers are stupid!"*
*--US Mars-program transport vessel pilot Guy Reisling,*
*aboard, referring to his counterparts in the Russian-Islamic*
*space-program*

If you can picture in your mind, three concentric circles, on a basically equal level with each other, or a plane: at the very center is the Sun, our solar-system's star. Disregarding the other planets, you place the Earth, and out from that is Mars. Now put them in motion, or orbiting heavenly bodies, at about their respective positions. Sol is at the center, two steps out is Earth-orbit, and one step out from that is Mars-orbit. Each orbit is a circle or ring, and the planets travel on that track. At the time of the US Mars Base takeover crisis in 2077, the planet Earth and planet Mars were of course in motion, and the speed of each was tracked by pre-launch navigators, so ships from Earth could transit to Mars with relative certainty. The distances are vast, not calculated in a straight-line, circle-to-circle, for obvious reasons.

Now try to conceive the space ships that had launched to Mars, from Earth. Guy's ship, the Penelope, was launched, or left Earth-orbit, at such-and-such a point in time-space; the five Russian-Islamic ships left at a second point, over ten days; a few weeks later, eight US ships, also in a series of launches, at a third point-in-time, even later. Think of these ships like a migration of large birds, or huge mechanical butterflies, a string of silver-wings and blasters, heading outward, or downwards and across, in an arch, separated by hundreds of thousands of miles, from group-to-group.

Each team traveled in a regular formation, also at a distance of many hundreds of miles at least, from ship-to-ship, not even within line-of-sight, for much of the travel.

Somewhere ahead, un-seen, was the planet Mars, always in their thoughts. The Molinari Space Dock station was there in the Earth-Mars corridor, also in orbit. If there had not been a war, the Russian teams might typically have planned to stop and rest at Molinari. The Eastern space-program had certainly been to the Mars Base before, and Molinari, on numerous trips in the past. They were at peace then, and like most space exploration, the shared facilities used for essential functions, were considered 'open research/free-access', based on mutual agreements. Those same agreements were now what the Russian-Islamic program leaders pointed to, when confronted by Earth government and global interest, about their intentions with these launches in 2077.

With the sessions and meetings in Europe on the question, news of the approach of BBB, was now known. And each side 'knew what they knew'. But over 100 years of international space-program development (1977 to 2077), the Russian-Islamic program leaders knew they had the 'right' to travel to Mars, under previous treaties and agreements. These established the basics of the programs as education and research, since no one really 'owns' Mars, space itself, or the Earth. The fact that so many ships, 14 in all, were headed for Mars at the same time, was very unusual. For those 'in the know', it was no real secret: they were headed for a war to control the Mars Base.

None of this was lost on the 260 or so people living on Mars at that time, feeling much like sitting ducks at a shooting gallery, without much else to do but prepare.

Perhaps 250,000 miles behind Guy's ship, the lead vessel carrying the Russian teams, was also propelled through the abyss at high speeds by hydrogen-fuel engines. The Russian ships were very similar to the US Mars ships, in design. Earth-science technology had reached a certain point of advancement, which was successful, practical and realistic. But they were not identical, and had different features. In general, the Eastern-block space program ships were even less comfortable than the US ships. They spent less money, cut corners by simplifying various functions, and trained the astronauts to cope. They were about the same size and equally fast, with similar cargo capacities, fuel-limits, sustainable life-systems, navigation-standards, and communications. By standardizing various functions and technology, any of the world's space programs felt more assured that in an emergency, the other space-program ships and men, could help, or find them with scanners, or open their hatches and doors, or dock, etc. Thus, co-operation was a life-saving approach for both sides, even now while flying towards war.

The lead ship for the Russians was called the 'Krenika', from an old Russian folk-song about a doll. The pilot was a husky man named Zolotny, who liked to eat pickles while at work on the helm. The Krenika was on-track too, now, and their routine was much like that on Guy's ship. His navigator was with him, for an hour-long shift in which they reviewed communications from Earth (the Ukrainian bases), with navigational instructions and up-dates.

"He is here, Zolotny," said the navigator. "Look at this map, these projections. We can scan ahead from radio telescopes back home. We have one ship ahead, about, uh, 300,000-kilometers. And behind us are eight more, the US ships."

Zolotny fingered another green pickle from a plastic tin. At the helm, they had data-monitors, controls for all the ships gear and decks, etc. As usual, it was 'steady-as-she-goes', nothing happening. The vinegar-cucumber snapped and crunched in his mouth.

"How far behind us are the other US ships?" he said.

The navigator had done his work ahead. "At the rate we are traveling, on a time-scale, without stopping or troubles, our formation would enter Mars-orbit 60-days ahead of theirs, roughly.

Maybe, 58 days. The ship ahead of us is a single transport. He will be at Mars 40 days ahead of that. So, we will have two months to deal with him, and put the battle into action," he said.

"What about his ship? What type is it? What is on-board? What do we know?"

The navigator brought up a data bank on a second unit-screen, with all sorts of research data or information easily available, including 'new' or recent expedites from the Ukrainian bases. He found what he wanted within a few minutes. "It is this one, a Dunlop. The Condrum 21. Very nice ship."

Zolotny yawned and scratched his beard. "Those have an exercise room," he said. "Reports from our side indicate he is only a transport, launched prior to the conflict. A cargo ship; let me see." He paused and read over the file. "Communications gear," he added now. "Crew of eight men. Hopefully accurate information." Zolotny now would lean back in his pilot-seat, where he had strapped in with a tether they used for the null-gravity. The navigator speaking with him was actually floating by his shoulder, prone, there at the helm of the Krenika. On-board Zolotny's ship was a very different cargo: about 40 men, armed and equipped for the anticipated attempt to win control of the Mars Base. Soldiers.

They all had Mars walker-suits and life-sustain gear, and weapons. Each of the five Russian ships had similar passengers. "He won't get in our way. One cargo ship with no soldiers or weapons. He's just ahead. But we'll also have to deal with the residents of the Mars Base itself. But that is not my problem. I am not a military planner. I am a flyer of airplanes and ships in space. I no nothing about battles," Zolotny commented. Now the pickles again were at his lips, sweet-and-sour, dripping. "We just kill them," said the navigator. "That is how wars are fought, yes?" "Eeehhh," groaned Zolotny. "Too bad, I'd say. But you may be right, they all must die, what do I care?"

"You have a big heart, but only a little brain, Zolotny," said the navigator. They laughed. The ship hummed and purred around them, full of energy and dull background sounds. The other flight-crew were each doing their jobs elsewhere. The 40 soldiers were only killing time, for now, away in the cargo-area, converted to keep them comfortable during the travel. It was very boring, close-quarters.

"I want you to see something else. This came over, two days ago.

It's a request from KK-F/Region Six. A man there, named Sarcasian, one of the council, but the Islamic side, not ours."

"What does he want? Why didn't someone tell me?" said Zolotny.

The navigator opened this file from his computer-kiosk. It was a voice recording. He worked a few buttons and keys, and the device re-played the sound of Sarcasians voice: "Greetings, Commander Zolotny and your crew about the glorious ship Krenika. I am Doctor Martin-Sarcasian, with the Central Planning Committee, presently assigned to the KK-F/Region Six launch-control site under General Terchenko. If you are hearing my voice on this recording, you and the other ships may be halfway to Mars by now. How I wish I could be with you."

"Wait! Stop to playback! Turn it off!" said Zolotny. His co-worker complied and punched a button. Zolotny seemed up-set. Sarcasians voice vanished suddenly.

"What?" said the navigator. "You need to hear what he has to say.

It is very---interesting."

"Then why haven't I been informed of this until now? Answer me that!"

"The message was low-priority on your communications-schedule, two days ago. You must have over-looked it. It is propaganda. Agenda. Visionary. Inspiration. Meaningless."

Zolotny paused. He took a few moments to quickly review

some of his normal-routine gauges and monitors, as far as the ship's functions. During their conversation at that point, the Krenika had traveled 10,000 kilometers.

"So I have a small brain, eh?" Zolotny said to his friend. "Look, just tell me. I don't want to listen to that right now."

"You should listen. This was sent to recording from KK-F! It must have taken these idiot weeks just to arrange it. He's an egghead. His research team feels they have some sort of mystical messages from other world aliens, extra-terrestrials, from many years. So he wants his point-of-view included when we take Mars. He feels if the meteor hits Earth, the aliens can still save mankind based on the survivors left on Mars. Our survivors, not theirs. But, oh, I can't recall everything. The main idea is that he needs special consideration, like equipment and experts." Now the navigator had pulled himself down into a chair, and strapped himself by a tether.

Zolotny, the pilot, was also strapped down. He toyed with a half-wet pickle hanging in the null-gravity in front of his face, which he could spin around like a top, in mid-air, weightless.

"I hate that crap," Zolotny said. "He's crazy. No one believes that crap."

The navigator laughed. "Yeah, very funny," he said. "Anyway, it was on your communications log. I thought it was interesting. He say they know about the meteor-strike and want to help us, they have known for hundreds of years."

"So what? If they did, why didn't they tell us? Why didn't they stop the thing?"

"They're aliens, Zolotny, that's why. They're evil!"

They laughed again. The Krenika cruised ahead. A red light began to blink on Zolotny's control dash, and there was a beeping sound, not alarming, but needing attention.

"The engines need re-grading," he said to himself. "You will excuse me, Penchka. This takes me an hour or so, with the engine man. Please be dismissed. We'll talk more later. I just wanted your information on the other ships. That is all. Dismissed. Thank you."

"As you wish," Penchka said. He released the tether from his seat, retrieved his binder-files and minor gear, shut down his computer monitor and closed his files, then pushed away down towards the rear of the helm, where he could slip through a doorway hatch. After a moment, Zolotny found a clever way for the floating green pickle he was spinning around in the null-gravity to pop into his mouth without using his hands, by bouncing it off a pencil. He crunched it down.

"Kill them all," he said to himself, with a bitter chuckle. "Seventeen years in the program and all the training. And this is all we know. Kill them all."

He sighed heavily.

CHAPTER 22: Who the hell is Terchenko?

*"How do you cook a goose? First you kill the damn thing. Once it's dead, it's much easier to cook, and you can be as creative as you want. They're delicious, with stuffing and you baste them with the best butter you can buy."*

*--Commander Rudolph Terchenko, leader of the Russian-Islamic space program, 2077, from the KK-F/Region-6 space-port in the Ukraine.*

And so it went, with the race to Mars, where Tweedle-Dee (the US side), and Tweedle-Dum (the Eastern-block) had agreed to have a battle. Mars, the barren, virtually dead world and stuff of Earth-legend for ages, sat in the sky like a fat brown-red marshmallow, half-roasted. And there on it, somewhere (a planet, after all, of considerable size) was a life-support research station deemed far more valuable than the 'marshmallow' itself. The reason, re-stated: God or whoever, had sent a very large rock through space towards home, looking very much like a collision, down the road some four years or so. Deep-thinkers on Earth thus now felt the battle to take Mars Base was worth the effort, or, an intelligent 'move', like an epic struggle for survival for human species survival, into eternity, or, however long the human species might survive. Despite the fact that no one seemed to get along anyway, as far as the human family was concerned.

There in the deep woods and mountains of the Ukraine, where KK-F/Region-6 was actually a very similar facility to the sought-after prize on Mars, Commander Terchenko was sweating the details, there in the frosty cold. The whole thing was making him feel taller, he thought idly. To move his thoughts skyward towards Mars, reddish star in Earth's night sky, and to plan, rehearse, study and commission Old Mother Russia's considerable resources 'out there', to do her will, it gave him a headache. Not to mention the whole idea of his Earth-home being hit by an asteroid the size of the British Isles, or larger, and what all that might mean. Standing tall might help. But maybe not that tall.

"I'm not God, Milana," he told his young secretary-assistant. "Despite what you think. I wish I was. Personally, I can only do my best. We are all in the hands of fate together."

Once again, Terchenko works at Region-6 meant many hours of labor, holding together the many-numbered threads that would give Russia and the Islamic-coalition the 'win'. This meant, in effect, that he had stepped into the role in direct opposition to that of the US Mars program leader, Lynn Rogers-Smith, the Commander of Angels. Had they spent any time together in-person, such as over a Lone-Starr longneck beer back home in her native Texas, it's true, the two would have likely been good friends, or 'muy sympatico', in any case. Smith could certainly be as tough as nails, like he, and was hella-smart. To kill a goose was not beyond her means and ways, at all. Or a space ship full of men, sad-but-true. Not her venue anyway, military. Terchenko, on the other hand, had very similar responsibilities. Launches, ships, men, women, fuel, navigation, orbit-paths, science-tech, politics. Both had power, and also during a critical juncture in history, supposedly. Each sensed their own mortality, and just how small they really were in the grand scheme of things. But a job's a job. So, they laughed. It was all they could do.

Yet, they would never meet at all, ever, these two. Terchenko was Tweedle-Dee, and Smith his complimentary

Tweedle-Dum? It hardly mattered. Earth's political and global space ambitions now meant that two people, a man and a woman, in such roles, would indeed stand like giants on either continent, squaring off to reach for a star. Smith was but five-foot, seven-inches tall. Terchenko was six-feet, two inches.

Milana, Terchenko's deeply devoted assistant, had worked with him for more than five years. Gruff, hard, cold at times, yet, he loved her. Who wouldn't? She was very healthy, quite young and fresh about herself. Her devotion to him was near absolute. Rich, powerful, mysterious, Terchenko was her father figure. She was always busy, his needs were demanding. Meals, drink, and also books, data-files, research, many phone-calls, urgent contacts with distant lands and hard-to-reach men or women who held one or three or ten-thousand of the many-numbered threads he needed to control his complex task and command.

Milana and the Region-6 Commander were now working with three space-path navigators, and others, to plot the course ahead for the ships on both sides, projected into Mars orbit, and how the military-battle planners and Generals would intend for the take-over to go ahead. Because this was now a 'war in heaven', the navigations for the ships were as critical as any other aspect. Terchenko knew enough about the methods to keep up. They all worked in one of the 'mission-control' centers, with many computers and monitors, and telemetry devices, and data-streams from various points around the world, pointed skyward.

"Now, here, Commander, you see?" said one of the navigators, a small man with a red face, very tech-savvy with his endless math and projections. "The one transport vessel, which is ahead of the rest. Our ships will over-take his position by less than 10,000-kilometers. That will take place in about two weeks, at the current speeds and paths. This may be an advantage for us."

"How so?" Terchenko asked.

"Well, it has to do with the launch-dates and the curvature of the telemetry. The transport is ahead, yes. But prior to the mid-point, ours start to reduce the distance because of the orbital movement. It's not much. But it will place them closer. Even, close enough for some kind of, uh, inter-activity."

The Commander frowned. A certain Colonel for the program, a military man they called 'Bowder' (for some reason), now offered an opinion, hearing this talk.

"Ship-to-ship," Bowder said, his voice was like a clarinet with a broken reed. "Destroy them!"

"At 10,000-kilometers, colonel?" said Terchenko.

"It can be done. There is a way. Our team's ships were re-invented for this with weapons. Bombs. Missiles. Of course we can. They could target the transport and..."

So, with this, they spent more time on the idea, with more experts and math-scientists working the space-telemetry, there in the Ukraine. Guy Reisling, of course, knew nothing about whatever they might decide, now some half-million miles away, sleeping comfortably in his bunk, dreaming of Lila Meetek, with a very nice erection to deal with, too. But it wasn't lost on the US Mars-team at Vandenberg, that at some point the transport would be vulnerable. And this was why Okman, the transport program leader for the US, was placing long-distance calls to Guy, with the specific instructions for him to have no contact whatsoever with the Russian ships, of his own initiative.

The sunset over the woods in Russia that night was hidden beneath gray clouds without rain. The trees and woods were still, and if one had taken the time, perhaps half-an-hour, it was possible to view the floating drifts of moisture-clouds, moving en-masse, like a huge gray pancake, just above the tops of the hills and mountains, and the green trees that seemed to shiver in the wind. Then it grew dark and you could not see the clouds at all, or the stars. As the world turns, within a few hours, the sunshine of the morning on the California coast revealed the far-away Vandenberg spaceport, inland a few miles off the sea. Sparks of blinding white reflection bounced off the many windows and shiny metals from the cranes and fuel tanks and hangars.

Rogers-Smith and some of the tech-planners, along with Branson Porter, Ibrahim Mehudi, and members of the Mars-Base Defense Task Force, took an early breakfast conference briefing; complete with all the tools they needed for their work. Muffins, hot coffee, eggs, orange-juice, vitamins. It was a working day, they wasted no time, so to eat together while they talked and reviewed the reports was just more efficient. They used one of the smaller VIP dining halls.

"Go ahead with what they've been working on at Mars, Mehudi, would you? Some of the task-force haven't been updated," said Rogers-Smith to Ibrahim Mehudi, the loyal and highly intelligent Middle-Eastern scientist.

"Shitting bricks, I'd say," Porter joked, with a few chuckles. Branson, also a Texan, was the security man, now recruited to plan various war-efforts and measures proposed for the battle ahead on Mars.

"Wouldn't that be convenient, Branson?" Smith replied. "Easier to dispose of. Conserve water."

Mehudi replaced his cup of hot coffee, following a swallow. He had a laptop connected to other computers within the facility, and tapped the keyboard for a moment. "The base on Mars has not been idle since all this started," he said. "They've gone ahead with many of the task-force plans, but maybe not all, and some have not been completed. By the time the ships enter orbit, with the Russians ahead of us by 50-days or so, the Mars Base will have considerable means to defend itself. All the men at the base on Mars are now being prepared with whatever weapons they had on-hand. They didn't have many weapons, but of course they had some."

"What kind of weapons?" asked Porter.

"I'm not quite sure, Branson. It's in the database for their permanent inventory. But, you can pretty well guess. Standard modern firearms, for one thing. They work fine; the Mars-atmosphere has no real effect. So, the large-bore shotguns, the military style multi-shot rifles like our Earth soldiers have, they were included only, well, I guess only for something like this. But there are not enough for every man. There are others..."

"All right," said Rogers-Smith, impatient. "Yes, yes. Guns, they have guns, and that sort. We knew that. I want to know about the external defenses. The air lock gates, and the, what did you call them? The Oxygen-Igloos on the perimeter, outside? And what about any way they might have to shoot down the damn Russians when they are in orbit?"

"None," Mehudi replied instantly. "None at all. Why would they? No one would plan to build a research base anticipating to shoot down our own ships as they arrive with goods or people? So, there are no missiles they can launch or target at the Russian ships once they are in orbit."

"Too bad," said one of the task-force men.

"Anyway, the plan on the external perimeter fox-holes has gone ahead. They've created about 50 of them, at various points in a circle around the airtight facility. They aren't quite ready, yet, from the reports we've gotten. But they will be. They used cargo containers, and other supplies. Men from inside can survive in one of these for about two days, on the outside. They can also re-charge the walker-suits they use, for oxygen. Each igloo has electricity, and some of them have communications back to the base. They're airtight once they enter. They used the plans and designs from before the base was built."

"That's what we wanted," said the same man, from the Mars Base Defense Task Force, who was a military planner. He was finishing his small meal. None of the military men worked at Vandenberg in uniform. "When they arrive, if things get into an attack situation---and that is what we expect---the men inside the base take stations in the igloos, and defend the base, moving in and out, onto the surface, or into the fox-holes, if you will, and back into the well-guarded air locks. So, any progress there beforehand is excellent."

Lynn looked dour. In her thoughts, the idea of ruining the costly space suits and Mars Base resources they had worked so hard to build, over many years, not to mention the loss of lives--- it was hard to see the point of it all. Asteroid U2753b was just a dream, distant, un-real to them. Almost not a worry anyway, and she well knew of the plans to stop or divert the asteroid, as it grew closer, from other Earth-global resources. Shotguns? God! My Mars surface suits are worth more than a million credit-units each!! She cast an eye across the faces at the long table.

"All right, thank you," she said tersely.

They paused. People were going over paper work and reports, or gazing at their computers. Rogers-Smith also had hers. Two or three minutes.

"Who the hell is Rudolph Terchenko?" said Rogers-Smith,

after another moment. "Commander Rudolph Terchenko, at one of the Russian launch-sites. Does anyone know?"

Dull stares and ignorance. No one seemed to know off-hand. Some shook their heads.

"No one knows him? That figures. They're very secretive. Well, just FYI, everyone. According to my desktop, this commander or whoever he is, is the responsible party on their side for a series of intercepted radio communications and instructions in the past week to the Russian ships. Translations say they want their ship that is closest to the single transport we have---the 'Penelope', under Captain Guy Reisling, I believe--- they are moving to encounter him. I guess because they are in range at some point in the near future. Did we know about this?"

Okman, Guy's immediate superior, was not at this breakfast work-session, or he could have told her. But the Mars Base Defense Task Force man, named John Williams, was privy to that aspect.

"Yes, we know," he said. "We can probably get something on this Terchenko."

"What the hell do we tell the transport pilot? He's defenseless, less than halfway to Mars," said Lynn.

"Not sure," said Williams. "We're working on it. Depends what the Russians have planned for him."

She breathed a deep sigh, heavy with the entirety of the effort, the emptiness of it all. For Lynn Rogers-Smith, a career space-program leader with a ton of experience, the ships and technology, the gear, the science applications---these were things of great beauty to her, near-perfection in their advanced functionality, very rare birds indeed, valued for the causes of research and exploration. Each ship took years to build and prepare, and even ages of learning and discoveries, from the past, for them to even be possible. And they were expensive to society, if that meant anything. And the cargo, and the people working at mission-control, and the planners---and the men on the ships. Human beings, real people, they wanted to come back alive.

"Well, whoever the son-of-a-bitch is, I think I hate him," she said, half-joking, still reading more of her reports and data. Some laughed, some didn't. "If they shoot him down, his goose is cooked, know-what-I-mean? Helluva' way to die, floating away in space. Have somebody find out about this rat, this Rudolph the Red-Nosed Russian mother-ducker. And connect directly to the sources here that are picking up their radios to their ships. I want every detail every hour, to keep ahead of them, and maybe save this transport."

"You got it, Lynn," answered Branson Porter, a bit stunned, even for a Texan.

## CHAPTER 23: Much Too Close

Thinking of rings, and worlds, and ways between them. One, Mankind's eternal home, the other (Mars), virtually dead. And between, as they spin in distant, silent orbit, the small moons, asteroids in clusters or fields, like scattered stones, very small. After a million years of human evolution and progress, the Molinari space dock was also there, a monument to technology and the space-program's long-term success. All these, and more, in spherical paths, like grace, stately, slow, vast.

As the battle dance now moved forward, early in August of 2077, the space ships were an added factor, including Guy's Monsanto-Dunlop Condrum 21 Local-Planetary Cruiser (transport). And, just a few hundred thousand miles behind him, the Russian-Islamic ships, and the US Mars teams in their ships, led by Winton Berle, the Old-School astronaut with a lot of hours in the Abyss, and other pertinent experience. Who was the Lord of this dance, this 'war-in-heaven', if any? Perhaps again, as usual, only human pride, ego and vanity.

The 'Penelope' was now in-transit, at her normal speed of about 10,000-kilometers per hour. Molinari was Guy's next stop, for re-fuel and re-charge, rest, and easier communication with Earth. Lila Meetek, his one-and-only true love, (which was certainly disputable for either of them at this point), waited patiently, day-after-day, knowing her white knight was on his way swiftly to her side. Sex with Tommy, the external repair space-worker, or space walker at the space dock, was---well, fun? Healthy? Sleazy? Or for Lila, all of the above. But Tommy also had the unpleasant habit of sleeping with nearly every other sexually active female on the Molinari facility, and the gossip was terrible. Besides which, he ignored her when she needed more than mere orgasms, such as heart-to-heart knowingness or relatedness---talking about things, sharing. Which was why she really did love Guy Reisling. He gave her so much more.

Despite the brewing situation ahead on Mars, Lila managed to link with Guy's radio-desk on-board his ship, at one point in the slow dance. He was still many thousands of miles away, even hundreds of thousands of miles, or kilometers. In her role as one of the Earth-Mars Corridor Environmental Conditions Monitors, she had all kinds of communications gear at-hand. So she could set up a radio-call to her boyfriend fairly easily, and also somewhat privately. For the two lovers, it was a naughty moment their superiors may have frowned on. After all, they were now 'almost at war', and the enemy could intercept the communications-link or signal. But all they would hear, if they did, was their mindless lover's prattle, and both Lila and Guy knew this, too.

"Be clear with me, Guy," Lila said. "Ambiguity right now is not working. You told me before, we're not exclusive, and we're not---married. So if you intend to be angry about my sex-life, I suggest you go ahead and 'let-her-rip' right now. That way, when you dock, you'll be finished with all that, and you can just enjoy the pleasure of my company. Know what I mean?"

Guy was again on the helm, or operations-deck, aboard the Penelope, halfway through his shift. Nothing unusual was happening, shipside. Engines were trimmed clear-and-clean, perfect running. Earth had no urgent instructions, commands, or information. The pilot's cabin-deck looked to Guy much like a rather large flight-cabin on a traditional jetliner back on Earth, or 'jumbo-jet'. There were seats, numerous controls, view-screens and view-ports (sealed during deep-space travel). There were computers and radio-gear, and various hatches and levers. It was large enough for five men, or even six, to work at once. But, at the moment, he was alone. There was no video-link to Lila, at that time, which he regretted. But her voice was clear, through the radio-link. Guy was in his shipboard working 'jump-suit' or 'flight-suit', a single-piece, efficient pullover, elastic and warm. Once again, at the ship's controls, he was barefoot, which he preferred.

"Jane, you ignorant slut," he replied to her comment. The signal carried his words. "Saturday Night Live, 1982, Jane Curtain and Dan Aykroyd. Remember?"

"No, not really. And it was Chevy Chase," she said. "Those old-time TV shows are in the past. I don't waste my time on that sort, I guess. But this is now. I am not ignorant. My slutiness is my bliss. You ought to know."

"Well, I remember, yes," said Guy sheepishly. "But the memory is fading fast. Need to re-boot that one."

There was a pause. "I'm tracking you for about 10-days out, about 123,000-miles," she told him. In his thoughts, he confirmed. Just about right.

"Right on track," he said. "And into your loving arms. Or your pants. You know." He chuckled.

"Don't expect too much, hero," she said. "Molinari is not exactly a love-nest hide-away for private space-romps with horny astronauts. There's a lot of gossip, it's a closed society, and there are no secrets. And we still have all the daily work-tasks. The Life-Sustain recycle has been losing power-integrity and re-charge purity. Translation: stale air."

"No smoking, right," said Guy.

Another pause. "What about those bad-ass Russians following you to Mars by about a week's worth of absolute nothingness? What are you going to do?"

Her voice over the radio was sweet and clear, unlike the signal from Earth, now much more distant. Maybe it was only because he liked her so much, wanted to be with her. Guy took a moment to answer. There was a minor energy-gauge measurement irregularity on one of the ship's solar-panel arrays, almost like a shadow had passed over them. He looked again. The levels returned to normal. He assumed it was meaningless.

"Guy? Are you there?" the signal penetrated again through space and into his pilot's deck.

"I have your signal, Molinari," he responded. "To answer your question, I have command from US-Earth on that. I'm to proceed with the transport as normal, as if nothing was unusual. The ship is loaded with communications tech-gear, for the Mars Base. A gal named---uh, Karen. Karen Tutturo, a communications-science analyst, and current resident of Mars, is waiting for my load. They had a communications problem. So, basically, I could give a crap about the Russians. My ship has no weapons. I'm defenseless. So it doesn't matter anyway, much. There are minor preparations I can make in case of a conflict."

"But---what if they attack?' said Lila, legitimately concerned.

"Then I'll be floating home, into the abyss with my crew, like a quick-freeze popsicle," Guy smirked. "And all my lustful dreams of rolling around with you in your private null-gravity bunk at the dock will be over for us both."

"Not necessarily. They might not kill you. Why don't the Russians just turn around and wipe out the US ships anyway, before anyone even reaches Mars? It's Winton Berle's ships they have to be afraid of. Not you." Lila was keeping track of the whole thing like everyone else.

"Totally impractical, not efficient at all. Deep-space dog-fight battles are just in the movies," said Guy. "In reality, it's essentially not even possible. The distances are too great. They have five ships, loaded with soldiers. Maybe 20 or 30 armed men on each ship. Plus bombs and missiles, and other weapons. So—if they attack the Penelope---what do you think they'll do? Ask me directions to Jupiter? Very funny. No, dear. They could easily decide to blow me to hell. But personally, I doubt that's their plan."

Back on Molinari, in the work-area where Lila was assigned, also surrounded by the cold-sterile façade of technology and LCD's, she winced a little privately, thinking about Guy as a frozen-solid corpse floating away from her forever, dead. "Just---drive carefully, Guy, okay? When we get home it will be different. We can start over; have a barbecue at your place near Santa Barbara. Travel around in a motor home. Or take the bullet-train somewhere, just you and me."

She now lowered her voice to a more personal tone. "What are we, Guy? What are we really? Just two star-crossed lovers floating around in circles? Is that all there is?"

"Standard-issue human beings, male-female, life span of 80 or 90 years if we're lucky," he answered quickly. "Personally, I enjoy floating around in circles. It's---a lifestyle."

There was a muted beeping-alarm in the inner-workings of the radio-link, telling them their time was up. It meant the system had other uses at that time, and they had to let go of their connection.

"Data-feed on the corridor for environmental," Lila told him, so far away. "Have to let go of this link.'

"All right," Guy said. "Not very satisfying anyway. Reisling out, Penelope transport-vessel, 1215-hours, day 61."

"Confirmed. Molinari out," Lila said in her official dispatcher-voice. "See you soon, Guy."

Another series of alarm-beeps and a few control-buttons later, the link went dead.

Twelve hours passed. Guy's shift at the controls lapsed, Rob Cowan took the deck, things were stable. Other crewmembers kept their posts---the navigation, communications, engines, life-support, and ship-systems. Right about the time Rob was ready to pass off his station-post back to Guy, who was getting some sleep, one of the two navigators, whose name was Tom McGee, had an urgent matter for the attention of Rob, or Guy. Tom was a mature space-man skilled in highly technical ship tracking and star-positions, such that he could get them home, or to Mars, or to Molinari, by studying the relative position of the ship and her destination-points. Without his work, if anything went even slightly amiss, they could literally be 'lost in space'. And since the Penelope could only sustain them for so long, getting to their destination-points safely was a life-or-death matter. Since Rob was the ship commander at the moment, Tom brought it to him, and appeared on the flight-deck or pilot's helm through the entry-way, floating by hand-pulls like they all did, then settling down into a seat with the magnetic strips.

"Rob, uh, you need to look at something right away," he said. "Is Guy on deck?"

Rob soured. "No," he said. "Guy's not on for another hour. What do you have?"

McGee had a small data-computer in his hands, the sort that could move complex information from one place to another quickly and easily, like a portable drive or hand-held unit for data-transfer. He booted up a page-file, which came to life on a small screen. Rob took the unit and studied the screen.

"I was running my usual plotting scans, in this case, pointed backwards towards home. You can see, the other ships behind us are still tracking. We already know this much. But look what happens when I magnify on the Russians, and then compare to our position," McGee said. Rob handed him back the data-unit, and he worked some controls for a moment.

"See?" he said. "Two of the Russian lead ships are now close enough to the Penelope to spit at us and not miss by much. I mean, they snuck up on us, over the past 24-hours, I guess. They must have increased their speed, and a lot, not a little. The fuckers could hit us with a rock. I mean---well, from this, at this hour---one is about three kilometers away---the other is maybe ten. See? That's not good, Rob. They want something. We need an all-hands alert, right now."

Rob scowled and viewed the data-unit. "How old is this file?" he said.

"Less than an hour," McGee answered.

"Mother-fucker," Rob said. He knew, as they all did, that any ship-to-ship contact was forbidden by US-Earth command, and also that ships in transit rarely if ever got that close, in terms of material proximity of the actual ships. Anything closer than about 5,000-kilometers was considered dangerous, considering the speeds at which they traveled, especially for ships from different launch-sources.

"All right," Rob said. "Good work, thank you, Tom. I'll alert the captain. Tell the men to expect a general alarm within 15-minutes. Use the ship's intercom. Then I want you to track every move they make to a gnat's ass, back at your plotter-scanner."

McGee sighed heavily. "What do you think they're up to, Rob?" he asked.

Rob was working to punch-up his voice-alert to Guy's sleeping berth. "Not sure. It's a fucking war, Tom. US Mars command warned us. But, you know---we should have been notified by Earth-tracking that they were moving in. It's bull-shit---"

"I'll get on the ship's intercom to the crew, from my station," McGee said. "Then I'll just track their ships." He quickly released the tether that held him down to the seat where they were working on the flight-helm deck, then pulled himself out of the room, down the hatch and into one of the level-to-level tunnels.

Now Rob had his alert-system connected to where Guy was sleeping. There was a loud beeping.

"This is the captain. What the hell—Rob? What's going on?" Guy said. He was already awake and getting ready for his shift at the command deck. But he was resting and hoping to enjoy a few moments of peace and a hot coffee in a plastic tube.

Rob was terse. "We have a problem, Houston," he said.

"Like what?" Guy answered.

"Uh---Klingons off the starboard bow, captain. No shit. Tom McGee just confirmed two of the Russian ships have snuck up fast behind us and are less then ten miles off from our position each. One of them is only three miles out. Right on top of us. He just figured it out an hour ago."

He could hear Guy huffing and puffing on the other end, in his sleep-cabin, living-quarters, like he was dressing quickly.

"God dammit," Guy said. "All right. I'll be on-deck in 10-minutes. Alert the crew and dispatch an emergency-alert to Earth-base, Molinari, and Mars, with the essential information."

"Yes, sir," Rob said.

They were not smiling.

## CHAPTER 24: Ship-to-Ship

To conduct an emergency situation or circumstance, from a command position, is difficult for anyone. Command, leadership, control, office or power, in emergency, or danger, is an urgency and trouble. Anyone who really enjoys it, or who seeks it, or thrives thereon, is probably un-reliable to begin with. With lives on-the-line, people who are depending on clear and successful decision-making, and with the dynamic and changing urgency of any dangerous emergency, a leader such as Guy Reisling is hard-pressed to win the day, even unto death. And yet his normal job was only to transport goods across the sky to Mars---easy and non-threatening, food, water, oxygen, and supplies. If they all died, in a conflict now with the Russian-Islamic space-ships having approached un-seen, to a dangerously near place, as the ships were speeding along---that is, if Guy and his crew were all killed, for some reason, for it was a war, now---they would not even know, or realize, or understand, that Guy had failed them, and they had lost everything under his command, for the aspect of the ignorance of being dead. In a flash of fire and freeze, if the Russians for example fired missiles at the Penelope, destroying their ship, killing them all, a moment of intense pain and anguish---maybe, ten or 20 minutes of panic, in which all hands were to be lost---as they left the Universe of the living, for the Beyond, it might approach awareness, in a miniscule, tiny, group or individual nervous-system of human consciousness they shared as the crew of the Penelope, that the hour had been lost to failure and fate. Guy, as the commander, was driven in a passion of perfect decision, to save them, his men, and not let this happen. And deep inside, he knew, he would prevail, somehow.

Guy arrived on the flight deck or command-control helm of his ship, within only 15-minutes of Rob's alert. Their discovery of the very-near approach of the Russian lead-ships was indeed quite alarming. A layman, or one who may know nothing at all of space-travel in the future year of 2077, might not understand. After 15 years or more of the establishment of the US Mars Base, called the Snikta Ridge Volcanic Basin Research Mars Facility, space flight back and forth was almost entirely educational and research-oriented. Now, with the advent of an approaching meteor, which may strike the Earth with overwhelming destruction, an act of God, if you will, Guy and his crew, and the teams of leadership and science-leaders managing the US Mars program, now were suddenly at war. This alone was a shock. For Guy, Rob Cowan (with his anti-biotic issue concerning his testicular loss), and the eight other men working this particular flight aboard the beloved Penelope---a grand and noble ship they all loved more than life---now, they faced uncertainty, that was beyond doubt. The common procedure for Earth-Mars corridor travel, was well-established for any of the ships and launches, either US Mars, Russian-Islamic, or even others, that ships never traveled the Abyss, so close or near, by miles or kilometers, as they presently knew and had confirmed, that these two Russian ships, were now 'at their side'. So without a doubt, they were alarmed and in a trouble about it. Again, the reason was, that aside from docking procedures, because space flight was dangerous and un-predictable, that distances of at least 10,000 or 20,000 kilometers—more than the distance on Earth from California to Australia, or the Horn of Africa---these safety zones of separation, were standard. It was only good practice. The ships flew at great speed, conditions were fatal if there were any failure, and also for purposes of navigation, variations might arise at any time. So with Rob's call to Guy's sleeping bunk, and the navigator's confirmed data that the two lead ships for the Russian-Islamic voyagers, were now ten-miles off, or even three or four miles off, from their ship's position, it was an urgent circumstance that Guy took very seriously. It could only mean they wanted something, or were 'up-to-something', and it was his problem. He wasn't happy about it, to say the least.

The entire crew of the Penelope was now at all-hands alert. Guy and Rob, along with the flight-path navigator Tom McGee, and the Flight Specialist in charge of Vandenberg-to-ship communications, a thin, brown-skinned man from Indonesia named Raza Brahman, were on-deck. The goal, or orientation to the situation, was to gather accurate information, and respond, within the structure of the instructions and current-moment advise from Vandenberg. To over-react was an error. Each man took his work at his own level of performance-skill and duty, knowing it would affect the outcome, either way.

Brahman was on his radio-link to Vandenberg within half-an-hour. He needed to transmit their codes and connection, and establish the link, and also to contact the appropriate Vandenberg leadership, who might not be instantly available, or even asleep, or off-duty, rather than the 24/7 US Mars Earth-base communication dispatch monitor underling, who would have no authority or information. This took time.

"We'll have Okman within 20-minutes," Brahman said. "That's the best I can do, he's the transport-cargo commander for this ship."

"Okman won't know anything," Guy said. "He's still hurting from my competency trail last year. All he knows about is our cargo and general orders. I need to speak with command. We have a general order forbidding any contact with the enemy. I have a feeling it won't be long before that becomes inevitable. I need to know what Rogers-Smith wants from me. Maybe we can get out of this."

"Working on it," said Brahman. He continued to ply his skill with the ship's communication set-up, mostly waiting for

Vandenberg to realize they had a situation, from their previous message. So, he simply watched the in-coming monitors for Vandenberg's response, across half a million miles. He also repeated their signal-code emission, a 'cry for help', that would not be mistaken, by the Vandenberg 24/7 monitor for their flight. All the flights were in perpetual contact with 'mission-control', but it was usually entirely routine, even dull, boring.

Guy confronted Tom McGee, again. "Give me the current details," he said. "Show me on a piece-of-paper if you have to. Who are they, which ships, and what are the relative positions from us to them."

Of course, McGee was tracking the Russians moment-to-moment. The Penelope had scanners, mostly for far-distant objects like Earth, star-positions for navigation, Mars, Molinari space dock, and the every-now-and-then meteor shower or heat-flare from the Sun. And also the Mars moons, and other celestial features.

"Hasn't changed much," McGee said. "Our ship is traveling at 10,000-kilometers per minute. They're matching us perfectly. I have to say, their pilots must be skilled. We're in formation, like it or not. One ship is roughly 4-kilometers below us and back. The other is ten kilometers, or I guess about nine kilometers, to the other side, above, like two o-clock starboard, left. These fuckers know what they're doing. They know we know. It's bull-shit, Guy."

There was a pause. Rob was looking at their forward motion, to plot a variety of control-commands that might be useful if they wanted to change position. Guy was looking at life-support, and preparations they had already made for any confrontation. For instance, he urgently needed the ship's four engines to be adjusted for a longer-term burn, without clean-and-trim cycle, in case they could not rest the engines for restoration, as they usually would. So, the rocket-engine team was working with him to make sure they would not fail to have full thruster-power for all engines, for example, to increase the ship's speed and escape trouble. Rob's momentary task was to look at how they might track the Penelope---down, up, across, loop-the-loop---even within the hour, to avoid the Russian ship's advances, should they attack. None of them had any idea what the Russians might do next.

Three hours passed in this way. Each team-member did what they could, and Transport-Cargo Commander Okman was on the line to the Penelope after the first hour. But he didn't have much to offer.

"You have your orders," Okman told guy, once Brahman had set up the connection. "No contact. No hostile action. No negotiation."

"Eat my shorts, Okman," Guy told him. "For all I know, they'll fire a bomb at us without warning. I need to ship-to-ship their pilot, find out their terms. They're right on top of us. It's totally abnormal. It's not like they're sleeping over there. The bastards are just waiting. It's been ten hours. They moved on us at high-speed, probably three or four times our speed. It was intentional. That alone is a hostile act."

"Don't interpret the situation, Guy," Okman told him. The radio-link was compressed and jerky. "Not your job. Hold your position, fly silent. We'll wait for command. Rogers and the others are being alerted. Just sit tight, it won't be long. I think that Jew, Ibrahim, the science-guy, was barhopping in Santa Barbara late last night. It's only going to get more and more complicated. Winton Berle's squad of ships is a month or two behind you. That has to be considered, too."

"Yeah, sure," Guy said. "And three other Russian ships between our position and Berle. Like he's going to get me out of this. He's too far back. 700,000-miles or so. They won't help me."

The link was silent for about five minutes. Guy and Rob, with Tom and Raza Brahman, chatted about it all.

"The signal to Earth is compressed," Brahman said. "He'll be back on-line in a few minutes."

"Even Rogers-Smith or Earth-tracking can't really give us anything," Rob said dismally. "We're on our own."

"I say we run for it," McGee offered. "Hit our maximum speed, maneuver away. Why not? The Penelope is probably superior to their ships for speed. We can hit 100,000 kilometers per minute. Why not? Run for it. Out-maneuver them, make it a race. All we have ahead is either Mars or Molinari. At least it would save us from a blind attack, like a missile."

Guy was cold, thinking, structuring his options sub-consciously. A missile attack was maybe un-likely, but from a strategic point-of-view, it might be possible, or it might even be their goal. Ten hours had passed; at-speed, and the Russians had been silent. No attack, which they certainly could have accomplished at such close range, if that had been their idea. On the other hand, Guy didn't really have any confirmation that the Russian ships were equipped with such ship-to-ship destructive armaments. After all, it had never happened before, in the history of space-travel or space-exploration. But this was---different.

Okman came back on-line with the link. He had nothing else to say of any value. What Guy really wanted, was permission or authority, from Vandenberg command. He was handcuffed by the standing orders. Okman couldn't change that.

Also, command-authority would have other information he could apply to their crisis. Or, even, as Guy might wish, some kind of plan, or back up. The radio-hookup with Okman finally went dead, until Rogers-Smith and her people were advised and could respond intelligently. Estimated time for that was an hour or so. Meanwhile, the crew of the Penelope started to sweat. The prospect of death and disaster, floating away into the Abyss like corpses of stone, was not welcome to their minds. The Russians were laughing at them, behind the silence. The Penelope thrust forward on her course, and they along, with the other ships at-pace, only to vary their relative positions by a few hundred feet. Thousands of miles of nothingness went by, and yet, was like stillness, or even seeming motionless.

Guy was seated in his command-chair, looking at ship's systems. Oddly enough, he had on the normal foot-ware they all used, the magnetic slippers that held them down in the null-gravity, on his feet, instead of bare-footed. Raza, the tele-radio expert, floated upside down above, working on a hand-held computer that kept track of his radio-link, to assure the connection from Earth when ready. Rob Cowan worked on the engine status and other tasks, by inter-ship link to the engine crew. McGee, the path-plotter and star-guide, seemed to have fallen asleep, his head slumped over into his hands at his post. Then without warning, a radio-link monitor started to bleep loudly with an alarm. Raza lurched to his desk-top kiosk, part of the helm-deck work-area, to respond. They all knew what that alarm meant, from the sound and desk-of-origin, there on the helm-command: an in-coming signature-coded radio communication signal, indicating and outside message or link to a responder. It could be Earth, Molinari, Mars, Winton Berle's armada---or 'them'.

"Go ahead with it," Guy told Raza. "Proceed. I'll take responsibility."

Raza activated the link, with the standard salute. "This is US Mars transport vessel Penelope, under command of Captain Guy Reisling and US Space Authority, in Vandenberg, California," he said. "We have your signal. Go ahead, please."

He pumped the volume onto the helm-command deck, so they could all hear. There was static for a tense moment.

"This is Colonel Robat Zolotny, of the Krenika, originating from Ukrainian space-port KK-F/Region Six, on Earth, commanded by Russian authority under General Rudolph Terchenko. Do you have my radio? Please respond?"

Static, dead air. "Handle it," Guy told Raza Brahman tersely, who was at the radio-monitor.

"We have your signal, Krenika. Go ahead," he said. The flight deck on the Penelope was now all-ears. A moment.

"We are apparently at war, Penelope. Who is your commander?", came the voice on the other end of the radio-link. Tom McGee, the plotter-navigator, quickly attempted to track the radio-signal to figure out which of the two Russian ships was the Krenika, which was not clear.

Guy was steaming. "All right, dammit," he told Brahman. "I'll talk to him. Give me the phone."

He moved over to Raza's station, floating and pulling himself by chairs and handholds, then settling down. Another moment, static.

"This is Captain Reisling of the Penelope," Guy said, using the radio-microphone. "The Krenika is too close to our ship. We are in danger of a potential collision at flight-speed, as per protocol. I'm aware of your position. I'm requesting the immediate withdrawal of your vessel to a safe distance of at least 5,000-kilometers. What the fuck are you idiots doing? This is non-standard, and you know it. Get the hell off my ass, Krenika."

Now, from the other side, on-board the Krenika, they could hear laughter, over the radio. The crew on the Krenika flight deck found Guy's request very humorous.

"We are at war now, Penelope," came the reply, apparently the pilot, Zolotny. His voice had an Eastern-European accent and dialect. "Surely you realize this?"

"Please withdraw the Krenika to a safe distance," Guy responded. "Surely you realize standard practices. I am requesting your vessel to comply for the safety of all concerned. Withdraw the Krenika to 10,000-kilometers immediately."

A long pause. Static. "We shall see, Penelope. We shall see." There was now more laughter and rude remarks from the other side on the radio.

## CHAPTER 25: Run Silent, Run Deep

*"You are old, Father William, and your hair has grown most white. And yet you incessantly stand on your head, do you think at your age, it is right?"*

*-Father William, Lewis Carroll ('Alice in Wonderland')*

At Vandenberg, California, the Western US Mars Base program headquarters and launch-command, transport crew Commander Okman was correct. Within an hour of the radio-message from The Penelope, the crisis team organizing the moves on the empty, black, blank, and airless chessboard where the East-West contest was playing out, a million miles away, was quickly brought together. Okman was the point man as far as details.

"It's our lead transport, launched before the conflict," he said. "The captain is Guy Reisling, a very capable pilot. The ship is within two weeks of Molinari. As you know, there was a secret launch, the Russian ships, and five in all. Up until about 20 hours ago, those ships were far behind, I'd say eight weeks of flight time. Now, however, I guess the root of the crisis of the moment, two of those ships have moved ahead quickly, and are very close, even much too close for safety, to our transport. That's it in a nutshell. Reisling wants to know how to respond, or authority to act on his own. I should warn you though, I know this pilot. He'll eventually take action if he feels his ship is in danger."

Lynn Rogers-Smith, the program head-honcho from Texas, along with Ibrahim Mehudi, the Science Lead, Branson Porter, also a Texan, the current program Security Chief, along with some of the men connected to the data-flow that tracked the ships moment-to-moment, now quickly met together in one of the 'mission rooms', used for each series of flights to Mars. By now, the teams following the conflict, and the approach of Asteroid U2753b, were exhausted. The discovery of the asteroid, and the release of the information to the public, had set so many things in motion. They had scrambled, first to learn how the meteor would impact the Mars program overall, thinking it would be an environmental issue, such as the delay of ships or disruption of launches, etc. It wasn't long until the two competing Eastern and Western space programs, escalated their reaction until it was becoming the world's first war in space.

Smith regarded the past year of work, as a personal challenge, in particular to outwit her counterpart, General Rudolph Terchenko, the KK-F/Region 6 commander in Russia's Ukraine. In a year's time, the asteroid had moved closer to a disastrous collision with Earth, and much more was now known about that. The year was 2077, mid-August.

"Thank you, Okman. Well done. Obviously our orders to your pilot, now, depend on what we feel the two Russian ships want, or why they've moved in on him, or what they plan to do? Opinions? Anyone?"

"They're taking him out, pure and simple," said Branson quickly. "The Penelope will be their first kill. They can't turn back to face our ships behind them, they'll wait until they reach Mars. So, they're moving in, probably just shoot him down. Make their intentions clear."

Rogers-Smith winced at his idea. Her face was normally somewhat feminine for her age, a rustic, busty Texas woman, with luscious blonde-reddish hair, trim-cut neatly. But as her Security Chief mentioned shooting down her ship, her features were very unpleasant. "All right," she said, grim. "That's one view. Doctor Ibrahim? Is there another way to look at this?"

Mehudi was a dark-skinned man, thin but also athletic, with dark freckles, and a typically Eastern face and smile. "Porter may certainly be right, of course," he said. "It probably wouldn't be difficult to destroy our transport, at that range, especially given they've prepared for a battle on Mars, so they have those kind of weapons. So, please, I am not military---don't hold me responsible if those men are killed, in terms of my opinion here. But---I think maybe the Russians are testing to see if it is really war, or not. It is a threat, and they are challenging us. But, even if they destroy the Penelope, from a science point-of-view, the debris and parts of the ship, will be in the path of the other three Russian ships. A single part of an engine-rocket, a hatch-door, an antenna---"

"Or a body," added Porter sarcastically.

"Yes, that's why the ships fly so far apart, on long voyages. If a large piece of metal or something strikes a Russian ship behind them, they surely know it could be a disaster. Even a small piece."

"Then Guy and the men would have their revenge," Porter said, staunchly consistent in his role.

Lynn Rogers-Smith, who was basically in command of the entire US Mars program (although most of the leadership work was done in groups), felt at this point that she was here in charge of a zealous football coach and his rebellious quarterback, who would rather be out chasing girls, instead of mature men who were responsible for a modern space-program. Porter never cared for Ibrahim, but the egghead intellectual class held sway in many program decisions, if only because the science was so critical. For most of his career, Branson Porter was a fifth-wheel, and not very busy, because as Security Chief for the program, there was never much threat, being an educational and research-based effort. A big, hunky Texan who enjoyed carrying a hand-gun for personal use, in his hydrogen-fueled rotor-electric car, and an Eastern-born scientist, with vast knowledge of what was needed to travel is space, thin, dark-skinned, and with an accent, who liked to go clubbing and pleasurably consume drinks at Santa Barbara night-clubs.

"You are begging the question, Mr. Porter," Ibrahim said. "We only need to decide what to order the pilot to do. We cannot read their minds, as far as the Russians. If they kill our men, then that is what they will do. Nothing we can command from Vandenberg will save the Penelope if the Islamist simply chooses to attack and destroy his ship. But the better course would be to avoid the war-like presumption and give them a chance."

Branson huffed. "Then order Reisling to fight back!" he said loudly.

"He has no weapons, Branson! The Penelope has no weapons at all! It is for transport!"

"All right, enough," said Smith. "Thank you. It's been 20 hours. They might have acted by now, if all they wanted was to attack. I still don't know what message to send our pilot, and we're running out of time."

There was a pause.

"He has orders to avoid all contact or action against the Russian-Islamic ships, and to move his cargo to Mars as a normal transport mission," said Okman.

Another pause. Ibrahim was gazing at a computer-screen, with one of the flight-path tracking experts, who then offered to comment.

"There has already been ship-to-ship radio between our Penelope transport, and their ship, the—uh—the Krenika, I think it's called, according to this," he said. "The logs show a communication a couple of hours ago. The Krenika is their spearhead vessel, ahead of the other Russian ships. He's only three kilometers from the Penelope. A stone's throw."

"We have to act, our pilot is waiting," said Smith. "Get a radio message to the Penelope. Tell him to relay our message from US Mars at Vandenberg, to the Krenika, that the United States demands full compliance with all space-corridor standard safety practices for travel distances, and that they withdraw the Krenika immediately to a safe distance. And also, that we will hold them fully accountable for any action against the Penelope whatsoever."

Another helper, a mission-tech specialist, went to work then on her order, first jotting down her words on a pad of paper.

"Lynn, please, also this," said Ibrahim. "Have our pilot relay to the Krenika that we intend to diffuse the situation without loss of life or ships. Let's not escalate to violence and lose our men. Why the hell can't we just agree to share the Mars Base, after all, eventually I mean, if the meteor hits Earth? Did anyone even ask that question?"

Branson Porter laughed sarcastically. "Tell your Russian pilot there are eight fully-armed US space-ships behind him by about three months of flight time who will gladly obliterate every single Russian astronaut who gives us a problem up there," he said. "Tell them that."

"Secretary, hold on that, from Branson," Smith said. She nodded towards Mehudi. "Send the first two. We demand compliance. We hope to diffuse the situation. Draw up a message, and let me review it quickly first. I want the same message to go out to all Earth-bound related Mars-program technical workers, Vandenberg, and other launch-sites and commands and the Presidential White House. Sorry, Porter. Your machismo isn't cutting it right now. Let's get this done in the next few minutes. Go."

Within a few minutes, the message was prepared, as they considered other options, and more instructions for Guy Reisling, as things developed.

The night sky above California, there on the blustery coast, held no clue of the drama being played out above, far into the darkness. The stars like tiny fireflies as a cold or lonely gray cloud from the beach may drift between, in the all-too-thin Earth atmosphere they lived and breathed. Inside Vandenberg, and elsewhere, the scenario was understood well enough, on their monitor-screens, radio-links, on computer files, telescope records, radio logs, and in the halls of power and in their thoughts.

Back aboard Guy's ship, things had not changed much at all, as they waited for their instructions. Guy knew he was mostly helpless. If they fired a missile, it would probably hit his ship, and they would all die. Even at-speed, some 10,000-kilomteres per minute, with all three ships in motion, it was much as thought they were stationary, not moving at all. Such was space travel, hard at times to visualize. There had been no further radio with the Krenika and Zolotny. The radio-sphere was monitored, and Guy was hoping they could pick up any of the enemy's transmissions, perhaps to give him a clue as to their plans. *Run silent, run deep,* he thought.

After an hour from the time they had been in contact with Okman, and were waiting for Vandenberg's reply, the second of the two Russian ships that had snuck up on them, changed position, and moved closer to the Penelope, with no announcement or warning. So by the time Vandenberg's message reached the radio-monitor desk on the helm of the Penelope, as directed by Lynn Rogers-Smith at their meeting, the Krenika was at three kilometers, less than three miles, and the other was at about five miles off. Guy felt like a bird in the mouth of a cat. He was disappointed. They were only really repeating what he had already told the Krenika, more-or-less adding their authority to his own. Whereas for him, he wanted out of the situation, and his ship and men, out of danger completely.

So they worked the message relay from the ship's helm, and all systems were functioning well enough. Rob Cowan, Guy's co-pilot, had now been on-duty more than 20 hours, and needed a brief rest. He took his break, as things seemed to be at a stalemate, and retired to a nearby resting berth, off the main cabin of the pilot's helm, where men could sleep briefly, in a comfortable area, no larger than a closet.

He fell asleep, dreaming of his Montana family ranch, his wife and children, and his relatives back on Earth. In his thoughts, hardly really asleep and yet half-awake, he fancied the meteor hitting the Earth, and the damage, wondering if Montana, in the deep interior of the continental US, would somehow be safe. He dimly sensed that his mind was not well. The images and ideas he was creating were mixed up, dark, fluid, grasping, like ice-and-fire, then flaring, burning, and swirling like a vortex. An asteroid streaking through space, very huge. Sudden impact, no hope to turn it away beforehand. A huge cloud of darkness over half the Earth, in Biblical proportions. Cities, civilization, Vandenberg, lands and regions, wiped out, under a toxic plume of water, dirt and dust, fire, cloud and who-knows-what---tidal waves, and earthquakes. Men, women and children, crying out for a moment's panic, then lost. All lost. Home no more. Apocalypse. But in his mind's eye, placing himself far away in space, his attachment to Earth, and his role as an astronaut, made it impossible for him to justify his beliefs, or to process the ideas, as if he were gazing down a tunnel, within himself, with the dying Earth at one end, and he there on the ship, so far away from family and friends, at the other. The up-and-down un-reality of space, the darkness and near-death airless vacuum of cold all around, worked on his psyche. In his blood, the genetic-triggered antibiotics he was using for his testicular loss, melted into the brain-chemistry of the deep-tissue nervous-system awareness and consciousness. Tossing and turning and trying to sleep, a small panic occurred to him, as he realized he might be losing his grip, or even going mad. It was an ill feeling, dread, and then fear. He tried to shake it off, snap back. But it was still there. His face grew reddish, he strained, feeling alone, beginning to sweat.

Now he told himself in firm commitment. "Don't hide it away, tell Guy if you have a problem, it's okay," he said to himself, sitting up on the bed, where the null-gravity straps held him down for rest. If he went insane for whatever reason, he had to take responsibility anyway, to protect the others, and then see what would happen. "It's the antibiotics, that's all."

## CHAPTER 26: Peace Betrayed

The transmission from Vandenberg was prepared then sent aloft on radio waves. The communications towers at the California base hummed and vibrated with specific energy, directed carefully, at just the right angle and intensity, to reach Guy's ship, so far away. Yet, invisible, with no indication anything was happening at all, outwardly. Within an hour, the content of the message was available to Guy.

Guy complied, working with his communications man. This time, the link to the Krenika, and the second ship, both too hot on his tail for comfort, which had created the crisis---this time, Guy sent the message as a packaged recording, without direct contact between himself and the Russian-Islamic pilot. He well understood that Zolotny would mock, lie, stall---and then possibly attack. So it was sensible to avoid talking with him. The Penelope's communications man confirmed the transmission, and receipt, then again all they could do was wait and prepare as they were able. All three ships were speeding towards the Molinari Space Dock at very high-speeds through the Abyss.

He considered his realistic options. Running was beginning to look rather appealing, or out-running his opponents, as his navigator had suggested. It would take a few hours, but in theory they could increase the ship's speed to maximum, or close to maximum. And in theory, the Penelope was a faster ship---the Monsanto-Dunlop Condrum 21 Deep-Space Local Planetary Cruiser was a hell of a ship, and the Russian-Islamic designers had never quite matched the superior Western technology for deep-space travel, mostly for cutting corners on cost and materials. At 50,000-miles per minute or so, and given his flight-path navigator's pre-planned escape-maneuvers, they might leave the two Russian intruders far behind with two or three days---far enough to exclude any dangerous collision or attack. But, Guy had to consider the remainder of the voyage to Mars, and the high-stress engine burn that would be needed, and the use of fuel, as well as Vandenberg's instructions. If things went wrong, as they were trying to out-run these two ships, they could end up really screwed for fuel and engine-function, even badly enough to leave them in serious trouble. And a high-speed run could not be sustained for long, anyway.

Meanwhile, aboard the Krenika, Zolotny and his flight-deck crew were also waiting. The reason was similar to Guy's delay---he couldn't act without mother Russia's decision on his choices. Unknown to Guy, or Vandenberg, weapons were ready to take out the Penelope. Or, to try. Although the Krenika was also basically a transport ship, a hastily rigged missile-launch system had been installed prior to launch. It was fixed into the cargo-bay and an air lock opening (door)---it was targeted by computer, and would be launched, and then navigate as a short-flight rocket, directly at the Penelope, only three miles off from the Krenika, and thus a dead-shot certainty for the kill, tracking to the radar-scan of the hull of Guy's ship. The ship had twelve such missiles. Two were ready for launch at any moment. It took half an hour to prepare another launch. The explosives were of a typical Earth-side wartime design, and functioned just fine in space. The Penelope would be obliterated, shattered into a million bits. The other Russian-Islamic ships also were equipped in this way.

"We shall see," Zolotny repeated himself. "I don't care, either way. I don't enjoy killing my fellow astronauts. I am not a soldier. We shall wait. We shall see. This recording from Vandenberg means nothing. They have not said anything new, have they?"

"They want directions to Jupiter, in case they get lost!" his co-pilot joked. They both laughed. But the mood was grim. Zolotny had of course considered the possibility that Guy would turn-the-tables, and shoot him down, given that he really had no idea whether or not the Penelope had any weapons similar to his own. But, KKF-Region Six had enough information to understand that Guy's ship was only a transport, launched prior to the advent of hostilities, and thus unlikely to have been equipped for a war.

"Start-up whatever you need to do to increase ship's speed to 90-percent maximum, on command," Guy told McGee. "Work with the engines-crew. Get it ready, just in case. If we need to, I want to be able to initiate high-speed quickly."

McGee was happy to oblige. It was his idea, from their previous conversation. "We're also ready with escape-maneuvers," he told Guy. He gathered his gear and logs, hand-held computer, and alerted the Engines Crew what they were doing. Like anything on the ship during a flight, they had to prepare things carefully---it was a complex task. By working out the details ahead of time, when the order came, they would be able to increase speed quickly, and with safety. "You need to rest, Guy," McGee said, on his way from the flight-deck helm. "When is Rob's shift up?"

"Forget it, for now," Guy said. "He was off-deck four hours ago. I'm good."

"You okay?" McGee added.

"Get the fucking engines ready for top-speed. I'm not dying out here, and neither are you. We'll get this done. Just do it," Guy responded.

"What about me?" said Raza, the on-deck communications

guy. "I'm not fond of the idea of floating home. Ha-ha. I know you'll kill me if I screw up this link to home. Ha-ha."

"Never," Guy said. They were getting a thrill-kick out the whole thing anyway, but crew-cheer and stamina was essential, now. *Run silent, run deep,* he thought. The phrase had now become a mantra.

Back on Earth, communications were also flying from continent-to-continent. As the situation reached the US White House Presidential Council, and its four-member decision-making group, a 'hotline' message to Russian leaders trickled across the wires and radio-waves, through the Russian Kremlin leadership---disconnected though they were from the real power and control of the Russian-Islamic space-program command.

The US Presidential Council Seat now included Renolds, a southerner named Boline Bouvier (a Black educator from the University of Texas), Martha Hazlett, age 45, a popular swimmer, with a law-degree, and a Hispanic man with a business background, named Martinez Jeses-Gaurrerao (age 38 years). All four of them were fully informed from Vandenberg. The decision was made to try to link directly with Terchenko in the Ukraine, the mystery-man calling the shots for his ships. The basic logic was simple: if there was an attack on the Penelope, it was now officially a war, a real one, with all that implied both on space, and on Earth. Somewhat of a Cuban-Missile Crisis circumstance, or a 'Pueblo-incident' type, from the standpoint of a military war footing (the 'Pueblo' was a small US war-ship that was attacked in South-Asia without warning during the 1970's, and was cited as a reason for warlike advances at that time). So, from the White House, and with their military planners, it was now to where they must rescue their astronauts, and avoid a war, or else, fight the Russian-Islamics in a full-on battle-royale, with no pulled punches. If the Krenika attacked.

More hours passed. Then word reached Guy and his crew. A radio pow-wow would hook-up in half-an-hour. Online were to be Terchenko, Lynn Rogers-Smith, Zolotny, and Guy. The arrangement was rather Mickey-Mouse, rigged for the moment, as far as the technology. Rudolph Terchenko was summoned by his commanders, and had to comply---otherwise he would never have agreed. This took time. But Smith, and Guy, had a lot to say, and welcomed the opportunity. A video set-up was also prepared. The Vandenberg teams were ecstatic---they would get Russia on record, and hold them accountable.

"Christ," Guy said. "You've got to be kidding."

Rob Cowan had now returned from his rest to the flight deck, again to resume his post as co-pilot.

"Let me get this straight," Rob said. "Twenty minutes from now, Lynn Rogers-Smith, the Russian launch commander, you, and the pilot of the Krenika, are going to work this all out by radio link?"

"While you were sleeping," Guy answered. "But, yeah, that's it. What's going on, Rob? You don't look well. You okay? Get some rest?"

Rob stalled. "I don't know," he said. "I had a bad dream or something. I feel woozy, not thinking right. They told me in the pre-flight, it might be the antibiotics I'm taking."

Guy frowned. "Don't freak, pal," he said. "I need you. You start feeling nuts or something; I'll tie you up with straps in the cargo bay. Seriously, it's okay. Tell somebody. You're all right. You get feeling really out-of-it, just checkout. We can handle it. It's not your fault."

Rob seemed to sink in the chest, somewhat humbled. *Not the right stuff,* he thought. Astronauts have a certain amount of pride, they all knew. "I'm good for now," Rob said. "I'm okay. Anything else with me, standard-procedure medical flight practices, that's all. I'll be good."

Guy dismissed the topic for the moment, scamming out his logic for the pow-wow. The situation needed to resolve. It was getting old fast, their career as a sitting duck.

"Ten minutes," said Raza. "We have the signal from Earth, rock-solid, audio and video both. The system is online. Whatever they got on the other ship, we'll find out in a minute."

"What machine do I talk at?" Guy asked him.

"This one," said Raza. "This screen will show you Rogers-Smith. Talk into the hand-mic. We won't see the Russian commander on Earth. But they will. The White House, too."

"Can they see me?"

"Nope, sorry. System wouldn't handle it. Just put a smile in your voice. I will be able to switch from image-to-image, and Vandenberg will change the images, too."

"I look like shit anyway," Guy said. Within a moment or two, there was an alert-sound, and the various parties came online: Rudolph Terchenko from the base in the Ukraine, Lynn Rogers-Smith from Vandenberg, Robat Zolotny, pilot of the Krenika, and Guy Reisling, pilot of the Penelope. No one doubted many other ears and eyes were listening and watching, even among the most important and powerful ears-and-eyes on Earth. Signal cues and ship-ID's were quickly exchanged as a standard-practice. Rogers-Smith took the lead to get things out-in-the-open.

"We meet at last, Commander Terchenko, after a fashion," said Smith.

"Greetings from the Ukraine and my program team, Smith," Terchenko said tersely. He was not being very friendly at all.

"All right," said Smith. "You have our message from three hours ago regarding these ships in the Earth-Mars corridor. Your pilot is too close. This is standard for safety. Let's not bullshit, commander. Are you going to destroy our ship and men, or not? If the answer is no, as we certainly hope, then have your pilots withdraw to a safe-distance. Any other answer is un-acceptable, and your people have started the world's first full-on war in space, which we all understand has always been a peaceful, educational and research area of exploration. In any case, please state your case plainly."

On Guy's screen, he could see Rogers-Smith, in a kiosk at Vandenberg. Terchenko's voice was like that of a lifetime tobacco-smoker with a harsh throat infection. He coughed and cleared his throat. "These ships are visiting Mars under Earth Space-Program authority agreements from 30 years ago," he said. "We have every right to visit the planet, on our own terms, under these treaties. Any Earth-based space-travel programs are entitled to visit any planet at any time. The United States does not OWN Mars, woman. Our pilot has made an error by moving too close to your ship, due to a mix-up about the intentions of your other US ships, which are now far behind mine, and full of soldiers and bombs, too, as we know. Don't deny it. The United States is every bit as much responsible for any hostilities. Admit it."

"I am a woman, sir, but there are plenty of big strong men over here who may disagree with your assessment of things, not to mention your transparency as far as with-holding your truthful intentions," Rogers shot back. "Please have your two ships withdraw immediately. That is the only question for now. Our pilot and crew are operating under standard practices."

"Get the hell off my ass, Krenika!" Guy suddenly said into his hand-mic. "I dump my toilet tanks out there, and we could all be killed. You know this. Pull back both ships to 10,000 kilometers before somebody fucks up. The ships are too close for deep-space flight at this speed."

Terchenko stalled. "My flight-track monitors are telling me the Krenika is having technical difficulties," he said. "Russia and the Islamic-Hindu Alliance are operating under common treaties for this space-flight. We are not in violation. Your ship, and your men---"

Now, without warning, the radio-link started to break up badly. There was a wave of very loud static-disruption; the screen-images began to fade, then blinked off. Raza Brahman, Guy's radioman, worked on the problem.

"It's dead. We lost them,' he said. "There's a power-surge now---geez, the Krenika is transmitting a huge EMP wave—they're frying our systems. Shut down your pilot controls!"

Alarms started to sound all over the Penelope. The Electro-Magnetic Pulse from the Krenika was nothing more than a massive release of radio-wave energy, trillions of volts of magnetic radio-waves that now suddenly and without fair warning, washed over the Penelope, shutting down much of their electronics, even essential gear.

"God-dammit! Life-support! Life-support! Go to batteries, now! Protect your systems, we're under attack! Back up all systems!" Guy broadcast on ship's communication to the below-decks crews.

They all scrambled. The peaceful radio pow-wow from Earth had failed, to say the least, doubtless part of Terchenko's plan. Not a rocket-launch or missile with explosives, but a sneak-attack with a technology-destroying EMP. Later, Terchenko could claim they had nothing to do with it, and the radio-systems had failed for some other reason. And at the same time, the Russian ships would proceed ahead to Mars, while the Penelope floated nearly dead in space, her electronics now toast.

"Communications down," Raza cried out. "All dead. Nothing. Radar. Tracking. Telescope. Holy fuck."

"We're slowing, engines shutting down on safety-sequence, power-loss," said McGee. "You got no spark. Losing speed. They'll pass us within hours."

Guy punched up the link to his cargo crew, from his command station, while they still had power. "Dump it, Arron! Dump it all! You have five minutes of power to open your bays and release our gift, like we planned. Do it now, use battery-power!"

The Penelope now slowed, but because there is no natural friction or resistance in space, her momentum would keep them in motion, slower, even for a day or so. Aaron Munoz, the cargo bay chief for the flight, had previously worked out their plan, between he and Guy, hours before. He was to open the cargo doors and release about a half ton of trash and un-wanted items, including small beds, desks, computer-gear, mirrors, bottles, and sewage, including heavy objects such as glass and metal, that would drift off quickly behind the Penelope, in the direction of the Krenika and the other ship. Guy obviously was hoping they would be hit by the debris.

"You got it, captain," Aaron's voice replied. The rest of his teammates did whatever they could to sustain the ship's systems.

But it was really no use. The massive EMP pulse had ruined their electronics, as intended. Lighting went down, computer-controls shut off, alarms sounds, engine systems such as air-circulation, plumbing, doors, air locks, all went to safety stand-by. They had back-up batteries, but only at the survival level. Unless they could restore their power-systems, Guy's ship would thrust ahead for another few thousand miles on momentum only, and then begin to drift.

"Life-support is on back-up only," said Rob, looking at whatever monitors and gauges were still working. "Your maneuvering jet-thrusters are off-line totally. All engines now in final shutdown sequence. Ship-to-ship and Earth-Mars radio gone, dead, no power. Lighting is out ship-wide except for the flight-helm stand-by. Computer systems for navigation---dead, to re-boot status, unless our data-logs and programming was fried, too. We're screwed, Guy. They must have hit us with a million volts."

Within a few hours, the Penelope had slowed significantly. Both the Russian ships that had been too close for comfort, at one point passed Guy's now seriously disabled ship, not even pausing to help. Their ships were at full power, of course. As the second ship moved through, still at high-speed, a large window-port beam, about the size of a curtain rod, but much heavier, from among the debris released by the cargo chief Arron, smashed into part of that vessel's hull, creating a dangerous rupture.

Now, it truly was a war.

## CHAPTER 27: Electro-Magnetic Pulse

Within about 30 hours, the Penelope had slowed to a virtual standstill. The ship's momentum, in the frictionless, gravity-free void of space, diminished gradually according to the laws of physics. Without power, or forward-thrust from the normally powerful engines, now silent or dead, the ship lost movement, so carefully tended and directed. Without her engines, Guy's Dunlop-Monsanto dream-flyer transport ship finally slowed even more, ending her encounter with the Krenika, in a state that resembled complete motionlessness. A physics or rocketry science-expert might have argued that she was still actually in motion, and indeed, even at high-speeds, any deep-space ship would appear to be motionless, from an offside observer. But from a space-man's perspective, it hardly mattered. They had stalled, completely adrift now, a space-pilot's worst nightmare, or close to it.

The Electro-Magnetic Pulse emission from the Krenika was a masterful and devastating choice for the Russian ship's pilot Zolotny, directed, no doubt, by the KK-F/Region-6 commanders. EMP's are a common usage for any high-tech application, especially radio communications. Any radio, and also microwaves, use EMP's, but at a much, much lower level of power. Your FM, and AM radio, receives the same kind of energy, attained by frequency modulation, or amplitude modulation, to carry forms of information. Much like a normal magnet, waves of invisible electricity-energy, from power-sources, using antenna, move in waves or ripples of energy. By the Turn of the Century, science had learned to use this in other, more intense ways. Microwaves operate the same way, for cooking, at higher energy vibrations. True-to-form, by increasing the source-energy to extreme voltage levels, using specific directional antenna-projectors, an EMP wave could produce remarkable effects. At billions of volts, the type of EMP that had been directed at the Penelope, would disrupt most electronics. By overwhelming the sensitive and delicate computers, wires and circuits, scanners and cameras, radios and radar, with huge amounts of energy, as electro-magnetic waves, an EMP of sufficient energy was great way to attack an enemy who might depend on high-tech gear. Thus, a very modern weapon, and interestingly enough, somewhat non-violent, or non-lethal. Apparently the Krenika was equipped to generate this kind of effect, and her Earth-side masters knew they could easily disable the Penelope in this way, with no real trace of more physical damage, such as explosive missiles, guns, bombs, etc. Whatever their technique, such as battery-stored voltage, Tesla-coil type energy-amplifiers, or some other means of generating high-levels of energy for single-flash release, it was enough, and more than enough.

Although dangerous, an EMP like this would not immediately end human life, or actually kill a living creature. So Guy and his crew might have been thankful they were still alive, or had been spared. But they were not happy at all.

The first two Russian-Islamic ships, which had been so close to the Penelope, at that point then for days, when the communications-link radio hook-up with Vandenberg, KK-F/Region Six, and Guy and his counterpart Zolotny, was interrupted by their surprise-attack: those ships quickly over-took and passed Guy's Earth-Mars transport. Guy and his cargo-chief Arron, figured that a well-timed release of selected waste and debris, might result in some hurt or damage to their opponents. It was prepared ahead of time, during hours of waiting. A half-ton of stuff from all over the Penelope was placed at one of the cargo-bay main doors. The bay could operate as a sealed air lock area, and the materials could be ejected automatically. When Guy knew they had been betrayed, he gave the order, and Arron caused the machine-systems to do the deed. With the Krenika and the second ship behind them by only a few miles, in motion, he was hopeful that some bit of debris, hopefully a heavy piece, would hit their ships at high-speed. And he was right.

So as the Penelope's electronic systems were shutting down, a 3-foot beam that was part of a port view assembly, smashed into a spot on the hull of the second, unidentified Russian ship, causing a small rupture or fracture (though Guy had no way to confirm this). As his ship lost all power (many of the systems were programmed to shut down harmlessly in such a case, automatically protecting various components), the two Russian ships then passed him at higher speeds. And then there were the three other Russian ships approaching from far behind, at about four or six weeks Earthward in the planetary passage-corridor, 'that-aways'. A small victory, then, by the chance arrangement of the positions of the three ships at-speed, and the motion of a certain three-foot metal beam, that Guy could have his revenge, depending on how much trouble he caused his enemy. Even a very small hull-fracture, in space, could be a big problem. When Arron released the load of junk, it dispersed quickly into space, like a fan spreading behind, outward and expanding, a small explosion of space-crap, small items, junk and trash, un-wanted tools, hinges, plastics, and glass, until it had spread over an area of ten miles or more. This was what they expected. Revenge had come randomly, then, and with any luck the second Russian ship would leak precious life-sustaining oxygen or fluids, by the resulting hull fracture or rupture, until the ship's crew was dead, suffocating, or exposed to the deadly ice-cold vacuum of space.

It was by means of this incident, then, that Guy, the transport pilot, whose ship had no weapons, realized in no un-certain terms, that it was indeed war, and him in the middle of it. He and his crew were still in grave danger even deadly danger.

A long, long hour, there within Guy's ship, so small against the backdrop of the infinity all around, as the harsh reality of it all settled into their minds and thoughts. Damage report: ship's systems were screwed royally. Most, but not all, of his 24-7 electrical, had been seriously interrupted, and of course their functions. The good news was that many if not all of these, could be re-booted. Running lights, computers, doors and air locks, seals, tracking-data and radio-radar, solar-panel energy collectors, oxygen-air circulation, main engines and maneuvering, water, food-service/storage, cargo, navigation and maps, were all electric in the essential operations. But circuitry disruption was not permanent. The EMP voltage created a system-shock. Normal power levels, such as required amperage-voltage, or circuitry flow, very delicate and precise for data-streams and hard-drives, were over-powered and then failed for a hundred different reasons. But as the EMP passed, which took about ten minutes, the same systems were simply 'off-line' or 'down', and in theory would function again once re-booted properly. In other words, the EMP did not melt wires or plastic, fuse glass, or scorch metals, or cause fires or explosions. So the urgently compelling problem they now faced was blood simple: essential life-support systems must-needs be properly re-booted for electric, or repaired, in a short time, before their air-supply was exhausted, and they all died. The estimated time-period for this was about a week, and they could be as precise as they wished there. After that, the crew would seriously lack for air. They might somehow sustain their lives for a while, at that point, or even be rescued by one of Winton Berle's ships, months behind them, or a lifeboat vessel from Molinari, but in the interim, they could only struggle to correct the situation as quickly as possible. So, although they knew the systems and methods, it was no simple task. Dozens of systems, dozens of computers, dozens of machines, dozens of controls and interface inputs, some hard to reach, or requiring special tools. They had already started on basic work and getting organized by priority, nine men, working together at a furious pace to save their own lives. Their only relief, as they worked, or morale-booster to keep up their spirits, was cursing the Russians who had attacked them.

"I'm pretty sure we had a hit on the second ship with some of your space-junk, Guy," said Tom McGee, the navigator. "We had just over ten minutes before I lost all radar, when the wave hit us. We logged a scan on a shock-point, or flash, on the underside of the second one, the one farther out. I was tracking them the whole time until we went down, so I could see the data-image on my radar, like a small vibration or peak, or like the radar would indicate for a tiny meteor hit. You could hardly see it, but it wouldn't have read back that way for much of anything else, if you know how to read these. I could also read the even smaller scan-points on your space-junk. So, I feel we may have had a hit on those fuckers. Just FYI."

"Just fucking fantastic," Guy replied. "I am so agonizingly happy about that. Okay, next. Keep on your re-boot for that station-level air-circulation. Get it back up. Any idea how long there?"

McGee was on the flight-helm when most of the crisis had taken place. "Not sure, time-wise," he said. Guy, McGee, Rob Cowan, and Raza Brahman, were still on the cockpit deck, and had only retired for various forms of rest, such as bowel elimination and clean-up, minor food and rest, in rotation, a few minutes at a time each. The other men---Arron in cargo, Bryan

Price and Hugo Mortinger in the very large main engine control room, and two Ship's Systems men, a younger astronaut named Peter, and a more experienced flight-worker, Charlie Ripa. Herbie French was not with them. Adrift now and powerless, each work-area was dark, cold, still and soundless. They could only communicate by using the space-suit helmet radios, from level-to-level, or some of the hand-held radios. Emergency and battery lights were all they had. Hushed tones and essential information was about all they bothered to share, as saving batteries was also important.

"How not-sure are you? Come on, Tom. Guess. How long?"

"I'm not very familiar with the station-level modules for that," McGee said. "I need procedure manuals, hard-copy."

"Each ship's level has three station-level air-circulation points. They pump oxygen. Four ship levels equals twelve circulation pumps. Ripa and Pete are re-building the system program for the main air-scrubbers and oxygen tanks. By the time they're ready, we still won't have good air without at least one functioning station-level air pump on each deck. So, that's four, we can repair the others later, the air will function with even one pump for each level, it's designed that way. Find the fucking procedure on it. The pumps need to re-boot, you need to get into them, one-by-one, re-set the power-supply levels and controls to proper calibration, the proper settings, then check for function, or any serious damage. Ten-hour job."

McGee huffed. "Well, dammit. Nine hours, then," he said, determined.

"Do it," Guy said.

And so he did. The station-level oxygen circulation pumps were also somewhat hard to reach. Guy was busy with a main-panel flight-control on all ship's systems. It was his most essential interface to control what was happening with the Penelope. There were back-ups, the one his co-pilot Rob used, and a third emergency control interface, on another deck. Guy wouldn't need this right away for any maneuvers, docking, speed, radar-navigation, engines, and so on. But the main control panel, aside from simply being an emotional comfort-zone for himself as ship's pilot, also linked directly to life-support monitors and over-ride controls there. Life-support, now, was Job-One, so the effort made sense. But like the rest of the ship's electrical---and it was all electrical---Guy's pilot controls were toast. It was very technical, complex and demanding for a proper re-boot and repair, re-setting the intricate power-supply and programming demands, for which there was simply no substitute for perfection. So, Guy was busy, too.

His co-pilot Rob, worked with him on the problem. 'Eighteen main motherboard circuits, each one handles multiple systems to our controls and meters," Rob said.

"I thought it was twenty," Guy said.

"Well, you're in luck. Only eighteen," Rob answered. "And they're all fried."

Ship's pilot control screens and panels were dead, blank, no power. Guy was looking over a lifeless, powerless machine that all their lives depended on. The loss of decision-power was frustrating, to say the least. "Call me a control-freak," Guy said. "I want most or all of these re-booted. Job one for you and me, for now. Geez, where do we fucking start?"

"Start at the beginning, go on through the middle, until you reach the end," Rob joked. He now was reviewing a technical procedure hard-copy binder that diagrammed the entire main control panel system, using a battery-flashlight. The schematics were awesome.

"How's that whole thing with your antibiotics?" Guy asked his co-pilot. He hadn't forgotten what Rob had told him--- bad dreams, fuzzy thinking, fears, even visions or hallucinations, He had a tool-set from one of the storage-compartments on the flight deck, with electric testing prongs, small tools, and so on.

Rob was flipping pages in the hard-copy control panel schematics and procedures binder. "Uh---well, let's just say that unless my worst fears are true, and those amp-voltage test-probes in that repair kit you have there are seriously intended to be used as cerebral implants for you to analyze my dreams and brain-chemistry, as I fear---short of that, I can probably handle things, for now," he said.

Guy was a little shocked by his comment. Paranoia? He braced.

"If you have serious fears or un-healthy thought processes or delusions and any hallucinations or apparent psychosis, you will inform myself or another crew member immediately," he said sternly. "That's a fucking order."

Rob turned, humiliated, now more serious.

"Yes, sir," he said. "Sorry."

## CHAPTER 28: My Boyfriend Started a War

*"They screwed us blue. Damn."*

*--Lynn Rogers-Smith, General-Authority lead, US Mars Program, Vandenberg, California, 2077, aka, 'the Commander of Angels'*

Of course it was only a few minutes, after the carefully arranged radio-link hook-up had gone suddenly blank, that the Vandenberg team recognized that they had been betrayed by the Russian forces, in all probability. Ibrahim Mehudi, the Program Science Authority-Lead, had been sorely wronged, humiliated by his idea that they could still make peace. The radio-visual link was intended to stall, delay, or correct any hostilities, and possibly save the crew of the Penelope. Lynn and others felt they could simply demand the other ships to withdraw, give them a chance, under common space-travel laws. She was no fool, she knew the stakes. On the other hand, Asteroid U2753b was still three years and eight months off, from even a potential collision with Earth, and other teams were working on a solution there, and would solve it, in time. So there was no real urgency. But, on the other hand, the Russian-Islamic may have seen the situation as simply a good time to try to take Mars-Base Snikta anyway, whether there was any real need or danger at all. Why pass up the opportunity?

"They screwed us blue. Damn!" Rogers-Smith said, shortly after the attempted radio pow-wow. "The fuckers knew we'd be ducks for a surprise-attack. I should have seen it coming. Mother of god. What the hell has happened?"

"Hate to say I-told-you-so," offered Branson Porter, the Vandenberg Security lead. "They're just not like us. Negotiations were a bad idea. The Penelope should have run, out-run them at top-speed, while they had a chance. We're the good guys, for Christ's sake, but we can't deal with them from weakness. And now you see the results, don't you?"

It was half-a-day following the collapse of the pow-wow radio hook-up. As the screens and monitors and radio-speakers and mics at Vandenberg suddenly went dead, blank, with a few moments of heavy-noisy static and signal fluctuation, those on-hand could only guess what had happened. But tracking-and-telescope scans that were following the Penelope and the other ships, made it at least somewhat clear: Guy's ship started to slow, there was a release of some debris, and then the Russian ships speeded past. Big-picture scan-maps placed those three ships perhaps six or eight weeks away from the Molinari Space Dock. Then Guy's ship slowed and slowed more, finally stalling. Radio-workers scrambled to re-connect with him by standard radio, and emergency-radio, but no luck, which was really frightening. Complete loss of radio communication with an in-flight Mars-transport was very serious, and meant something was quite wrong. So their first assumption was a missile or rocket hit. At the same time, US channels to Russian and other space-authorities, as well as global and world space-authority, and US White House and military, rushed to figure out what had happened and to respond. It started to look a lot like hostilities, in space, and on Earth, was ahead.

Rogers-Smith, Porter, Mehudi, and now a larger team of experts and technicians, continued to manage things from various work-areas, and launch-flight command center, in California. At one point, a radio-link connected Rogers and others to Winton Berle, the experienced space-program veteran, leading eight US ships through space to Mars, just over half-a-year behind Guy's ship, and of course the Russians. The compressed-data signal flickered across the distances, but Berle's raggy-sounding voice was distinct and clear. His ship, leading the other seven in formation, was called the 'Understandable' (he had named it himself, years before, somewhat of a personal joke). Like most of the others, this was a very capable vessel, about the size of a large oceanic cargo ship on the seas of Earth, perhaps several thousand feet long, much larger than the largest Jumbo-Jet ever built.

"Wait---wait," he said. "Just slow down making choices, study the situation. You still can track Guy's ship, so that means they didn't blow them to hell. It's still there. They lost power for some reason. I'd guess an EMP, they work great for that they leave no trace, hard to confirm, but no-kill technology. Your scans would have a log with a big energy flare, or the tracking from Molinari. My group can't get to him in time, calculate a rescue from Molinari, he's only two months out. A high-speed lifeboat might make it. Main thing is if they lose air. But don't lose hope," Berle consulted with Rogers-Smith, as the link was stabilized and they could communicate.

"Yeah, thanks, we thought of those options, of course. It's underway. We felt an EMP might have been used, too, but we scanned debris from Guy's ship, so we weren't sure. But we still have a solid scan on the Penelope, so it doesn't match," said Smith, on a radio-link, now from the launch-command center, in Vandenberg, complete with teams of mission monitors and aspect-controllers, all working together, especially with so many ships in flight at once, probably 40 men and women and helpers.

Berle was now in-residence as the commander of the Understandable, on that ship's flight deck. All-systems were fully functional, and there was no danger at all, given the three Russian-Islamic ships nearest them were thousands of miles away, even more. Space is big, the distances are vast. So Berle was very comfortable, at cruise-status, watching over his other ships in formation behind and around him. Of course, each of these was full of weapons and soldiers, anticipating all they had planned for to protect the Mars Base, so far ahead.

"You did what you could, I guess it was a crisis-mode choice," Berle comforted her. "It's a hostile action, without a doubt, so I'll be ready for more of the same when we reach Mars. You also need to protect Molinari, Lynn, we planned for that, too, but now they've advanced two ships ahead-of-schedule, closer to the space dock."

"Yes, yes, underway," she replied. "Be informed, at this point, for your teams and pilots. It looks like there will be no way to prevent a battle, and all that means, damages to ships, killed astronauts, damage to Snikta. It's just inevitable. Please have your communications ship-to-ship with all details as you receive information from us here. We're initiating a regular all-ships-in-space dispatch. And one other thing, for yourself personally. Your brother in Utah says to remember to eat your fish-oil pills for your heart, I guess you knew. He says hello."

"Fish oil is an excellent source of blood-thinning and amino-acids that have a wide variety of healing applications," Berle said in a lighter tone. "Entirely within program dietary specs."

Lynn's voice seemed droll and languid, stuffy, but they were laughing, just a bit. "I know, Winton. I take them, too," she said. They talked more briefly on specific details, and Berle's call was turned over to a specific flight-command desk, where other data was exchanged. Work continued.

All of these events reached both Mars and Molinari, and staff-tech workers at both facilities had been advised previously, on-going. Of course, interest was intense. Mars Base Commander Bojji-Than, and his people, were very sure they were sitting ducks for whatever was ahead. How could they really fight off five orbiting Russian trans-planetary ships, full of soldiers and weapons, with only what sufficiency they already had at the Snikta-Ridge Volcanic-Basin US Mars Base, home of more than 200 people for years at a time, and a completely peaceful research and educational outpost for American-US and global Earth interests? It seemed hopeless, and whatever they could learn on Mars about the approaching ships, and what had happened to Guy Reisling's transport, only seemed to build worry. The tech-staff, tracking and navigation-mappers, or star-trackers, obviously kept very close track on whatever was confirmed, and also data from Earth, even hour-by-hour. And they did what they could to implement the various security-measures ahead-of-time, as directed by the Mars Base Defensive Task Force. Major work included preparation of the very few ships that could actually enter orbit from Mars, only two, called Orbital Lifts, because that was their only function, and they had no weapons. The gates and entryways of the base-facility were fortified, and some were sealed; committees of staffers prepared oxygen-suits and short-term life-sustain arrangements for everyone at the Mars Base---every single person, and extras. This was in the event the base itself was breached or entered, such as with explosives that would blow off the gates and air locks, creating a sudden loss of air-pressure. And then there was the work on the perimeter Life-Sustain Igloo 'fox-holes', planned by Earth commanders, for US Mars soldiers to survive in as they supposedly fought the descending Russians, out on the cold, airless, lifeless planet-surface. But Bojji-Than was intelligent enough to realize just how unlikely it was that either side would do actual hand-to-hand combat, or fight each other with guns and grenades, successfully, on the surface of Mars. The environment was simply too alien to human survival and hostile in general, extremely harsh, without a properly life-sustaining atmosphere. Even in the suits, or using the igloos, it would never work. There was no use arguing the point, the battle-planners had to do something. As Mars Base commander, Bojji-Than was worried about the whole thing, you might say. From one of the social view-port garden patios, he looked over reports until he was exhausted, then found himself gazing out at the Tharsis Montes mountains, four times higher than Earth's Mount Everest. The woman technician from Earth, Karen Tutturro, entered, and was enjoying a portable meal she carried in a tray.

"Inspirational, aren't they?" said Bojji-Than.

"What? Oh, the mountains? I hadn't noticed. Is that them?" She also looked out the window. The commander knew she was joking. "I'm kidding, Commander Bojii. In all my life, I never dreamed I would see mountains as high and extreme, as the Tharsis Montes. It's overwhelming, they---they aren't hardly even like mountains at all, if you get my meaning. More like--- immense tidal waves of rock, or a roof, or a stone grave we are all laying in together. You can't even see the top."

"Where did they come from? In your opinion? Do you mind me asking?" Bojji-Than said.

"Is that a trick question?"

"Well, yes, but that's not my intention. It's sort of a Buddhist thing, or, spiritual. Where did the planet come from?"

"Uh, from God, I guess," Karen replied weakly. "Is that the Buddhist answer?"

Bojji-Than, who was an Asian man, assigned to run the Mars Base, now for the past five years, as the cap-stone of a wonderful and thrilling career in space-sciences, grinned at her, for Karen was a beautiful and exciting woman, very intelligent, with frizzy-style hair, a pinched-looking cute nose, puffy cheeks, and a smile of her own. "You are god, aren't you?" he said, joking with her.

"No, no, sorry," she said sweetly.

"That's okay," he said. "I am only joking with you. But if you were, or if you were goddess, I would say, you have created an amazingly beautiful and stunning planet, this world we call Mars. And these Tharsis Montes mountains in particular."

She giggled girlishly, and the dark mood of impending war and doom lifted a bit.

The red planet turned in the slender rays of sunshine.

Halfway home, when home is but a death away, or seeming so, the dark, cold, empty, airless void to carry who would make the journey, and certainly would, for Earth is a better place by far---halfway home, a home-away-from-home, the Molinari Space Dock, which we have described, was likewise a buzzing hive of activity, following the incident with Guy's ship, still drifting powerless, and seeming dead, out-there, somewhere. Molinari had many similar tools and tech-gear for one of its main functions, to monitor conditions and traffic in the circular-orbital pathways between the two worlds. Lila Meetek worked at this role, and was well trained and personally endowed for her job. So, when word passed from Earth-sources such as the Vandenberg launch-facility, to Molinari and the Mars Base, there she was, like the bride of Christ who never returns, alone, afraid, and angry, that her one true love, Guy Reisling, was in such danger, Lila along with the other Molinari workers, scrambled to figure out how to save him, as a team-member assigned the task, grim though it was.

Lila was again at her work-area kiosk, surrounded by the computers, screens and radar, communications, various data-logs, scanners, maps and controls, rather a dark little world of her own, that she alone operated. Molinari's environment was null-gravity, but she was held to her seat by a small, comfortable strap, and at the moment, wore only a flimsy bath-robe, which was a colorful fabric, warm and cozy (many of the long-term space workers would sometimes 'dress-down' for their shifts, for comfort's sake, since their work was demanding in terms of personal accommodations). Guy would have gobbled her up.

"Son-of-a-bitch," she said to herself. "He started a war. I knew it. That man. Christ. What next?"

A long pause. She quickly reviewed what they all already knew, with a one-sheet read on her VDT, showing the essential truth, Guy's ship, dead and lifeless, a pinpoint on the maps, and he inside with his men, either well, or not well.

"He'll be fine," she mused. And then, back to work.

## CHAPTER 29: Wind Chimes, Bells, Music and Wood

The most direct form of assistance and rescue for Guy and his crew was immediately recognized and implemented. A short-distance space shuttle, essentially a space flight 'life-boat', would make the 17-day trip from the Molinari Space Dock. Given that the crew of the Penelope could repair their electrical systems within that time frame, or sooner (they had only about five days of actually 'fresh' air left to survive, at that point), they might have it 'easy', in that sense. Or, they could make repairs piece-meal, working first on essentials, and then wait, and work together with a fully equipped ship, once the Molinari shuttle arrived. But it was obviously their best hope, the rescue ship, quickly undertaken as a flight-mission of urgent importance, under orders from Vandenberg.

It was a strange dichotomy of beliefs and practices that distinguished between the Russian space-program, and the US Western view. In the West, as in warfare, or deep-ocean exploration, or air-force activities, rescue workers or firemen, and other high-stress jobs, the foot-soldier, or astronaut, and so on, was almost coddled, or nursed along most cautiously, by comparison. And in an emergency, the West would spare no expense or effort, to save or rescue an injured man, or a stranded or dangerously disabled flight. This was thought to be compassionate, and of course it was. But, as per Islamic-Hindu tradition, and Russian bravado, the lives of their astronauts, were lesser valued when things got tough, at least in terms of rescue efforts, and sometimes in terms of medical or 'a-priori' safety provisions. They would just let the men die. This is true in much of the split between the two worlds, spiritually---the Western and Eastern. In the West, death is the absolute worst fate that can be imagined, the end, without exception, though fear of death might vary; in the East, death was often viewed as a release and escape from suffering, part of an endless cycle. But, the result was that the Eastern had a low-value on human life and survival---that is, 'life is cheap', while in the West, life was preserved as a sort of favor to the living, who may wish to go shopping, or out for dinner, or visit with family. And the pain, the cannon-fodder poor, the slavery, and all that. So, many millions of dollars (no longer in use by 2077), in terms of simple cost for a rescue, were not even a consideration, where Guy's crew was concerned.

McGee, and the man from life-support named Peter, whose surname was Thomason, a bold young adventurer, but shy and tense, trained with all there was to know about the Penelope's ability to keep them alive during transport---the two of them spent more than 24-hours, just to restore function to the air-pumps, circulation, oxygen tanks and release-mix, $CO_2$-scrubbers, etc. They had help from other crewmembers, too. Fortunately, all that was really needed was to start-over those systems, as far as computer control and management-operations. But those were demanding, those machines, for high-end quality programming, and some of the hard-wired applications had burned out, and had to be re-invented from other sources---back-up control programming. So they had to re-program how the life-support worked, while waiting to see if they would all suffocate, and also in radio-silence to any sources of help and guidance from Earth, or Molinari. Yet, remarkable to their skills and determination, and also courage, they fixed the damn thing for air-circulation within that amount of time, about 30-hours of work. A shout and round of cheers went up from every man on the ship, when McGee gave word, and the pumps started pushing their most vital resource---air to breathe.

"Take a deep breath," said McGee. "We did it. Back online."

"All air-circulation and oxygen systems should be fine for the remainder of ship's service for this flight," said Thomason. McGee grabbed the younger man's shoulders and hugged him roughly. Thomason smiled.

It was the same for main-batteries and solar, and the radio, scanning, radar, antenna, etc. What remained after those were the engines, and flight-deck controls. Other systems could wait, but it was only a matter of time, and a certain level of exhaustion for them all. It was fitting during their ordeal, after about a week drifting, that none of them had the slightest idea what was happening back on Earth, or with the Russian ships, or the US ships, or any info of the so-called 'battle for Mars". There were a few indications---they knew, for instance, that even at best-possible speed, the other three Russian ships were unlikely to reach them in so short a time. And they could estimate the travel-time for a rescue from Molinari. In many ways, they didn't know, or care, what the rest of the Universe was doing. It just didn't matter.

They rested in short shifts, and continued working. It started to make sense. Each critical ship's system had gone down with an overwhelming EMP jolt, but was not destroyed. So assuming they had main battery power lines, each system could simply be re-booted, one-by-one. But, computer also operated the main batteries, even if their stores of energy were not depleted by the same EMP action, which mostly they were not. More time-and-effort, before they could even begin, and before things started to returned to normal. The only other choice was death.

Emergency lights made the ship's interior seem dim and dreary. The air had been stuffy, even moist. Food was in cold-packets. No computers for any service, including music, or news. They started going around half-naked, too, much like Lila in her colorful robe, on the space dock. A little known fact-of-life for veteran space-travelers in 2077; the regular suits and out-fits worn while working, inside, were not much real comfort or use, or protection. Bathrobes and a few helpful items like handbags and tether-straps, and the magnetic-footwear, to hold them down in the null-gravity---much more common. Newbie or novice astronauts were often shocked, and it was an un-spoken rule not to report this as a violation. If women were part of any crew, however, it was very popular.

"I got the fucking mother-board re-installed on the mains for the co-pilot's computer," Rob Cowan told Guy, at one point. "That way, you won't need your mains up first. We can work those controls from mine."

"Great," said Guy. "Good job. My task is just on the maneuvering controls, for now. But it's coming along. I'll get it. Complicated. You got all the controllers, which are fine. But the second-tier relays are all down, the programming that moves my signals to the actual parts. I can't even test anything to see if they work until those are restored. And every inch of it is hard to reach, and needs special tools. And the fucking schematics manual is printed in itty-bitty fucking type so you need a fucking magnifying glass to read it. Who the hell thought that up?" He paused, and sighed heavily. "I'll get it, I'll get it."

"That's show-biz," Rob countered with humor. He was feeling better. Maybe it was the focus of pressure.

"Not a TV-show. Not a movie. Say it often," Guy said. They both laughed, it was a fraternal thing.

Another two days, and Guy was on the radio with Lila. Once the radio communications system was up, the first order of business was the official, perpetual link to Earth, orders, re-affirmation of their survival. Vandenberg was transmitting May-Day constantly, to them, and even as the Penelope's radios came online, within an instant, Guy's ship was receiving from Earth. Another group-hug round of applause and cheers at Vandenberg, which spread to the White House, Mars, Molinari, Winton Berle's ships, and then Russia, and even around the world (for those 'in-touch'). Life on the Penelope started to seem more hopeful. The engines, other mains, and the approaching rescue 'life-boat' from the Space Dock, which now would probably be only an escort, it all looked good. Somewhat like a typical space dock harbor guide-ship, or re-entry 'orbital tug'. After all the work and labor, and a few hours of dead-solid sleep, Lila and Guy could 'chat'.

"So---how's it going?" Lila said.

"Working it out. How about you?"

"Okay. Not bad."

They paused. The distance between them was only 250,000 miles, now, as far as the Earth is from the moon. The radio signals were compressed and slow as usual, but clear. The Penelope's batteries were full-charge. Within ten or twelve hours, main engines would be back up. In another 30 hours, the rescue ship from Molinari, where Lila was at her work-kiosk, would arrive. Life was good.

"So---your ship fucked up, huh?" Lila said casually. Guy had to laugh. She was acting all calm and bored on purpose. That was Lila. He knew better than to get angry.

"No," he said. "My ship didn't fuck up. I was attacked. We took a trillion volts of EMP. Shut down all our system-electrics. Like---you know---air?"

"So you couldn't breathe? Wow, that sucks. Oh well, you seem okay now."

"Better, yeah. Can you breathe where you are?"

"Yeah, sure. How could I talk if I couldn't breathe? What do you think, the dock was attacked and we're all dead and I'm a ghost? Grow up, Guy. Stop fantasizing. I'm fine."

"Just checking."

Lila made an effort to pant sarcastically and heavily on her end of the call, into her mic. Guy was confused for a moment, then listened more closely---Lila was panting, in-and-out, huffing, obviously so he could hear.

"Stop it, okay, okay," said Guy. "You sound like a Beagle."

There was a pause. "Now you," Lila said. "Do it."

"Do what?"

"Breathe for me," she said. "Do it. I want to be sure."

The data-compression slowed down for about a minute. When the real-time recovered, Guy responded by whistling a tune, an old American song called, 'Dixieland'. He wasn't a very good whistler, and after a few moments, he started singing to her.

"Oh, I wish I was in the land of cotton, old times there are not forgotten, look-away, look-away, look-away, Dixieland. I wish I was in Dixie, awaaaay, awaaaay, in Dixlieland I'll take my stand, to live and die in Dixie---awaaaay---awaaay---" He sang better than he whistled.

He could hear Lila laughing. "Guy! Guy! Stop!"

So he did. "What? It's a sing-along. You-no-like?"

She giggled. "Oh god. They let you drive that ship and everything? Why doesn't the space-program get a grown-up?"

"Hah, ha, very funny. With any luck, my aging process will continue for quite some time. I like growth, except between my toes, those little funky skin-things that peel off when you pull on it with your fingers."

"Calluses."

"Yeah. Calluses. Not my idea of personal growth."

"Too much information, space-man. So anyway, when does the rescue ship arrive where you are? I have my scanner-maps. They're pretty accurate. They're getting close. Why don't you just crank your engines and move out? By the way, that stupid meteor thing, the one they figure is hitting the Earth? You may not have heard. It's now absolutely confirmed. It will hit Earth for sure, no doubt-about-it. ETA is now 36 months. How's that for science?"

Guy tried to relax and just let go of all his fears and concerns. He liked his home on Earth, too, even much more than billions of people who had never left. Absence makes the heart grow fonder. "Tough, I guess," he said. "Sucks."

"Well, massive destruction, no joke. Same as those old movies, but worse, and real, too. Darkness falls over the entire human race, that's all. Maybe a few property damage issues. Tidal waves, earthquakes of impossible intensity, vast and un-bearable weather-system changes, cities and coastal regions utterly destroyed, dust-clouds of debris and gasses, half the planet in silent-death, bodies and corpses without end. You know."

"Have you thought about this much?" Guy asked dryly.

"No, I try not to. Just standard government visualization. Sorry. But don't worry. With guys like you on the job, not a problem, huh?"

A long pause. "What, me worry?" Guy said. "Hey, pant like a dog some more, okay? Turns me on. I like that."

They chatted more for a few minutes, but behind the mask of their jokes and light-hearted bravado, it was exactly what it seemed. Just like a movie, or TV show. Then they ended the call, by required schedule. He knew he would probably see her soon, and it gave him something to live for. Sex.

As for Lila, she eventually retired from that night's shift, with 'night' being a relative artificiality. From the moment they all knew that Guy's ship was in trouble, she had followed the crisis, hour-by-hour, with every scan or radar and radio-transmission from Earth, and with every authority available to make the right choices from Molinari. She also prayed, for what it was worth, as she went to her private quarters to sleep. Then as she slept, she dreamed, and it seemed that she was conjuring up some sort of 'new world', or a 'new Earth', in the slumbering sub-conscious of the mind's eye. A beautiful world where a meteor-hit was less likely, for many millions of years even, and where there were all kinds of new and healthy environmental aspects, green grasses, trees, rivers---and people, clouds and blue-green skies, food and fresh water. Even kids. And she and Guy lived in a small hut, which they had decorated with fish-bones and stones they made into wind chimes and bells, and wood. There was music.

## CHAPTER 30: The Understandable

The lonely planet, Earth, like any stellar or cosmic object, has its physicality in the material realm of globes and spinning, orbiting moons, and yes, stars, along with all of heaven. One had called us, 'the silent planet', but to live on Earth, with it cities and eras, people and children, and most prevalent and persistent quality of ignorance, if only for our dead, it can be very noisy and deafening, loud with every din and crying infant ever born, seeming without end. Wounded, old and treacherous, our eternal Mother, without a doubt. So old, so very, very ancient that even an elementary-level student, gazing upon the Earth's moon, through telescopes or satellites today or in 2077, far superior to those of the past—any student of the moon would observe its innumerable craters. And by comparison to the verdant and more-or-less successful appearance of life forms, ecology, and green-things-in-general, here on Earth, macro-biology, or common geography, it's fairly easy to see that the planet has far fewer meteor-craters, than its moon. Exactly why this is so, might be a lifetime study. So, though many, many millions and millions of years might pass, with the Earth and moon so near in space-time, isn't it inevitable that a large asteroid or meteor would eventually hit the Earth? And modern science in planet-review would tell us this has happened in ages past, with results such as the Ice-Age, or the extinction of the dinosaur, even Noah's Ark, for all we know.

Earth sciences such as those established by NASA, at mid-20th Century, had by curiosity, or prudent stewardship, tracked and followed all sorts of cosmic objects. Cataloged and recorded by type and size, also by pathway and speed, researchers yawned and sipped hot coffee, at their observatory posts, including many amateurs, jotting down their notes, sharing and comparing, to learn what they could. Somewhat like a man might examine his own body and find he has a cancer or dangerous bleeding, or spreading rash or infection, we set forth to find out what we never really wanted to know at all, that is, how and when and if, the end would come. That is, if total Earth-destruction was guaranteed, now in 2077, by Asteroid-U2753b, 'Big Baby Bertha'. The end, or not, more optimistic folks might recommend that we could somehow prevent disaster by predicting it, once the emotional shock passed and we could organize a way to do so, if we could.

This wasn't Guy Reisling's job, or Winton Berle's or Lynn Rogers-Smith's, or Rudolph Terchenko's, or that of any individual. But, as Guy's struggling ship was escorted at last, to a docking point at the Molinari Space-Port, halfway to Mars, and the rest of the Mars Base drama was playing out, the task of tracking Big Baby Bertha, and figuring out what to do about the damn thing, was now a major effort among Earth's science and space forces, as well as military and political-government. Fiction writers had long speculated, even in the Bible, about such a meteor, and how a strike would effect Earth. It didn't look good. Really just a heck of a lot of major property-damage, with no really effective liability insurance in place. No way Earth's funeral industry, or cemeteries, could handle the load, at all. Plus, many people might experience sneezing with all the dust, and for those who were not strong swimmers, in coastal areas, there were not enough pleasure boats and yachts. And there might be a loud noise that would wake up a lot of people who would want perhaps to be sleeping in, on that particular day. So, yeah, end-of-the-world, all that.

Shoot it down with a nuke and send it off course? Send up a ship, or a probe, land on the meteor if possible; install some sort of action-reaction thruster-rocket or large explosive, to change the path of the object? Melt it, or dissolve it, or break it up into a million smaller pieces? Speed the progress of the Earth's solar orbit somehow enough at that time so that the meteor would miss us? Mass exodus of the Earths population, to off world? From the point-of-view of the common man---they were working on it. From the point-of-view of the worlds elite class, the very wealthy or powerful, or the especially educated or skilled, the off-world exodus option looked better than it ever had. By 2078, comfortable bases on the moon, orbiting space stations, and such as the Mars Base, were a home-away-from-home in any case. So, even if the exodus were rather small in number, even only a few hundred people, not to mention a selfish option, it made sense.

Thus as the world turns, people really were 'working on it'. Thousands of University-level scientists were set to task, organized under more potent authorities, military, space programs, municipal-emergency, industry, ICBM, financial, survival organizations, and so on. The spectre and phantom menace of panic and anarchy must-needs be controlled, so true information, or the actual facts, were kept mostly 'under wraps', as far as the general public. Rumors, half-truths, speculative guess-work, filtered out into society, however, and most people looked at the topic with the view that there was nothing they could really do, it was all beyond their control, beyond any passionate prayers. It's probably just another well-timed and calculated official false alarm, to improve stock-market indexes.

Ibrahim Mehudi and his wife took dinner a week or so later, one balmy evening, along with Lynn Rogers-Smith and her husband. They had a fine meal and short respite from all their feverish work, at Mehudi's home, a pleasant abode in the California hills, by Santa Ynez. They had a dinner catered by a local Chinese restaurant. It was an indigo evening with many hot and burning stars, gazing down from above, and of course, any of them could recognize the 'star' that was actually the planet Mars, standing on the patio with drinks (coffee). Rogers-Smith's husband was a mature industrial engineer, gaunt and thin, with old-world grace. Mehudi's wife was an attractive Eastern woman.

"They'll find a way," Mehudi said. "Personally, I do not feel it is the end, with the meteor, for many reasons, that I have studied, from the best information. The meteor is big, but not that big, only about a few hundred miles wide and a thousand miles long or so. By comparison, the Earth is more than 30,000 times as large. Also, as the meteor travels through space, it tumbles and rolls and moves, so the actual point of collision that we fear, is at least somewhat variable. And the Earth is moving, too. So the collision-point is very, very narrow. It could miss. It might be only very, very close. But it might certainly miss the planet completely, even with our best math and predictions showing otherwise. So, for this reason, I still have hope."

"I know you do," said Rogers-Smith. "And I'm glad."

Space-program shoptalk was eventually pretty dull for their company, but both spouses enjoyed their association with the 'insider' aspect, and all the drama. They talked a bit more about the incident with the Penelope, closing that chapter of the struggle for the Mars Base with a sigh of relief. They also were talking about how important it was for the communications-system on Mars to be fully functional, as things progressed.

They enjoyed the evening and some rest with other talk, and some great food they all found delightful.

A deep hum and vibration, like a steady thunder unending, low and slow, was the sound that Winton Berle and his team heard and felt, hour-after-after, there aboard the 'Understandable'. The ship's engines were at best speed. Berle was on the flight deck, with the co-pilot and one of the navigators. His ship was very similar to Guys, but the Understandable, and seven other ships with him, each held all they needed to fight the Russian-Islamic teams at Mars, including some 30 fully armed soldiers, each. Berle, the husky space-veteran, who knew he was possibly on his last and even most demanding mission, led the defense-formation, and was very respected. He was well aware of what had happened with the Penelope.

"Smart," he told his co-pilot, an African-American man named Ben. "They were damned smart about the whole thing, the Russians, and the commander of the Krenika. An EMP was a way-better way to get what they wanted, instead of a missile."

"Did anyone ever confirm any damage to their second ship, from the release of the space-junk?" Ben asked, as they chatted away the long hours, the distances passing.

"Not to my knowledge," Berle said. "Too bad. I hope they fucking croaked. It's sad but true. Have to say, not a bad move on Guy's part. May as well have thrown rocks at them."

Ben chuckled. Berle's team of defensive ships and men were now more than halfway to the Molinari, approaching that region of space quickly. All eight ships were spread out in a stable formation of about 1,000 cubic miles, and of course at-speed. Only two of the ships would actually stop at the Molinari, the other five would go straight on towards Mars, since time was a factor. The Krenika, her sister-ship, and the other three Russian-Islamic ships, had been tracked passing the Molinari region, days ago, with much relief, since an attack on the space dock was feared. Instead, it was clear the Russians wanted to be in orbit at Mars well ahead of Berle, to take advantage of the defenseless Mars Base. So Berle would have two ships dock, and collect goods and needed materials (fuel, oxygen), and then join up later. His plan included sending ships back to Molinari, from Mars, for more supplies, in the event the battle continued over a long period.

"Smart, though. Lets not forget. Fuckers are smart," Berle concluded.

"So are we," the co-pilot said.

"Yeah, but maybe not quite as mean," Berle said. "I was talking with the Conflict Commander. They've planned the smallest detail. But in a real way, they really don't know what they're going to do, or have to do, on Mars. It's a learning-curve for everyone, because its never been done."

"It should never have been done, never even been needed," Ben said. "There are other solutions. One of the science-leads was saying, why not just share the Mars Base with the Eastern-block if the meteor hits, or gets too close? Instead, we're likely to destroy it, and kill a lot of people."

"Well, that's not the main idea, but, yeah," Berle replied. "We're supposed to save the place, and the residents. They're going to try to take control of the whole joint. The Mars Base Commander, Bojji-whatever-his-name-is, will be led off in chains, if not killed. Any managers, authorities, tech-workers, same thing. By the time we get there, it might even be over, and the Russians are running the place, if they attack quickly, and the Mars Base men cant fight them off. Then all we could do is try to take it back."

A pause. Berle was hovering over his control-mains, viewing the monitors every now and then by habit. Ben, the Black man, was to one side, also at a work-kiosk, where he was tracking the flights and known objects, and doing a series of calculations regarding their Mars-entry and orbits. Not easy, since he had to consider eight ships that would reach Mars about the same time, and would need to establish stable orbits. He had been working on that navigation for days. A moment more, and an alarm sounded on Berle's console, from the radio-desk.

"Commander Berle here, go ahead," he said into his hand mic.

"In-coming radio from Molinari," said a voice, his radioman. "Please use channel eight."

Berle switched channels deftly. "This is Commander Berle, of US Mars fleet vessel ID 42UBK-Vandenberg. Please identify."

"Winton, good to hear your voice! Its Garth, at Molinari," came the fuzzy-liquid sound of the compressed-date radio-link from the space dock. Garth Frakesteen was the space dock commander. They knew each other.

"Hey Frankenstein," Berle joked. "How's life at your floating casino and deep-space beer-joint?"

"The usual, Milton, same 24-hour cycle, same shit," he said. "Your pilot docked two days ago. They were fine, but I had to send a lifeboat more than two weeks out. Used a lot of fuel."

"Fuel? Rocket-juice? You got any?"

"A little," the space dock commander answered. "Anyway, his ship is basically back to square, no lasting damage. They were sweating it for about a week, though. No power at all, for anything, had to re-boot from zero. They need their orders."

Berle paused. "The Penelope is ordered to R-n-R the crew, restore all systems to perfect, and proceed on their original

mission, with the cargo for the base, the communications-tech,"
Berle said. "Nothing has changed about that. The Mars Base
communications-system needs that tech, badly, and with the
conflict, even more so. Reisling needs to co-ordinate with my
group, and fall into line, and basically stay out-of-the-way.
That's about it."

"That's what I thought," Frakesteen said. The radio link
created long pauses between their talk. "One of the men on the
Penelope may be ill. The co-pilot. Apparently a form of
psychosis, something about a gene-trigger anti-biotic he was
assigned."

"Fine," Berle said. "That's too bad. Psychosis in space-travel
can be bad, I'm sorry. Have your medical do an exam, and have
him dismissed from duty at my order, regardless of your
findings. Just do it. No way I need that risk. Standard medical
protocol, but I want him dismissed anyway. My authority on that.
And I suggest you follow the same rule-of-thumb with any of
yours."

They talked more on other matters then exchanged radio
signature sign-off codes, and the call ended.

## CHAPTER 31: The Million-Mile High Club

The Molinari Space Dock, in that year of 2078, was not necessarily much of a fun place, at all. It wasn't even very beautiful, no glittering palace or fancy wheel-in-space, with smooth curves and arches and shiny-pure surfaces of alloyed wonder-metals, not at all. Of course, the floating research-and-rest-refueling 'station' was very impressive. But by 2078, technological improvements were not so advanced as to be Olympian or such as the gods would build for themselves to live in. Delusions-of-grandeur were an occupational hazard for spacemen and women, dimmed somewhat for more practical reasons, that made life at the Molinari half-way station a hardship, demanding, and often dreary and lonely.

The Penelope was escorted to dock safely, with most or even all of her electrical system restored. 20 days or more passed between the time Guy's crew was finally able to reprogram and reboot first their air circulation, then lights and computer control, then communications and navigation, and finally engines. The lifeboat found them, but they cruised the distance to Molinari on their own 'steam', something Guy was proud of. Docking was standard, always a nervous time for a pilot. Things settled down, the men passed through the air locks as they had been trained. Technicians could now go over the Penelope's systems more carefully, and Guy and his crew were 'safe'. Beyond, 'out-there', Berle's ships were now increasing speed, per schedule, and Terchenko's squad of ships were about 120 days from Mars-orbit. But for Guy, and Lila Meetek, there at the Molinari, it meant only more work ahead, more long hours of labor and boredom, and a potentially ruinous effect on many lives, if they grew lazy or failed. Which was why 'R-and-R' was a standing order for them all, even there in the depths-of-space, cold, airless, deadly, and a million-miles from home.

They met again, face-to-face, now almost a year apart from their time in Santa Barbara, in one of two main cafeterias. Lila, ever the femme-seducer, had died her hair reddish, with a new cut, and was careful to wear a revealing and colorful robe outfit, with dance-leggings, as per the declining space dock dress code. Guy had somewhat recovered from the ordeal of the passage, and the pressure of the crisis. He didn't exactly dress for the occasion, but had shaved and cleaned up, and was in his Captain's jump suit.

"Hey, space-man, new in town?" Lila said, entering the cafeteria then stopping next to Guy, at his shoulder. Everyone used the magnetic handrails and straps or slippers, which made for quite a scene. Some floated as they pulled themselves along to fetch another Happy Abyss Meal, others found they could suck down nutrition drinks as they floated in a quiet spot, held by a tether, reading a book, cross-legged like a Buddha. Many were seated, like Guy, by an arrangement of magnets, strips similar to Teflon, on their seats or uniforms. The food was not bad at all, served in plastic trays, tubes, bags or as free items like apples or pears, or beef-sticks, candy-bars and nutrition items and drinks.

Guy turned toward her voice. They had arranged to meet in the cafeteria, after Guy's disembark and base-admissions process. As he turned, his chair pushed away, and he sort of floated gingerly into her arms, like a husky ballerina-astronaut with a boner and the relative mass-weight of a balloon.

"Is that a beef-stick, or are you just glad to see me?" Lila said. They kissed, long and wet. Guy always felt she was an exceptional kisser, with full, wide lips like the slippery valves of some delightful sea-creature or mollusk. Their Eskimo-moment came and went.

"Both," Guy said. "Of course I'm glad to see you. Hey, how was I to know Molinari served these? You know—they're okay."

His hand then found the food she had joked about, a foot-long, spicy-processed meat 'stick', chewy and flavorful. They preserved well and were nutritional and the crews could eat them any time. She took his hand, and the beef-stick found its way to her mouth. She tore off an inch with her pearly whites.

Guy laughed. "It's for YOU," he joked.

"Nice," she said. "I'm SO excited. Weee—eell, good to see you. I thought you were a goner. You came through like a champion."

"Fuckers zapped my ship with a trillion volts of radio," Guy said. "Could have killed us all."

Now they embraced emotionally, and he held her tightly, beginning to caress. Yes, he was glad to see her. They were a well-known couple, among the workers and crews. People felt they were 'cute', and could 'make it'. So, they settled down and ate together, happy they could finally share their moments together. In the same way that being alone was one thing, but to be with a friend could break the spell of separation, reuniting these two so close, after a time apart, and vast distances, it was unique and intimate, warm and cozy, there is the quiet hush of space.

After a few hours, things settled down about the new arrivals. Guy's ship was now the work and labor of a team of technicians, bringing her 'back to perfect'. Just like any flying machine, 'imperfect' did not exist, given the rigorous demands of space. Guy's co-pilot, Rob Cowan, got the news---he was relieved of service, on a private medical. Psychosis wasn't indicated directly, but Berle's wisdom was time-tested for the program. For the sake of everyone else who depended on Rob, and each other, there could be no chances taken, that a pilot would flip out. They all now could begin to 'rest'. The other crew found ways to amuse themselves, enjoying food, music, video and news, card games, and each other. Many had friends at Molinari.

Guy and Lila were determined to revive their romance, in the usual way. No one bothered to gossip much with Guy about Tommy, the external-repair space walker, who had been keeping Lila busy. Lila wasn't going to mention it, it meant very little to her. But Guy had few illusions about their relationship; they weren't married, after all.

"Let's do it," Lila said. "Come on." Her smile said it all, her eyes.

It may be of some interest to the eternally Earth-bound, that as the space program had matured, by 2078, sexual activity and protocol, was a serious consideration. Aside from health aspects, and various rules and restrictions, weightlessness, or the inevitable null-gravity environment, was 'challenging'.

The two lovers made their nest in Lila's private quarters. Like any woman, she made her personal living space into a special, sacred zone, where her heart could freely dwell, with her music, arts, reading, clothing, sundries and perfumes, hair-color and make-up, etc. As a semi-permanent worker at the Molinari, doing half-year shifts or longer, Lila was entitled to a room about the size of a master bedroom in a large house. Some of the short-term guests had much smaller rooms, and for all of them, the essential features were yet rather practical, Spartan, and utilitarian. It was a space dock in the depths of the Abyss halfway to Mars, after all. A bed (sleeping bunk, large enough for two), personal desk and computer with communications, toilet, storage, private electronic cooking, and amusements like music-player and video, etc. Of course, all in a weightless, or null-gravity setting.

"I missed you so much since Santa Barbara," Guy told her. He had then crept up on her, she laying sultry in a corner of the sleeping bunk, though hanging by a strap, as they all did, her legs and rear lifted, so she seemed to be almost inverted towards him. Guy's momentum moved his entire mass of about 170-pounds into her gently, a mobile love-bomb, a human balloon, as amorous as he ever had been on Earth.

Lila giggled. "Well, what are you going to do about it, that's what I want to know?"

He wrapped his arms around her, under her own arms, above the tender white belly, fully appreciating her exposed breasts. She was very beautiful, athletic, though not husky or hairy, more lithe and supple, perhaps as a swimmer's body, and also somewhat shorter than Guy. She responded, releasing her hand-hold, and in a miracle of that age of technology and advanced science, something wonderful began, which only a very few had experienced by 2078, but which almost any sexually aware adult might find exciting to contemplate, a rare dance of sexual love-making in null-gravity.

Isaac Newton would have been thrilled. With each attempt to kiss, caress, tickle, hold, lick, fondle, probe, hump, suck, or actual coitus, the laws of physics concerning action-and-reaction seemed a certainty for failure. A kiss created either her or him to float away again, and fondling the breasts or his man-root, went like a twin-backed ball of legs and arms, their necks crooked together by their chins (Guy had shaved his beard), and they would hold on in a passion, suspended over her desk or bed, or by the door, either high at one end, near the ceiling, or rolling slowly some other direction, hardly aware of the joyous absurdity of it all.

Lila was maybe more experienced than Guy, as a healthy resident of the space station. Guy had tried lovemaking in null-gravity before, but Lila had perhaps been at it more often. Often true of women, they know 'what to do'.

"Damn it, hold still," he said, now upside down and stretched out, directly beneath her, she maybe two feet above him, with her colorful robe wisping about them both, in the air-conditioned room-space. Guy of course was stripped naked except for a long towel he could wrap easily around his shoulders or waist. They laughed again.

"Does the null-gravity make it easier to get an erection, or more difficult?" Lila joked with him.

"Consult your flight-manual," he said. "And drop your leg this way, over here."

"Yes, commander," she said with a hot breath.

Now he had her left leg in the elbow-crook of his arm, winnowing his face and head upwards between, towards the goal, but once again, for all his efforts and lust, without gravity, the effect was to push her away, instead of to draw her closer.

"We use the harnessed nets, Guy, in the bed," she said at last. For another minute or two they would chase from one part of the room to another that way, chatting wryly as lovers do, like two bare-ass butterflies in a very unique bottle. Lila's bed was equipped with harnessed netting, nylon cords and straps and expanding elastic. Designed for just such an occasion, they could crawl in together, with pillows and sheets, lower the netting over themselves until the bed was like an intimate trap, but holding them together, such that sex was possible almost like it would be in Earth's loving gravitational embrace of all mankind.

The quivering and shuddering passion continued, and it couldn't have been sweeter. Almost a million miles into space, love found a way. And then found it again, and after a half-hour break and a cool sponge bath, found it a third time. Guy was a happy astronaut.

And again, and again, through the next four 24-hour cycles. Ahead lay Mars, also like a woman, maybe fatter, and redder, and far more cruel. Guy needed to take the time to establish his part in the up-coming assault, or rather, defense, of the Mars Base. His team also had to re-invent their mission, and re-learn or re-calculate their flight-path, entry and orbit, the position of the planet Mars relative to the Molinari, and also to retrieve specific commands and directions from Berle's leadership aboard the 'Understandable'. The Penelope's cargo of high-tech communications gear would eventually prove to be an essential component, as far as the Mars Base survival. So it was no small matter for them to succeed, even though they were not intended for significant battle, or killing any Russians, or shooting down any ships.

"Why does Berle call that ship the 'Understandable', Guy?" asked his navigator, McGee, as they were now able to work together in one of the observation decks, with various maps and navigational tools, and radio-links, computers. "That's an odd name."

"With Winton Berle, you just never know," Guy said. "He's one hell of a space-man, I'll tell you that. Maybe he wanted to remind himself just how difficult the high-tech data and programs can be to deal with in an emergency, if no one truly understands how they operate."

McGee grinned. "No one knows it all."

"Yeah, true," Guy replied. They continued working on the road ahead, a road made of less than nothing, but full of every potential, and every challenge or danger, either of them could imagine.

## CHAPTER 32: On The Road Again

Mars Base Commander Bojji-Than was waiting. Life at the Tharsis Montes/Snikta Volcanic Basin US Mars Research Facility in 2078 could not, of course, cease at a dead stop just because a war had broken out on Earth, the Mars Base itself being the golden poker-chip to bargain their survival, and perhaps many tomorrows. Bojji-Than could indeed project his own fate at the hands of the Russian-Islamic forces that were ready to invade his rather small and fragile kingdom. If they killed him, that might not be so bad. Imprisonment, torture, interrogation, mind-altering use of force or drugs to pry non-existent secrets from his brains---scary, painful. Maybe it would go another way. As he was responsible for 263 lives on Mars, Bojji-Than had taken whatever measures he could to prepare. But, he knew they were no match on their own for the Russians in their ships, and with their weapons. They could fight, but they would be over-run, out-gunned, and out-numbered (for fighting soldiers, anyway). In the hostile environment on the Martian surface-world, their chances were even slimmer. So, they waited, watching the skies.

The base had a Command-Control Center, designed to rest at a high-level, atop the various structures, not a tower, but a series of large rooms. Here, they could do it all, though many base-functions were operated from other tech-centers. But the Mars Base Operations Command-Center was certainly a 'do-it-all' button-pusher's realm of dedicated, direct-line computers, tracking telescopes, communications, doors and airlock control, life-support, launch and orbital processes, external surface mapping and weather conditions, staff and crew positions, schedules and monitoring, data-records archives, and many other needful functions.

On the same solar-day that found these other Earth-Mars transit corridor activities happening, notably the now approximately 1.2 year-long approach of 14 inter-planetary ships (five from the Eastern-block program, eight from the US Mars program defense, and Guy's ship, the Penelope), Commander Bojji-Than, together with base-security officer Juno Amorrossi, Vinces Grant, the base science-lead, Charley Barron, the life-sustain systems man, and other helpers, were busy for weeks trying to fathom the complexities of what was ahead. They could view the maps of the planets and space-regions, and also plot the Estimated-Time-of-Arrival of the ships, and receive data and info from Earth and Molinari. But communications to off world were not reliable, and were intermittent. Karen Tutturro, the radio-specialist, had been able to affect some of her repair work, but needed the tech and gear in Guy's ship, so far away. With all this, the Mars Base management was able to reach some specific conclusions.

"So, just less than four months from now, correct?" Bojji-Than said to Grant, who was on top of the navigational astronomy. "At that time, the first Russian ship will enter orbit, am I right?"

"Yes, sir, that's right," Vinces replied. Grant was a husky, 50-year old man, the mind behind the science and research for the base, with a habit for lame one-liner jokes. "I'm figuring all five will enter Mars orbit over a period of ten days, about 95 days from now. As the crow flies."

A pause. "How far behind are the US ships? Winton Berle's team?"

"30 days, only a month, maybe less. That is, 30 days back from the Russians, roughly. The position and speed of the Russian ships."

Juno, the good-looking athletic security-chief, laughed. "As the crow flies," he said. "Funny man. Ha-ha. No crows in space! Ha-ha!"

"None on Mars, either," Vinces joked.

"So, for at least a month, roughly, we will have to face them on our own, the Russian-Islamic ships and soldiers," said Bojji-Than, the 65-year-old Asian leader for the base.

"About right, yes, from these maps. Yes, sad but true," said Vinces.

Beyond the airtight walls of the base, 14 small 'oxygen-igloo foxholes' had been completed by this time, and another 10 were still being worked on. This was the concept the Mars-Defense Task Force had endorsed. Each position was ingeniously designed from available materials, such as cargo-crates. They were airtight when entered, and prepared with life-sustaining temporary, breathable air, water, toilet, food, radio, medical, surface-walker gear and air-recharge, small computer-link, and weapons and ammunition. When and if the Russian-Islamic forces descended to the Mars surface to take the base, the men could fight from these. The base's eight main air lock doorway-entrances would also be defended, and other forms of resistance. The orbital launchers on Mars, included only two ships, which were very small orbiters, intended as space-port 'tug-boats' for incoming transports only---these were also defended and prepared for any conflict. But Bojji-Than never wanted to use them. The launch pads, orbital ships and 'lifters', were extremely important, the only way off-planet without external assistance. Although it had never been done. In theory, one or both of the orbiters could even reach Molinari, or connect to a rescue ship from Molinari, and make the journey safely with people on-board. This was a significant comfort for the residents of Mars.

There wasn't much idea to actually launch the orbiters or 'lifters', into orbit, to attack the Russian ships, when they entered orbit themselves. Not practical, they would be completely over-powered by the superior number of larger, faster, better-armed ships. Bojji-Than's team did seriously consider how they might hope to shoot down, or damage the invader ships, with surface-to-orbit missiles. It was an insider secret that the Mars Base had a collection of six such missiles, computer-targeted, basically small, un-manned rockets. They had never even been un-packed or programmed to launch, or fueled and prepped, etc. The base was more than 15-years old, and Bojji-Than was only five years in command. There were all kinds of mysteries he had discovered, as he was able to take the time to explore. At some point in the past, someone had shipped over the six rockets. They were the size of typical small modern rockets, about 100-feet tall. The lesser gravity of Mars made the launches into orbit easier, and computer targeting was deadly accurate. The rockets had plenty of explosive power to destroy an orbiting ship, with a hull-breach or other. So, Bojji-Than directed two launch specialists to prepare to go ahead if they needed to. Matt Curisonn Van Templar was the launch-specialist in charge of all Mars launches. Bojji-Than also had standing orders from Earth, and had no real choice but to fight if he had to, or if he felt he could, prior to the arrival of Berle's ships and soldiers, and probably during and after as well.

And so it went, like digging their own graves, or preparing their own funerals, or, from a more positive view, confirming in their hearts that they would avoid both, with every effort. Mars Base was not a busy place, and could be quite sedate, calm and peaceful, quiet, with only a regular flow of research activities and astronomy, daily chores, and not much else. What else was there? They weren't 'up-to-something', or learning anything Earth-shattering that would bestow immortality on Mankind back home, or a cure for cancer, or finding tons of pure gold. They were explorers, first-and-foremost. But now they were trying to learn something new: how to deal with a war, on the planet ancient man had named for the so-called God of War. Earth's first war-in-space was just not looking like much fun after all, from the point-of-view of the residents on Mars.

Time passed. Guy's ship was restored to perfect, and after more romance and dove-talk with Lila, he and his men shipped out again, headed for Mars, on the last leg of their original journey. Co-pilot Rob Cowan was left behind on Molinari, and would catch whatever transport he could next, back to Earth. The gene-triggered antibiotics had done him in, his career was over. Psychosis in space was the end of any astronaut's career, because of the un-predictable nature of the human mind, and the high level of responsibility and stress. Rob accepted this as best he could.

"They should have known," he told Guy, prior to the Penelope's departure. "The doctors felt it would be okay. If I had either dropped the meds, or not done the flight, I'd still have a career. Now, it's over. I'll never fly again."

"It's the right thing," Guy said. They were good friends, after all. "It's not your fault. I mean, how bad was it? Seriously? Just between you and me?"

"Not that bad, but not that good. A little goes a long ways. A bad dream, fears that don't go away, irrational ideas, a sort of black hole of worry that gathers and dissipates, then gathers again. Like a swirling puke," he replied.

Guy was grim. "Yeah," he said. "You'd think the human mind was something the human family would finally have figured out by this date in history. But we never do, do we?"

"We are what we are," Rob answered. But Guy could tell, it had broken his co-pilot's spirit, to be forced into retirement, and the black-mark of mental issues on his record, when it was never really his own shortcoming or personal lack of responsibility. They parted as friends, and they both knew it was for the best.

Guy and his crew took their places again in his ship: Tom McGee, the navigator, Raz Brahman, communications, Arron in cargo, Peter, the young life-support man, and the others. He was now down to a crew of six, but it was sufficient, they knew their jobs. He decided to make the rest of the journey without a co-pilot. McGee would step in as needed.

The Penelope was docked, attached by a long tube-like walk or tunnel, from her main air lock doorway, to the appropriate air lock hatch on the space dock (Molinari). So the ship appeared to be perfectly upside down, against the side of Molinari, which was like a giant Christmas-ornament, with a large top, bulbous and ovoid, and then structures much like towers or buildings, below, and along the sides. Right-side up, down-side down, left or right, in space, it meant nothing, or very little.

"Thank you, Molinari traffic. This is captain Guy Reisling, US Mars transport, 'Penelope', Mission-ID, Vandenberg. Ready to disembark. Please advise for your air lock safety crew and clearance to engines." Guy was again in charge, thankful things had not been worse, working his hand-held radio-mic to speak with the Molinari launch-command. A few moments of radio-silence.

"Thank you, Penelope. All docking connections are cleared. Molinari traffic-control wishing you a safe and happy voyage." The voice was that of a youngish male traffic-dispatch monitor.

Guy began to sequence the needed thruster-commands to move his ship away from the mothering form of the space dock.

Only the tiniest thruster-energy was needed, at first, just a 400-pound blast of compressed hydrogen-oxygen jet stream. The thrusters had to be directed very precisely so the ship moved away safely and properly, but McGee and Guy and worked this out ahead, and manipulated the proper controls to point the action-reaction jets exactly where they wanted them. The control lights and alerts blinked and hummed, then a loud 'beep-beep-beep'. From outside the ship, in the soundless world of space itself, a jet-puff stream like high-pressure boiling steam shot out from one of the small thruster ports. The ship began to move backwards, out and away from the hull of the space dock, quite slowly. From inside the ship, there was only a sort of hum or buzz, not really a hissing sound, as you would expect. But of course they could monitor what was going on at every inch. It was a big ship, but weightless. Isaac Newton was again the winner.

500-feet. 1,000-feet. 2,000-feet. More short thruster-blasts. The Penelope dropped away like a huge fat bird. Now 3,000 feet, into the black.

"Looking good, Penelope," came to traffic-monitor's radio-voice, again from the space station, also on top of the maneuver. Guy recognized Lila's voice, and smiled.

"Don't give a damn what it looks like," he said. "I only care that it performs properly. The first step in a long journey is always the hardest."

"Well, it looks good anyway," Lila said. "I mean, it's a beautiful ship. Who painted that ensignia on the side of the left wing-lifter assembly? The one that looks like a marshmallow with eyes?"

"That's not a marshmallow. That's the official Mars-program Transport Crew Emblem, and we wear it proudly. I have no idea who did the artwork. It's supposed to be like a food-service thing or something. You never noticed that before?"

At 5,000-feet off from the space dock itself, another thruster blast in the opposite direction slowed the ship's fall. Now they would begin a turn, and carefully point the ship in precisely the right direction to reach Mars, while prep-firing (warming) the main engines. This would take another hour. Once ready, they would build speed slowly, track to their flight-path, until within about 24-hours the ship would be at-speed, perhaps 10,000-miles per hour or more, with her destination still half-a million miles away, or more, a journey of months.

"Okay, but why is it yellow?" More radio chatter.

Guy paused. He was a little busy, right at that moment. It was for those reasons, that he really loved Lila Meetek, the way she would go on about something like that. He never really could tell if she was serious or not. The Transport-Department Artwork was totally meaningless to anything that was happening.

"I don't know, Lila. It just is," he said.

"I guess it shows up better," she said.

A pause. "Okay, well, Penelope is away from your space dock, Molinari Traffic. We will be doing our turn and directional flight-path for the next two hours. I will alert you on our final departure and engines. Reisling out, Vandenberg transport to Mars Base, Mission Day 202 and counting."

"Molinari out, standing by" came the reply.

Guy's heart ached a bit. He might never see her again.

## CHAPTER 33: Commander Prokov Keeje

The five Russian-Islamic ships entered Mars-orbit, 25-hours ahead of schedule. There was no way to do this secretly, though space was vast. The lead ship was still the Krenika (piloted by Zolotny); the others we called the Saint Peter, the Tolstoy, the Kamchatka, and the Sir Soviet. By comparison, each was very similar to the US Mars ships (such as the Penelope, or Berle's ship, the Understandable, and the others with him). The reason was that space-travel techniques in 2078 were fairly standardized, because successful methods had been developed that each of the various space-programs had adopted. The ships appeared in form-and-shape, much like the early US space-shuttle types (the 'Challenger', 'Endeavor', etc., with a main body and apparent 'wings'), but they were quite longer and larger, perhaps three times the size of a 747 jet-airplane. They were also much wider, not designed to lift in an atmosphere, or to glide down, on 'wings'. Perhaps it was a mere psychological comfort that they were designed this way, no wings were really needed in space at all, and they weren't 'real' wings. The long-throw engines were in the rear, with large converters that re-circulated fuel, so that very small amounts would operate the engine-thrusters for very long distances, by repeated flash-bursts that sustained a constant 'push', almost like a machine-gun, staccato-effect, instead of one endless stream of jet-push thrust, or like a blasting-off launcher, with fire and smoke. By the time they were to enter Mars-orbit, these engines were shut down. Plotting the needed navigation to place all five ships into orbit, over about a week's time, was challenging. But the Russian teams were quite skilled, indeed, among the very best.

Mars of course, though smaller than Earth, is yet vast, so the five ships could place their orbit-paths as far apart as they wished. The Mars Base location was northern hemisphere, fairly high above the Martian equator. Their descent-to-surface would place much smaller, manned 'lander-pods', by parachute and glider, near enough to the Snikta-Ridge Volcanic Basin, to make the assault of their soldiers easy enough. But from the Earth, at Vandenberg, Texas, Florida, Australia, Puerto Rico, Canada, and other Western sites, the KK-F ships were tracked and monitored closely, with certain bitterness.

The Russian-Islamic Command Leader was not the pilot Zolotny (Krenika). Terchenko, the KK-F/Region-Six commander, and other authorities for the East, had appointed a thick-headed, muscular and coyote-like, youngish man, named Prokov Keeje, only 38-years old. Keeje had a decent number of hours in command of his inter-planet ship. Most of his experience was the famous Earth-Moon run (both sides had moon-bases during this period, some of them known for excellent gourmet foods). But more than this, Keeje had military leadership and combat-strategy experience, as a Russian submarine lieutenant commander, for four years at sea, with nukes (yes). From Terchenko's point-of-view, this was far more valuable, and Keeje was his point man to take Mars for their side. KK-F and other Eastern-program sources were in-touch by radio, and US Mars made every effort to intercept those messages, especially since the incident with the Penelope. And their opponents, made every effort to conceal their communications, standard for any modern warfare. As scattered, technically oriented, and often simply as a ubiquitous electronic mess, as the wide spectrum of radio waves could be in the vicinity of the Earth, this was also challenging. But as Keeje took control of what he intended to be a swift, devastating and efficient takeover of the US Mars Base, he knew exactly what he would do next.

Days prior to the KK/F ships entering orbit, base commander Bojji-Than and his team were following the approach. It was like watching a clock ticking away their urgent response, all too soon ahead. All that they could have done by then, on Mars, had been done. It was somehow more of a matter of scientific curiosity, for them to wonder and observe, how the Russians were handling the task. Even as a man with a bad tooth, or a patient with in pre-op for a medical procedure, might want to know the details, Bojji-Than, Vinces Grant, Juno Amorrossi, and the others, kept a very close view. They didn't get a lot of visitors on Mars. The Russian navigators had plotted their ships into a rather elegant orbital entry, passing the trajectory of Mars' twin moons neatly as a group-formation, at a clean distance, and then adopting a planet-orbit one-by-one, separated by thousands of miles, until each ship was gently embraced by the mild hug of Mars-gravity. A fine dance, thought Bojji-Tan, considering what must surely lay ahead.

Commander Prokov Keeje took the helm on the ship, 'Tolstoy'. "Captain on deck," said his Second, an Islamic man called Amin.

"Report please on all ships for orbit positions, right away. Thank you," said Keeje, to no one in particular, knowing he would be instantly obeyed.

"Just a moment," Amin replied. Like any of the space-vessels, the crews functioned in a weightless circumstance, gliding from post-to-post or at data-control stations, then holding on by magnets or straps. Also in the flight-command area were three other men. They all tended to speak either Russian, or a Hindustan Arabic language. Amin quickly viewed his maps and data-screens

"All of our group has successfully adopted Mars-orbit,"

Amin said. "More than 24-hours ago. Three are in equatorial; two are in a polar orbit, including the Tolstoy, as you know, commander. Crews and soldiers await orders."

Keeje tugged thoughtfully at his long, reddish goatee-beard. He appeared somewhat the physical equal to Guy Reisling, who would sometimes grow a beard himself on long journeys.

"Show me the base," he said. Another man, with a very dark complexion like an African, but Eastern in his features, whose name was Thomas, responded.

"I bring up the view, give one five minutes, sir," he said. In the interval, as he worked, Amin and Keeje chatted.

"We have 30-day's grace," Amin said. "The Americans cannot arrive before then."

"We will easily take the base in seven days," Keeje said. "They are powerless, almost completely un-armed. No weapons, only research."

Amin chuckled. "Everything easy is better," he said.

"They will obviously already know we are here," Keeje said. "More anxiety for them, I guess." He smiled.

Thomas alerted them to his work. "Here," he said. "Tolstoy's telescope-camera can show maximum magnification. You can see the base now. There."

Keeje moved near the man's station, a computer-control and view-screen arrangement, linked directly to external telescopes and cameras. They didn't use it often. Amin moved aside.

Much like a Google-Map, or 'Map-Quest' image, what they could see was a very high-altitude downward view of the US

Mars Base. The Tolstoy was easily 100 miles above, but not directly over. The image showed the sandy, vast, desert-like regions surrounding, and then the red, crusty, scabbish-looking length of the humongous Tharsis Montes mountains, higher than Mount Everest by three or four times, and as long as the continental US. At one end, as the operator 'zoomed-in', it was clear enough, they could now see the Snikta-Ridge base. They had more detailed maps and technical specifications. But at this view, the base, like any structural facility, was seen laid out as square shapes, various features like towers, antenna-arrays, outer-tanks, or storage, small roads, etc.

"Not much of a prize," said Keeje "I can't really see why it is so precious."

"Well, given the meteor, you know," said Amin. "Maybe we will soon call this not-so precious base our home."

"Do the other ships have this view, or this image?" Keeje asked.

Thomas paused. "They may not, because of their positions in orbit. Or, they may not have taken the effort. But, I can transmit a digital image or images, with all details."

Keeje paused, again tugging his reddish goatee. "Study what you see," he told them both. "Get any details, learn what you see. We have time, for now. Use whatever tools you have. Tell me everything, any military aspects, like rockets, or launchers, ships, roads, trucks. And make a map. Do it quickly, you have 24-hours."

"Yes, commander," said Amin.

"Yes, sir," added Thomas. Each returned to their work. Mostly their hourly tasks were to monitor various data and flight controls. Keeje took his own post, as they eased their bodies back-and-forth, weightless, yet heavy with their purpose to attack, big men, military-types every one.

"Six days, not even seven days," boasted Keeje. "That would give us plenty of time to take control before the American ships arrive, and to propose a defense."

"They have nine ships in all," said Amin. "We have five. How can we compete?"

"By fighting our cause with savage, brutal passion and efficiency," Keeje said. Amin scowled a bit and grimly. He knew the tone of the commander's voice. This was war, true enough.

And it was true enough inside the Mars Base itself, and at the Molinari dock, and back on Earth. Space was like a winter for all, in any case, cold and forbidding, death, Mother Night. Two of Berle's ships would momentarily dock at the Molinari, for services and fuel, supplies, re-stock, not just for themselves, but for his entire group. Molinari was a huge advantage for the US side, and Berle would take complete access, with his ships going back-and-forth as a tag-team from Mars, once he arrived. The Russians only had what stocks and supplies they could carry, or steal from Mars, or a possible attack Molinari. The Russians did however have a month to take the Mars Base, if they could. Of course, for all the Mars-crews and ships, the men, soldiers, and the residents of the base, actually returning to Earth was ever a specter of doubt and possible failure that would end their lives. It was simply the nature of space-travel during that era; they knew they might never return to Earth at all. The distance was enormous, and many things could go wrong. They planned to come back, even in great detail, down to the very ounce of fuel. So, it was a lonely feeling, in that sense.

24-hours passed. Keeje later on, ordered his radioman, to set up a voice-link to the Mars Base itself. There was no need to stall, or hide their intentions. After some initial trouble making the two communications systems connect, and with frequency codes and opener ID's, Keeje could speak with US Mars Base Commander Bojji-Than, directly. There was no video-image, but the voice-link was clear, much more so than in deep space, being closer, a shorter distance for radio waves to travel.

"This is Commander Bojji-Than, director of the US Mars facility at Tharsis Montes. Go ahead, please," came the Asian man's voice, from the base. His team, in the command-center, surrounded Bojji-Than. They all wanted to know what was next.

"I am Captain Prokov Keeje, of the space-vessel Tolstoy, launched out of Russian Ukrainian authority, Earth," Keeje replied. His tone was stagnant and strong, precise. "I am requesting at this time a simple communication, you may wish to enter into your logs or data-base."

A pause. "That may not be necessary, Commander," said Bojji-Than, responding more jovially. "I already know what you want. You and your ships and soldiers and men have arrived with weapons, and orders to take control of this base, is that correct?"

"I prefer to inform you more formally, sir," Keeje said. "But, yes, that's about it. We have five ships in orbit. Landers will begin to descend to the surface near the base within a day. My men are armed and prepared to handle surface conditions. You will please to prepare for them to enter the base itself to the safety inside through all available air locks, without resistance of any kind. All weapons at the base will be surrendered and gathered before we arrive. Your control services, command centers, data and transmissions, and all base functions, will all be turned over directly to my technicians, as a transfer-of-power, with all access and information. All base personnel and residents will be gathered into your meals-areas or large rooms. There will be no resistance, no killing, no shooting, no fighting, or your people will be shot dead without mercy. Is this clear? I am asking you to transfer power and control of the base to me by laying down your arms and peacefully surrendering. I consider this an act of grace towards you, as fellow astronauts. I don't wish to see a great deal of bloodshed. We have the superior power, and my orders are very specific."

There was a lot of laughter just then from the men in the control room, or command center, at the Mars Base, who had heard what he said. They hooted loudly, clapping or blowing horns, shouting out jokes. Bojji-Than also was chuckling. It was all so simple.

"You are like god, then, Mr. Keeje, yes? You come from the sky and give everyone orders they must obey or die! Very good, very good!" Bojji-Than replied, via the radio link. "I truly don't know what to tell you, this is outside my experience or training. Are you sure you wouldn't wish to, maybe, just fly down here and have a nice hot dinner, or bath, with real gravity and solid ground under your feet? There are some very nice women who would love to meet such a dashing and strong-willed figure as yourself. You know, for photographs! Very rare, to have dinner with a man who can conquer an entire planet all by himself!"

More laughter, hoots and clapping, the men in the Mars Base Command Center were warming up to the affair. A long pause.

Keeje was grim, bitter, and venomous. "You have my message. Good day, sir. Tolstoy out." His voice echoed down the waves of electronic energy, transmitted far less perfectly for all the technology, than what they surely knew were his hot emotions and intentions. But they knew, of course, the Martians.

Like all soldiers, the men and residents on Mars, were prepared to die, maybe with a laugh. Or, even possibly, to win the day for their side, as they must.

"What a fuck-head," commented Juno Amorrossi, the Mars Base Security Lead. After a few more minutes messing with their high-tech, the base command center returned to normal, as they all began to sort out what they would do. Computers buzzed and the air-circulators bled cool air throughout. Someone tossed a paper-airplane across the control room. Chatter among them began, then gossip, speculation, orders and action.

The battle for Mars at Tharsis Montes had begun.

## CHAPTER 34: Cargo

*"Tweedle-dee and tweedle-dum agreed to have a battle, for tweedle-dee said tweedle-dum, had broken his brand new rattle."*

*--Lewis Carol, "Alice in Wonderland: Through the Looking Glass"*

Guy Reisling's ship, the Penelope, had departed from the Molinari space dock a week or so ahead of the arrival of the Russians at Mars. During this time, as the planets in their orbits were yet ever in motion, and by virtue of the appropriate navigation of the ships across the Void, Winton Berle's squadron of space-ships, coming to the rescue, had also moved much closer to the end-destination, the planet Mars. The planets are always slowly in motion, a stately dance of gigantic heavenly bodies. The ships can't move in a straight, direct line, from Earth to Mars, but are directed in a shorter curved path, according to the positions of the planets, at departure and arrival. It is a very narrow calculation for the pilots, like threading the eye of a needle. In order for them to make the passage in a reasonable amount of time, given the needed fuel, oxygen, water, etc., and the sustainability of the life-strength and stamina of the crews and men, and their ability to perform actively on demanding duties and tasks, the passage time needed to be reduced as much as possible by navigation. In other words, they left Earth, and by predicting the positions of the planets in motion, made the voyage as short as it could be, or, indeed, possible at all. There was no other way it could be done at that time, in 2078. This was standard, and it was never attempted for a ship to travel to Mars in any other way, which would be absolute disaster. For the position of the planets being at opposite points in the circles of their orbits, for example, the distances would be unimaginably far.

It might be thought by the layman, as a collection of spinning tops, on a surface drawn with circles, viewed godlike from above, each in different places, with various circumferences. And in-between, the tiny, tiny specks of consciousness and life, mere flecks of awareness and technology, being the Earth, of course, but also the ships, the space dock, the residents of Mars and Earth's moon, all also in motion. To pretend to really understand it, without doing the rocketry, physics, astro-navigational algebra, trajectory, is maybe arrogant. But most high-school students would be able to comprehend the difficult and even amazing challenge, to travel from Earth to Mars in these ships, approaching 2079. Two of Commander Berle's ships docked at Molinari, for the stated purpose of gathering supplies, for the long, long journey. Unlike the Russians, by connecting to the Molinari, Berle was assured as a wise commander, that he would be supplied during the journey home, and also for the battle at Tharsis Montes. His plan included ships from his formation of eight ships, that would be ready to move from Mars to the space dock, and back again, and even back to Earth, in defense of the US possession on Mars, the base where 260 people or so, lived and did their research. The Russians did not have this advantage, but they might pretend to achieve their victory and sustain their long voyage, by taking the Mars Base and its supplies by force, and although it had not happened, it was possible they intended to take over the Molinari-station as well. It all sounded very easy, like pieces on a chessboard, but in actual practice was highly improbable.

So, two of Berle's ships had already docked at Molinari by that time, while the other six moved forward through space, spread out over some 20,000-cubic miles, giving each ship plenty of room, at the high-speeds at which they were traveling. Guy's ship was later piloted to be part of this formation, so that group moved towards Mars, now seven ships to the rescue, the other two later, to do battle with the Russian-Islamic vessels, and defend the valued, functional, essentially self-sustaining base on Mars at that time. As it was observed, each ship was self-contained, and could operate on its own. But as Guy's ship was united with Winton's formation of ships, it was something of a meeting of friends, glad to connect and join, in safety and conduct for the same purposes. Even though Guy's ship was a transport vessel exclusively, and did not have weapons or soldiers. Berles' ships had been specifically out-fitted and prepared for the military purpose. Each ship had about 20 soldiers or more, and all the soldiers were prepared for their tasks. They had weapons, they had guns, and they had surface gear so they could walk on the Mars-surface, similar to the walker-suits used by the Mars-workers, like Karen Tutturro. Mars of course had no air, and the surface was hot and very cold, and the gravity was very light. So the soldiers each were equipped with at least one of these walker suits, which had the oxygen and radio, and so on. They were very cumbersome for a battle, because of the nature of the job. None of the soldiers were looking forward to risking their lives, shooting at other astronauts on a strange planet. The weapons, guns, grenades, launchers, small arms, were designed and modified, to function on Mars. But it really wasn't known how it would go, because there had never been warfare in space, or on an off-world surface, such as the moon, or any other planet, involving men, ever.

Winton Berle, the Commander of the Defense Formation, was alerted to the hook-up with the Penelope with his formation on its course. The trackers and monitors, who perpetually logged and recorded their progress, if only for safety, were informed ahead by radio and radar, that Guy was joining their formation. It was a fairly easy task, and as Commander Berle learned that the alignment of the Penelope was successful, he directed his radioman, Raz Brahman, to set up a radio-link, so he could welcome Guy and give him instructions and orders. This took some time, as the ships and captains and radio-staff exchanged identification and formalities, until they were ready to talk.

"Welcome to the new riders of the purple sage, Captain Reisling. This is Commander Winton Berle. How was your journey from the Molinari?" The radio-link was typically delayed, and the communications of the voices was compressed, sometimes sounding as though under water, echoes.

"This is Captain Guy Reisling of the Penelope, reporting. Hello Winton. Glad to be part of your squadron. The Penelope is in top shape; we were able to effect repairs. Ship's crew is in good condition, with the exception of co-pilot Rob Cowan, who was dismissed for medical reasons. We're now part of the defense-team formation and will be continuing with you on our way to the Mars Base. Thank you for letting me be a part of the----uh---new riders of the purple sage. Reisling standing by."

"Ghost riders in the sky, Reisling," came Berle's reply. "It's an old country-western tune. You like horses, back home? I know you have a home in California, plenty of horses out that way. How's your girl-friend, Lila What's-Her-Name, did you have a chance to see her at space dock?"

"I don't have horses back in Santa Barbara, but my co-pilot has a ranch in Nebraska, with horses. Of course Santa Barbara is famous for the Arabians and some very excellent thoroughbred horses, but I don't keep them myself. Yeah, I did see Lila, we had a great time. Molinari is looking good, quite pleasant," Guy responded. They were like old chums, the astronauts always held a fierce loyalty and fraternity.

"I'm sorry about your co-pilot, Guy, there was nothing we could do," said Berle. "If Rob has a condition that makes his service un-safe or a potential failure, then we had no choice but dismissal. He was a good man, and I'm sure he'll be fine. It's difficult when the space-workers experience the psychosis, and I think the gene-trigger antibiotics certainly played a part. I know you and Rob were friends," Berle replied.

"Yeah, I feel the same way. Rob was all right when I last saw him, he's handling it well," Guy responded. He continued on the radio, to Berle's ship, the 'Understandable'. "Winton, I'm going to need your orders and direction for the rest of the voyage. The Penelope is really only responsible for the transport of the communications gear that we had at Vandenberg, but when we arrive at Mars, once the battle begins, I'll need to know how to deal with my part of the show, know what I mean? You guys are going to be conducting warfare, and I'm supposed to be delivering crates of high-tech gear, and I need to know how to deal with it. I have no weapons, no arms, and, well, although we have our standing orders, I need that clarified."

The link between the two ships delayed their speaking by ten or 12 seconds, then Winton had a chance to respond. "Captain Reisling, tell me, at this time, what exactly is the cargo on-board with you? Do you have a basic run-down of what's in your cargo for Mars?"

Again, there was a pause. Guy responded from memory. "Yes I can Winton. The Penelope is carrying a shipment of four large crates, about the size of small trainloads, or rail-carriers, smaller---"

"I'm familiar with them," Berle interposed.

"These four crates, and three smaller crates, contain the communication gear requested by Vandenberg for the communications specialist---I guess, that gal Karen, at the Mars Base. The specific technology is not my venue, but I do know that it's mostly data processing machines, data-compression, which was the main failure at Mars. The radio-transmissions are routed and processed to digital, and those processors, those devices, were failing. Some of the transmissions back-and-forth are rather high-volume, vast amounts of compressed signals, all the way to Earth. So, we have those data processors, data-compression, that she will install, and they are prepped and ready for her. We have four large crates, and three smaller crates, bottom-line," came Guy's reply.

Winton Berle took Guy's communication, on the flight deck on his lead-ship, and was operating 'on-deck' as formation commander. He had plenty of help, including his staff on that ship, and then the other pilots, navigators and military men, on the others ships. They knew very much what they wanted to do. It was planned out in great detail. But as an experienced person, Berle knew full well, that things could always go wrong. He was guided by the idea, that "if anything can go wrong, it will", and that "mother-nature is a bitch". And in space, Mother Nature is an extreme bitch, to be sure. "The Universe is actively hostile to intelligence," was his other maxim, in his travels beyond the Earth. But he had faith in his men and his equipment and his knowledge, and he knew as they all did that they might not come home at all. So, in his order to Guy, he had also thought these out, and knew what he wanted to accomplish with the transport.

"Guy, this is Commander Berle. The goal here is very simple for your transport. You'll proceed with the formation to Mars. The ships will enter Mars orbit by scheduled navigation, including for yours. Your job is to get that stuff down to the base, as you normally would. The only difference is, is that there will be a battle going on, which we intend to win, while you're doing it. To that extent, I expect you to make judgments and decisions that will accommodate your goal. You will also communicate with my command, and the other commanders, to coordinate all efforts for all success," Berle said.

"That's fine, Winton," Guy replied, the two ships far apart in space by hundreds of miles and at speeds of 10,000-miles per hour. "What the hell do I do if they start shooting at me? It's not going to make it easy."

Winton chuckled to himself. "You'll have a defense team assigned to your cargo and your descent to the surface," he said. "We have enough men to have a small squad help you get that shit down there and conducted into the base safely. I have no idea what the battle is going to be like, except the way we've planned it. So, don't worry about carrying a bunch of guns or being shot at. Your cargo-mission will be defended, and you may have to just fight your way in, if you have to. I don't know. We've got to get that gear into the base, and installed. It is very important, because the base must be able to communicate with Earth. Those messages are important enough to the success of the entire operation---do you understand? You have to get those cargo crates down to the surface, and safely into the base, one way or another."

"I hope you're not expecting my cargo-crew to handle weapons, Berle. My men are not prepared for that. Transporting cargo is a standard procedure; we've used it at least a dozen times together in the past few years as a cargo team. But if we have to shoot at Russians and Islamics, in fucking walker-suits, it's not going to work---at all," Guy told his commander, somewhat forcing his opinion.

"Just be prepared, Guy. Make sure your men understand that this is not a normal circumstance," Berle's reply was terse, the radio-link warbled his voice through the emptiness. There was a pause again, and Guy reviewed the logged orders and communications from his side, which were recorded automatically onto the computers aboard the Penelope.

"That sounds fine, Commander. Is there anything else I need to be aware of? I mean, I'm sure my crew has something around here we could bash a Russian over the head with, a curtain-rod or something, or throw something at them. But beyond that, we'll follow standard procedure, including orbital entry, in coordination with your navigators."

"I'm also directing you to take it easy, and don't panic if things change suddenly," Berle added. "You need to get your people home, and you need to stay alive and safe. The defense of the base on Mars is for my soldiers and my teams. We've got it mapped out, and I think we're going to win. We have the advantage. The main thing with the Russians, is that they had a head start by about 30-days. We'll be there in, uh, another 20 or 23 days from right now. You and your cargo crew need to be flexible and on top of your game to survive, and that's an order too," Berle said.

"Logged and noted, Commander Berle. Will comply, thank you. The Penelope will continue in formation for the duration, along with the rest of the riders of the purple sage---we'll—uh--- see you at Mars. Ending transmission for now, with your permission. Vandenberg transport ID-Penelope cargo-vessel, captain Guy Reisling. Thank you, commander, more later. Peace out."

The radio transmission went dead, and Berle from his side, and the two radio-operators on both ships ended the link. Later on, Guy, as the pilot of his ship, took the time with his crew, to make these same standing orders clear, and to continue to prepare, to transport the goods to the surface of Mars. The cargo transport was standard, and he himself and some of the others had done those jobs at least a few times, in their careers, an essential part of the survival of the Martian community. He was a transport cargo-vessel pilot, just like Zolotny, of the Krenika, and his work was to supply the base. He was very popular with the Mars Base folks, for this reason. The materials were shuttled down to the surface, and moved inside with the help of the Mars Base staff. His crews returned to his ship, then, in orbit above. The Penelope was prepped and restored as needed, and then would leave orbit for the return trip, with a stop at Molinari.

It was pretty straightforward and had been worked out over the 15 or 20 years of the successful Mars Base operations. Guy had made his career out of it, and was very proud of his work. The most challenging part was when the cargo came down from the ships, to the Mars-surface. To do this, it must be was worked out, but this didn't mean anyone had ever really made it very easy, or particularly sophisticated. A small lander detached from the main ship, for the descent, it also had to carry the crates. Parachutes, gliders, lifters, winged-craft like jets, un-manned crash-landing pods with crates, were hard on the hardware, harder on the crates, and no lemon-pie and ice-cream for the crews. But it could be done again and again with regular and predictable success, and was survivable.

*Why?* Guy thought to himself. *Because it's there.*

## CHAPTER 35: Not A Movie, Not a TV Show

"We're going to try to delay the conflict with the Russian commander, by attempting negotiations, and I'm hoping that we can put off his assault, at least somewhat, to give our forces a chance to arrive and rescue us. I know some of you think this is a foolish course of action, but, it is the peaceful path, and since we can only wait for Commander Keeje's ships and men to make their assault over the next few days and even hours, what I want to do, is counter-propose to him a settlement, that might postpone the violence," said Bojji-Than, the Mars Base commander, told his staff, there in the facility Command Center.

This was following the communication with Keeje, and the ultimatum. It quickly became common knowledge, and with somewhat of a panic, that the Russians had arrived on Mars, among those living at the base. The life of a space-voyager in 2078-79, people who had dedicated their education and training, and many years at a time working on Mars program, for them, their lives were not always very comfortable and were often fraught with dangerous outcomes. In this case, the danger was a purely political and military venture, yet they had to face danger, attack, death and enslavement, or worse, because the approaching meteor, or 'Big Baby Bertha', had created a desperate and urgent attempt by competing Earth-side forces, to control the operation of the Mars Base. And they all knew this was basically true.

But it seemed to make sense, since the Russian-Islamics certainly had the advantage, prior to the arrival of Winton Berle's ships, and soldiers. And since the Russians had ships and men and weapons already in orbit around Mars, now, Bojji-Than and his advisors understood fairly well that they would probably lose any battle or armed-conflict, especially if they assaulted very soon, and with great force. He also really was a man of peace. He didn't want to see a lot of his people killed, and had to keep in mind, that it was not a normal situation at all. Any breach of the life-sustaining oxygen-air seals at the base facility, a small breach or a large breach, caused by a bullet or a grenade or a battering ram, would suck out the air from within, and probably kill them all, within only a few hours. Whether or not Commander Keeje would go that far, he didn't know. He was hoping they would be spared, given the necessity of the Russians to preserve the integrity of the base life-support, so they could use it themselves. And also the idea that it simply made no sense to destroy the base, which would make it useless for anyone.

So Bojji-Than and his staff and planners, had discussed things long enough to reach the conclusion that it was in their best interests to try to delay, stall, lie if they had to, and put them off. Keeje had already made it very clear, that he intended simply to over-whelm them, and had his orders to do so. Easier said than done, bold claims.

No one at the Mars Base or out on the surface, such as external workers, or within, could actually see the five orbiting ships that they all knew had arrived. Above, or somewhere far away, yes, but because just like one might search for a satellite on Earth in the night sky, and to the naked eye see only the usual star-formations, the ships seemed quite small, far away in their orbit-paths, like small stars themselves, if they could be seen at all. The orbits were perhaps 100 miles above the planet surface, variably. Mars is only one-third as big as planet Earth, but large enough that the new arrivals might hide on the other side of the planet, out-of-range. The only way they knew they were 'there', was by telemetry, telescopes, radar, and scans. On the opposite side of the planet, there was really no way at all for the base technicians to know.

For students of warfare and the history of warfare, it might be thought an interesting study, that being the assault of the surface-ground base on Mars, by orbiting ships. Even back on Earth, ships in orbit had never really been employed by attack vessels intended to overcome people, cities, lands or nations on the surface. In other words, the space-programs had always been for research, and for peaceful pursuits, resource gathering, telescope-photographs, communications satellites, and to some extent for travel, such as to the moon. They had 'never' been used for war. So it was a significant historical development, somewhat, those now on Mars, orbiting ships were making war against a ground facility and residents. It was true that the central philosophical debate about the whole Mars Base conflict in 2078, was the idea of war-fare as an Earth export to a barren and lifeless neighboring planet, as some sort of staging-ground for the acquisition of resources, and the sustainable facility. The reason was that the space-program, like the exploration of the deep oceans, and like Earth's historical explorers, and advances in science-technology, were universally regarded by the elite science-intellectual as peaceful, and non-military. The heart-and-soul of the space-program was always non-military in its essential nature, and never seriously adopted for conquest or the domination of peoples, especially the people of the Earth. So, both sides of the question, might have been interested in how things proceeded on Mars, during that season in 2078-79, and the battle at Tharsis Montes.

As the space-program had progressed during this era, there were many ships and satellites orbiting Earth. There was a successful base on the Earth's moon, with travel back-and-forth regularly and extensively. It was well known that the ships might someday be used for war, on Earth, so that to prevent this, there were old and current treaties, agreements, and laws-and-rules, established by space-authorities such as the by-then defunct NASA, and global Earth powers, and world international interests. To load orbiting ships with such as high explosives, weapons-of-mass-devastation, deadly devices, nuclear-bombs, this was a big boundary, or line to cross. Earth had many war-machines, air-planes and jet-aircraft, missiles, ICBM's, the ground warfare with tanks, scanning by radar, and the EMP's, and exotic weapons, but the use of the space-technology was perhaps thought to be 'pure' or free of all this ill-will. Or so it was hoped. It was not especially fearful to use the ships in space to attack the Earth, but without a doubt, military think-tanks and planners had thoroughly examined various potential ways this might happen, and if they felt it would be successful, and different ways they could do this, or someone else could. Nuclear weapons launched from spacecraft were just one particularly cruel and heinous scenario. It would likely be equally successful as the same weapons delivered by high-altitude aircraft, or by ICBM; similar, but different enough to have different dynamics during a conflict. But there were many other applications, such as spying, and the use of telescopes pointed back at your enemy, and other types of war from space. Men would never parachute down to the battle-field from a space-shuttle, or even glide down on smaller planes loaded with soldiers, but as human ingenuity would ever likely have it, all kinds of destructive capacities for ships in space were banned and banished and outlawed, from the beginning of the space-program in the US, and worldwide.

Those agreements were routinely broken, and when they were, they were either re-built, or not. But now, on Mars, things started to look like students of the Art of War, would have a high-quality case-study to review these policies, with a very destructive battle poised to begin, as many, many years of advancement and progress would be obliterated. It was different enough from a similar conflict on Earth, but definitely related, but as a matter of war-studies, was challenging for both sides. As the Mars Base commander Bojji-Than had observed, relating to his staff and crew, that many of the terms of the battle were highly unlikely to succeed, and even absurd, he also felt inwardly, that it must always have been so when men go to war in any case. There were only some 263 people at the Mars Base at that time. The entire area that the base occupied, including out-laying launch facilities, storage tanks and buildings and structures, and support facilities, was only just more than some ten square-miles, of actual acreage. So as a military objective from orbit, and from the transit across the Abyss, the base was quite a small target, even miniscule, a needle in a haystack. The airless surface of Mars, the cold and heat, the ultra-violet radiation, the great distances, the need to bring the men down from the ships to assault and occupy the base, and other factors, for experienced space-travelers, it was to laugh, believing that it would be as simple as a Hollywood movie.

"Not a movie, not a TV-show. Say it often," said Karen Tutturro, the communications-specialist assigned to repair the Mars Base radios that could reach Earth effectively. She was part of the Command Center team at this point, as well as Bojji-Than, the Security-Chief Juno Amorrossi, Vinces Grant, Charley Barron, and Matt Van Templar, and the numerous tech-specialists, who worked in shifts, under each department. It was now the third day since they were fully aware that the Russians had entered orbit, and since Keeje's ultimatum, with at that point was only 48-hours past. Juno, the Belgian security man, very muscular and husky, seemed to rise to the occasion, as the looming crisis grew tense. Perhaps it was there had never been a conflict at the Mars Base, and for the most part his employment as security chief was to settle arguments and disputes among the residents, over the years, or to handle any crowds or visitors, where personnel or people might be unfamiliar with the various processes and safety standards. He laughed at Karen's comment.

"I love movies," Juno said. "I love American movies, the old ones. I love action-adventure, I love movies with airplanes and guns and soldiers. I love the robot movies and the spy movies. I love the super-heroes. Everybody loves those. They're great!"

Vinces Grant, the mainstay science-officer at the base, familiar with almost all of the science-tech, especially external exploration, offered his view. "By the time they make a movie about my life, I won't be moving much," he said. "So is that still a movie? Or---a film?"

"The main idea is to try to be realistic," said the commander, Bojji-Than. Bojii was a wiry, thin Asian man, with dark-skin and a somewhat bald head, with only a little hair around his ears in the monk's pattern of hair-loss. "If we are arrogant, we could all end up dead. This is not a humorous situation at all. We have to keep our heads. We have to do the right thing at the right time. We have to make the right decisions if we possibly can. It won't be easy. During normal operations, we know what to do in any situation, and it's dangerous anyway. If we lost air-pressure, or life-sustain. We all know this. With a hostile force, it's a brand-new situation. Please, keep your head, and use good-judgment. Don't be a hero. Try to do what you need to do, when you need to do it. By all means, form groups and get support, and get a second opinion when you can. This is real-life, and suffering and death are a part of real life. I am here to protect you all, and I intend to do so to the best of my ability."

There was a long, somewhat morbid pause. Hour-after-hour, for two days, and prior to that, they had been keeping track. Work in space is slow, things take time. It was hard work as well, many long hours. They all knew their jobs, but none could really see the future.

"I wish we could just do whatever we wanted, and not be afraid," said Karen, musing for a moment. "You know what I mean? Like, just go exploring around, see what's out there. Instead of this. Instead of fighting. It's like fighting over an empty wasteland with a single giant diamond on it. Big deal. It's still a wasteland. If it wasn't for the giant diamond, at least we could just look around, take up our own ships and go out in the buggies, climb the mountains, take pictures."

"Karen, dear," said Vinces, "That has been all we've ever done, since we came to Mars 17-years ago. This is your first time. The rest of us have done plenty of painless exploration. A wasteland, yes, but full of mystery. Personally I have always hoped to find some new resource that would be valuable back home. Vast oceans of oil, perhaps, or huge mountains of gold. El-Dorado."

There was an alert-sound beeping on one of the radio-monitors---buzzzz-buzzzz-buzzzz. A large light blinked green. "In-bound radio from the space dock, Commander. Connecting now, through the Comm-Tower Control," said one of the techies. He started to manipulate his buttons and devices, working to route the signal from the base's main radio-group, the same system Karen was supposed to repair.

"Molinari," said Bojji-Than. "Good. I was wondering when we'd hear from them."

It was more than a few moments to connect this 'long-distance call'. Finally, they all could hear the voice of the Molinari commander, Frakesteen. But the signal was very weak, garbled, and they could hardly make it out. It may have been a solar flare, or other cosmic disturbance.

"Yes, Mars Base---this-is—Frakesteen---Molinari, Orbital Time-Frame---221.22---do you read me? We're having some---there is---but it's fine—"

His words were dropping out. "Yes, Molinari—Commander Frakesteen, this is Mars. Weak signal here. Bojji-Than at Snikta-base control. Are you there? Please respond?"

A long pause, static, then garbled sounds. After a few long moments of noise and buzzes, the link seemed to be lost. "I've lost the signal, Bojii, I'm sorry," the tech-operator said sadly. For all of them, those slender and fragile connections to home were sometimes more cheering and joyful than anything else that ever happened at the Mars Base at all. It meant far more. It meant they could still go home one day, and that the other teams were functional. Sometimes they would go months with almost no news or communications at all.

Bojii sighed. "See if you can get it back, or connect by out-bound from us," he said. "It's probably nothing."

A brief silence. For some reason, the whole room then exploded in laughter as he said this, then they all went back to work.

CHAPTER 36: Landers With Men


*"We are not terrorists. We are explorers. There's a difference."*

*--Commander Prokov Keeje, of the Space-Vessel 'Tolstoy' from Mars orbit, 2079, in a private communication with US Mars Base leader Bojji-Than*


Later, as it turned out, the radio-link to Molinari was re-established. True enough, it was not un-observed that the Mars Base communications-system and radio-signal long-distance processors were failing. Tutturro had earned their respect, there at the base on Mars, by determining the underlying cause; the signal compression data processors had been designed without sufficient UV-radiation protection for the way they were being used. Mars has a very high ambient UV radiation surface profile, because without a denser atmosphere, such as Earth's ozone layer, the Sun's rays, though fainter than on Earth, are yet harsher, un-filtered. The processors were installed externally to the main interior of the base, among the antenna and satellite dish power-supply towers and outbuildings. This turned out to be an error. The technical design of the processors had not considered the high-level UV radiation. Seals, plastic parts, some of the film-based circuitry, and some of the 'silicon chip' were drying up, and became brittle, fracturing prematurely, and then beginning to fail. They were repaired as they went along, but eventually the processors were getting worse. Karen, on her first trip to Mars, was called upon to surface-work and Mars-walker suit excursions, numerous times. But it paid off, and the problem had been recognized. The data processor designs had not intended the units to be exposed to high-UV levels for such long-periods, because they were inter-changeable with others used elsewhere, in other environments.

But they still worked, and the Molinari summons was reconnected, a few hours later. Commander Bojji-Than had by then retired to rest, and Matt Currison Von Templar handled the link, the Mars Base launch and astro-navigation specialist, working double-duty in the Command Center, like the rest of the lead-staff. The gist of what Frakesteen had to tell them was relayed from Vandenberg and the US White House and military. The message had also reached Berle's formation of ships, in-transit.

"The Eastern-bloc powers on Earth have declared a state of war with the West, back home," Von Templar told Vinces Grant and Charley Barron, and others on-duty in the Command-Center at that time. "So, congratulations. They're fighting over us. Or, gearing up for it, anyway. Relayed from Molinari."

Maybe Vinces, Charley, and those on hand, were visualizing their homes and families, people they knew or favorite places, childhood vistas, Mother Nature as it was intended for mankind, green hills, pleasant small towns lined with shade trees. They often did, it kept their spirits up, including photographs and video. There had been recent wars, it wasn't anything new. In this case, nations were anticipating the approaching meteor, that it might hit planet Earth. The moon-bases were also part of the equation, and were also disputed. Any off-world sustainable resources suddenly had a premium value for Earth's power brokers and elite planners. So it was no surprise that hostilities were at hand---back home. It was even somewhat routine.

"Have they attacked? Is it just diplomatic, or a declaration of war? Or has there been destruction?" Charley Barron asked, turning in his chair towards Von Templar. The Command-Center was arranged as a series of computer work-kiosks with all their monitors and controls, tracking and power-supplies, etc. Barron was the base life-sustain systems-man.

"No nine-eleven for now," Von Templar reported. "Apparently there has been a new panic. New details on the meteor, too. But it's official. The Islamic-Hindu alliance have made public demands, threats. War-footing on both sides."

So, the message was logged, and then would be released to all-hands at the base. After a while, Bojji-Than returned to the Command-Center, rested for a few hours. He also got the news from Molinari. There was nothing new on the ships in orbit. It had been more than three days. As long as he had been in charge on Mars, ('the king of Mars', they sometimes joked), it had always been obvious to him that almost all of his people were very bored and restless from day-to-day. Like a farm or homestead in the Old West, isolated and set apart by wilderness miles and snowstorms or distances, visitors, or 'mail', and the transports, were always a thrill. Now they seemed to have more than usual, to be sure. But the effect was the same.

"Any ideas about their next move? Let's predict our best and worst scenario. Vinces? Matt? We seem to have time for now. Please, go ahead," said Bojji-Than.

Another pause. Von Templar was still feeling he had not been heard. "You're not listening to my opinion, here," he said. "These men have just traveled across the Abyss. They want to, even very much need to, rest and refuel, repair. Fifty million miles away, a bunch of sweaty Russians and Mohammeds in close quarters. We've all made the same voyage. You were talking about negotiations; offer them peace. We don't have to surrender."

"He didn't sound very peaceful," said Vinces. "He has his orders. That type of man will do exactly as he has been told."

"What about just, just tell him to send down a team with no weapons. Invite them inside. Make things easy. Maybe the situation on Earth will change?"

A long, somewhat morbid pause. "Paint me a picture," said Bojji-Than. "How do you see that happening?"

Von Templar shrugged. "Radio his command-ship. They bring down no more than five men. Shuttle down or they use their pods. Escort inside. No weapons on either side, guarantee return to their ships. Offer them a small supply of fuel, or food-and-water as an olive branch. They can take it back, we use the transport dockers. Tell him we'll try to set up a link to our Earth-authorities and theirs as well, I guess that Ukrainian base. It might work."

"Wait a minute! Wait a minute, Matt. That's exactly what happened with the Penelope cargo-ship, Guy Reisling and his crew, stranded and powerless in deep space. They were almost killed! All ship's system-power and life-support lost, for two weeks! They barely restored function to reach the halfway station. Negotiations didn't work. They took it only as an opportunity to attack. And they did!! No way! No way!" Vinces Grant was uncharacteristically passionate, angry. Guy was a friend. They all knew the story, what had happened.

"True, true," said Bojji-Than. "But, it would buy us time."

"Commander Berle has nine ships," said Juno, the big Belgian man. "He can kick their asses right back to Russia, once they arrive. And they will. Three more, or even four more ships than they have. And supplies from Molinari."

"A sand-storm would buy us time as well," said Karen Tutturro, also there among them.

"Sure, just hard to arrange when needed," Von Templar said.

"How much time do we have to buy, exactly?" said Juno.

"At least 20 days," said Vinces. "No way they'll be here in any shorter time than that. Twenty days, even at high-speed."

"This Russian commander could attack any time. He's had a solid 48-hours in Mars-orbit," said Von Templar.

Another morbid pause. The Command-Center buzzed and hummed. It had become dark outside the walls of the sealed, air-safe base. Night on Mars, eerie, like a soft stillness and dark. Bojji-Than moved to the main mapping-screen computer-projection array they used to track the planet-region, skies, and moons. This linked to radio telescopes and radar, etc. But it was in no way complete. Mars had not been Google mapped. "There is no movement from their ships at all? No intercepted communications?"

"We have only one ship within tracking range at all, on these machines, in the past day. They're hiding on the other side. Realistically, those ships could move into attack within hours," offered Vinces, who seemed to have a bull-dog approach with the data and information, working alongside Charley Barron.

"Oh what a feeling," Karen joked.

"It's called anxiety," Juno replied. "Shake it off."

Karen had never been to Mars before, and now she was in a war on her maiden voyage. After more brief discussion, rather than waste the chance, they decided to go ahead with a version of what Van Templar had proposed. The opposition point-of-view

(Vinces and others) were mitigated by the idea that it probably wouldn't work anyway, and that the Mars Base team would certainly be telling lies, too, buying time. It took about an hour to figure out how to direct-radio the single ship they had spotted in-range, which turned out to be the Saint Peter (the ship that had actually taken the damage of the hit by the metallic bar released by the Penelope as debris, they had survived, however). The relay link connected to that pilot, and he was able to connect to their Command, Keeje, somewhere out-of-sight on the Tolstoy. The Mars Base counter-option to his ultimatum was delivered as a content, not as inter-active voice-to-voice, that is, basically, as 'text'. The Mars Base team carefully put together their terms: they were inviting a Russian-Islamic team of five men without weapons to drop down to the surface as they usually would, and to be escorted inside the base, to talk it over. They would provide fuel, food and supplies, and guarantee their return. By the time the planet had turned again towards the warming Sun, on the Mars Base side, the message had been related, and they were waiting for a reply.

Keeje had not been idle. Following his threat and ultimatum (demanding their surrender of control of the base, or else to die) he and his chiefs and pilots, and soldiers, were getting ready to attack, and enter the base by force. The five ships would co-ordinate an orbital pass near the proper region above the base, one-after-the-other. They would use the descent pods, which used both parachutes and small thrusters that reduced impact and descent speed to manageable. Pods would hold 10 armed soldiers each, and also pilots, or they overload them with more. Once on the surface, each descent pod had a return 'lifter', that could return to orbit by means of a rather rough ride by powerful rockets, in even small pods, and then re-dock with the main ships. The lower Mars-gravity made much of this possible. The five very large main ships each had four of the descent pods, and spares, or emergency back-ups. Once on the surface, soldiers in walker suits would attempt to take the Mars Base by force. The discarded descent vehicles would be used as shelters for their side on the surface. They had planned this out as well, including bombs, guns, grenades, life-support, and internal base take-over process. All that remained was the final order to begin the dance.

Keeje received the message from Bojji-Than, at this point presented to him as a hand-held mini-computer screen that could move messages and info. His seconds and technicians were also aware that this message had come over.

"They are lying. Stalling. It is so obvious," said Keeje to his co-pilot. "They're hoping their reinforcements can arrive and save them."

The overly long rays of the Sun, slim and skinny, colder than back home, reached the giant feet of the Tharsis Montes, sandy rust grains dusting across the rocks and stone near the base, where ageless frozen moisture glistened like silver. Snikta-Ridge Volcanic Basin facilities operated as usual, for a base on Mars, that is. There were always duties that kept them all alive. Re-circulated air and water were scrubbed or cleansed for usage again. Scheduled seal checks and confirmed safety standards were maintained. Hydroponic indoor farms, with vegetables and beans, fruits and various grains or rice, which were a very large part of the operational internal space, with many benefits, were tended. Research and educational, health and cafeteria, external vehicles and workers, suits, air locks and doors or cargo-area operations: all 'status-quo' for the Mars Base population. Also, now, the defenders, the external oxygen-igloo pits, the armed soldiers and their own walker-suits, all also prepped in process. Shifts in the Command-Center were on-rotation. The peace-proposal was calculated to have been in Commander Keeje's hands for about five hours.

"Word? Any reply?" Bojji-Than watched over their gambit like a hawk.

"Nothing," said Juno Amorrossi, who was monitoring any related communications. "They're not going for it."

"That figures," Bojji-Than responded.

Another three or four hours passed. Their effort seemed to have gained them nothing. At least they had tried. By about noon, Mars-time, there was yet another alarming development. Tracking-monitors could detect the first of at least three of the orbiting Russian ships, tracking a path directed in the region of the base. Estimated time-of-arrival to attack position or for descent, only about an hour.

"There they are!" said the technician on the tracking station. There was a huddle at his post, little more than a grid-map with tiny blips and signals indicating the ships and planet, and other navigational features.

"My men need to be directed now to alert-status," said Juno suddenly, as he realized what might soon happen. "Right now. By permission, Commander."

"Go ahead. It doesn't look good from this, that's for sure," said Bojji-Than. "Set yourself up for some kind of defense. Damn!"

"Yes sir," said Juno. He looked again at the grid-map, making a mental note of the general time frame, then moved off quickly.

"What about the other two ships?" Bojji-Than asked the technician. Vinces Grant was hovering nearby to watch what they had all feared start now into action.

"Not on this grid at all. Probably too far out. This covers 3,000 miles in a circle. The approach is from the Northeast. Three ships visible, 500-miles apart in a line-up, apparently, from this. The lead is maybe 1,500-miles out."

"What's happening?" Grant asked them both. Others were with them. The Command-Center was alive with all the activity.

"They may be dropping in on us soon, my friend," said Bojji-Than. "Take a deep breath."

"Told you so," Vinces responded sourly. "Shit. Fuckers. This base is, uh, fragile!!"

In another two hours, the first of the descent-pods were spotted, dropping from orbit by parachute and para-glide thrusters. Then more after those. They started to land-to-surface in a pattern about two miles or less from the base, and some farther out.

## CHAPTER 37: A First, a Second First, and a Third

*"Mars ain't the kind of place to raise your kids. In fact, it's cold as hell."*

*--Elton John, 'Rocket Man'*

Two of Commander Keeje's ships, the Sir-Soviet, and the Krenika, cruised at an ambling pace across the Martian horizon, about 200 miles apart by this time (plenty of room for each ship to position and drop the descent-pods). The height above the surface was about 150-miles, appropriate for their task. To maintain orbit, ship's speed was between 15,000 and 17,000-miles per hour, the Mar's gravity being less than Earth's, and the atmosphere much thinner. The process took several hours, for each of these two ships, to deploy the shuttle-pods to the surface near the base. The pods had to somehow land near enough to the Mars Base for the men inside, the soldiers, to realistically hope to be able to traverse the distance to the walls of the base facility itself. This was calculated closely and the navigators knew their business. So, at the proper moment, the first ship released two of the descent-pods. These were about the size of a hefty Earth-side RV-camper, or motor home. But they were rounded, looking much like the original Earth re-entry vehicles from the early Apollo missions. That is, rounded-circular, with view-ports and hatches, and for these, both para-glider chutes that popped open at a certain point, and also thruster-rockets (using a hydrogen-peroxide and chemical oxygen formula). Ten men could ride in one pod, and each had a pilot, or descent captain, who manipulated the para-glider chutes and thrusters. Of course they had life-sustain and other controls, but they were no fun to ride in. Cramped, bumpy, dangerous, and claustrophobic.

There wasn't much to see, if you happened to be walking on

the Mars-surface, at about that time on that day (Mars-day). The Mars Base used a calendar that was structured similar to a normal Earth calendar, but was based on the actual length of distance-time it took for Mars to make one solar-orbit, more than three times that of an Earth-year. Seasons on Mars were thus very long, very cold, and then hotter, and the sandstorm seasons, and so on. These were still being studied. But, from the surface, the Martian sky was a dingy sort of indigo-reddish, lightly blue like Earth, but distinct with other colors. The ships were too high to be seen with the naked eye, but a decent telescope would certainly capture them as clear images. Within the time-allotted, however, one might have observed the pods as streaking 'contrails', the friction burn, even as thin as Mar's atmosphere was. Then the para-glider chutes opened. These were designed to work on Mars, but resembled the type that have been successful among recreationists and athletes, with long, curved tubes of tough ribbed fabrics with thicker or more firm wings or vents, attached by diagonal strings of very strong new-material fibers. The pods slowed, and then tracked the way the pilots intended. The rockets began to fire, as they grew closer to the surface, ready for impact. This slowed them even more, and they acted like very awkward, heavy-moving air-craft, for just long enough for the pods themselves to hit the surface in such a way that they all survived in reasonable condition. It was an impact landing, and they tended to roll, or overturn. But a good lander-pilot could touch one down with a long skidding slide, raising Mars-dust into long-thin wispy clouds that had not been disturbed for perhaps millions of years. The pods had bottoms with various kinds of 'skids' or 'sleds', or even wheels.

The Krenika released two pods first, and then still in orbital motion, two more. More than 40 soldiers. The second ship, the Sir-Soviet, repeated the action, a few minutes behind. The 8 lander pods hit the Mars-surface in a pattern only about two miles out from the base, some more distant. They were scattered here and there, however they happened to hit, but far enough apart to avoid collisions on the way down. So, 85 soldiers in all, plus the pilots, also ready to fight. The pods were corrected for their landing-spots, meaning they were righted and secured so the men could get out. Inside, the soldiers were prepped with their walker-suits, gear, and weapons. Teams had leaders and captains with enough information to move them towards the goal: the gates of Snikta-base, seen to them all far ahead, in the wisteria of the Martian-moment, a glistening-shiny techno-fortress, with its walls and outer boundaries, towers and antenna, small roads, fuel tanks and launchers.

For these soldiers, it was a bit of culture shock. They had been in space for many months. Suddenly, the lower gravity of Mars was holding them firmly to the dirt under their boots, but even that seemed like a ton of bricks. Strong, tough men, they had all been in space before, mostly on the moon. But the stress was very hard to overcome, for the first few minutes and hours. Heart rate, muscle-motion, breathing, lifting and walking, some of the soldiers stumbled around like drunks, even falling over on flat surfaces, without a cause. And falling was hard on the walker-suits, if those failed, they would die, unable to breathe in the thin atmosphere. A fabric rips or tears, a torn glove or boot, a cracked helmet, the inner air leaked out, and they would die, and they knew it. They had weapons, packs, and more. Two of the shuttles had small, easily assembled electric truck-rigs, or 'golf-carts'. These were un-packed and cranked up. By mid-day, six hours later, they were all in a march to the base, 85 armed Islamic-Russian astronauts trained for battle, to take over control of the US Mars Base, the Snikta-Ridge Volcanic Basin facility at Tharsis Montes. It all seemed to happen in silence, but of course they all had radio-links, as did the ships above, and in one of those, Commander Prokov Keeje and his attack-management planners, viewed the big-picture progress and called the shots.

"I myself will enter the Mars Base on-foot within less than a week," Keeje told his crew.

The base-defense men were not so sure. Before the pods even hit the surface, the 24 outer 'life-sustain igloo stations' were manned and ready, fully operational. The Mars-teams had had plenty of time to build these, and did the best they could. Half underground, rigged-materials bio-domes could hold each about six to ten men. They also had communications, weapons, medical, food-and-water. They could hold out in these for days, re-stock, hide-and-seek, and fight. Mars Base had many other advantages, especially the use of far more numerous electric 'golf-carts', as many as 45 of these, some as large as big regular trucks. So Mars Base soldiers could use these to haul men directly to the approaching soldiers from the opposition-side, and attack from the carts, complete with extra gear and supplies. Mars Base also had a much better view of the battle, with in-place monitoring stations and equipment, long in-use for other purposes. Mars Base also had more men, who could fight (and women), in-total. Strictly by the numbers, the five Russian-Islamic ships could ultimately deploy about 120 men. Mars Base population was at 265 on that day, and only about 40 of those people would not be able to fight, for various reasons. So Commander Bojji-Than could count on about 225 fighting men and women (there were no children at all on mars at that time). Mars Base had more than twice the forces of the opposition, including every available person. Essentially, from a strategic viewpoint, the invaders were at a serious disadvantage and uphill struggle. The odds were against them; they had less to work with, and were in more danger throughout the battle until their goal was attained, if it was to be. So Juno Amorrssi and Bojji-Than, Vinces Grant and the others, could rest easier, confident they could hold out until the American forces arrived to put the thing to an end.

"All right, for Christ's sake, here they come," said Bojji-Than, working very long shifts now in the Command-Center, high up on the main structure. "I assume you have some kind of plan?"

"Well, yeah, but I've never run a war before, sorry. Hope this works," said Juno, the hefty-sexy Belgian man, with his huge arms and broad chest. Juno had managed to bed one of the base-women who liked his company enough to allow the once-in-a-while sexy-romp in private, the night before (on his shift-break). They made passionate red-hot love for hours, she was a medical worker. The result was that he felt refreshed and very healthy, confident.

"Yes? Go ahead," Bojii replied. "I've looked at your work-sheets anyway. It's simple, how far out are their first-wave on foot, and at what point do we deploy fighters?"

"You can see on my map here," said Juno. Now four men huddled near the maps, Juno, Vinces, Charley Barron, and Bojji-Than. Other task-team staff worked on other problems and plans elsewhere in the Command-Center, at the same time. Everything was happening all at once now. "The igloos are stationary, they just wait. But by the time their first men reach here, give them another three hours, and see? They'll be within small-arms range, that's like a quarter-mile, but still half that distance again to the first circle of igloos."

"So you have two hours or less to get teams of men into the electric trucks to meet them with heavy resistance?" Vinces said. "Right?"

"Yeah, I'd say only an hour, may as well be ahead of the

game," Juno said. "It will be hit-and-run in the trucks and carts, most of theirs will be on foot. In a way, it looks like we could even just win the whole thing that way, wipe them out, I mean, these men are in walker-suits. They'll fight back, but from the electric carts."

"They have carts too, the reports said they had several," said Charley Barron.

"Yeah, and what-the-hell else anyway, who fucking knows, Charley?" Juno replied. "If they choose to, their fucking orbiters could drop a damn bomb on our heads, within a couple of hours, on their next pass!"

"Never mind, never mind, one thing at a time," Bojji-Than interrupted. "Colonel Amorrossi, give the necessary orders and get that action underway. Do it immediately. And tell your men, go ahead and wipe them out. This is war now. They will kill us all. I'm sorry to be in this position, but---"

"Buddha might disagree, eh?" said Vinces.

"Take prisoners and spare any lives you can, Juno," Bojji-Than added. "But we will defend this base. Every single life on Mars at this hour is hanging by a thread anyway. This is madness."

So it began, and so it continued. A radio-link was set up to reach Earth, Molinari and Berle's ships, this also took time. Matt Currison Van Templar was getting ready to launch two of the Mars Base orbiters (used to assist in-coming transports until then). This would take hours as well, but would create an 'air-equality', though any specifics as far as counter-attack from orbit were not clear, the battle would be much too dynamic and constantly changing. Ship-to-ship conflict was highly unlikely, for the orbiters, against the Russian ships. The distances were too vast, the orbit-paths too specific, and the ships were not that maneuverable, not jet fighters at all. But by having their ships in orbit, Mars Base could deal with all kinds of contingencies, so it was essential.

The battle could only form to the flat-regions, East-and-North, not from the mountains against which the base rested its back. Now a mile-and-a-half in that direction from where Karen Tutturro had first viewed Mars up-close, months ago, the invaders trudged towards their fate. The men in walker-suits looked like hard-plastic and thick foam-fiber body-armor sealed robotic machine-men with shiny glass bowls on their heads and air-tanks. They each had large guns, a type that easily fired heavy-caliber repeating-action munitions, even in the thin air. Other weapons, too. The Martian sky turned just a bit, the distant Sun that supplied both Earth and Mars with life sustaining heat, glowing brightly and warming, almost smiling at them all.

"I have never been to Mars, have you?" said one of the Islamic men, walking-walking-walking, only rocks and stones and sandy dust to view. His radio was linked to his 'buddy', also Islamic. Other invading soldiers were around in squads or teams, also walking-walking-walking. In the distance behind them, the landing pods rested, still, the glider parachutes crest-fallen, flat on the ground, or rolled up.

"Are you a fool, Mohammed? Of course not," said his friend, Aminadab. "Neither of us has. No one has."

"Then we are the first men of faith to set foot here," Mohammed said. "We must bless this world for the glory of Allah and the Prophet."

"You go ahead," Aminadab replied. Their radio-helmets buzzed and hummed. "Damn! I can hardly walk in this! How far ahead?"

"Half a mile," his partner replied. "Or half a million miles. We have come far to kill these astronauts."

"It will be worth it to get inside the base and eat or drink normal food or take a hot bath!"

They went on. They could see the Tharsis Montes, of course, in fact it was impossible to avoid seeing them, they were so high and tall and towered over everything, more like a roof or ceiling than a mountain with a top that could be seen. The position of the base ahead was determined by Mars-adjusted compass, they could only assume the actual direction towards it. After a while they could see the first forms and shapes, outer buildings and a few structures.

There was an alert-signal suddenly inside every walker-suit radio-helmet---*beep-beep-beep*. Then the Surface Commander's voice over-rided all other communications. *"Forward units, hold your positions, take cover if possible, prepare weapons. We are under attack! Repeating: we are under attack! They are using electric trucks, ahead South and West, over the ridge! Prepare to engage! Prepare to engage! Here they come!!"*

And then they could see, from the direction he had said, a row of electric carts and small electric trucks, each loaded with the Mars-men soldiers, about six carts in all, scooting towards them as if following a golf-ball to the next hole-in-one. Dusty sand lifted behind them into the thin air and almost vaporous so-called 'wind'.

It was only minutes before the billion-year silence of Mars, was broken by the all-too-familiar sound of men shooting war-weapons at each other, there, for the first time ever: *bada-bada-bada-bada! Bada-bada-bada-pow-pow-pow!*

Fire, bullets, blood.

## CHAPTER 38: The U.S. Presidential Seated Council

The situation on Earth while the battle was taking shape on Mars, now escalated into an international scenario, by which the two sides in dispute concerning control of the Mars Base, were posturing to gain advantage. Typical of such things, there was no real center of control or rational or reasonable way in which wide-scale or broad-based disputes were to be settled or negotiated, or dealt with in the best interests of the largest number of people. Instead, the powers at work, the Eastern and the Western, sensed their necessity to initiate action and counter-action, attack and counter-attack, strategy and counter-strategy, in the super-power model that had dominated the last part of the Twentieth-Century. It made sense to one side, and it made sense to the other side, but in the broader view, it didn't make much sense at all, and even a uninterested person without much power or knowledge, could see a way out, when the entrenched and very powerful leaders and planners on either side could not. There really wasn't that much at stake. The base on Mars had been in operation at that time for about 20-years, but only represented about 260 people as a sustainable facility, and had very little power of its own. It was also very difficult and challenging to maintain in the longer-term. The Mars Base at Tharsis Montes required constant maintenance attention and repairs, because of the hostile Martian atmosphere and environment. There were no mountains of gold, there were no riches, and there were no serious military or positional advantages. So essentially, it was a very high-maintenance piece of real estate, with the only real attraction being that if the disaster came, that someone could live there, approximately 300 people at maximum. It may not have been known among the Eastern space-program elite, about many of the details of the function and operation of the Snikta Ridge Volcanic-Basin base, and all that was involved in the day-to-day activity there. The need to supply the base with services from the Earth, such as Guy Reisling's transport, was extensive, and the US-base operational assumption was that a regular flow of transports would sustain the Mars Base, so even in the event that control of the base was lost, it still needed to be supplied. No one seemed to have looked ahead that far when the East secretly launched the five vessels under the command of Rudolph Terchenko and others. But they certainly knew all this. But it didn't seem to matter in the panic of Big-Baby Bertha's approach. If they could control the base on Mars, it would be worth it, they felt.

By now the Earth-calendar for yearly date keeping, was turning towards 2079. Tracking of asteroid U2753b, had continued, and the anticipated destruction was also estimated in great detail. So although it was not incidental to the circumstance on Mars, the affair regarding the meteor, was another story in itself, and as one would expect a defense at that time, the people of the Earth with their powers and glorious science and technology, were set to work to overcome and prevent the disaster. Busy-busy-busy. *Not a movie, not a TV-show, say it often,* was the jocular by-word of some called on the US-side, to deal with the crisis. The end of the world had become a joke among them, but serious efforts were giving many hope, and it seemed possible they would survive, or avoid the collision with the asteroid.

But you never know. In the view of many including the Mars Base commander, the space-science on the meteor was merely an excuse and opportunity for the Eastern forces at work in the local space-vicinity to make war-like advances and steal the Mars Base, for whatever other reasons they may have had. So it made sense, but it didn't make sense, it could be solved rather simply, and yet it was utterly impossible to deal with. The notion of sharing the resources on the Martian planet, might have functioned without war, but in the heat and passion and panic of it all, the military chose to move quickly and lethally, as they usually do, and let the asteroid-chips fall where they may, right onto the heads of the people on Mars, and the hostage masses of the rich and poor on Earth, who may even have known nothing if anything about it all.

So with all this, between the nations and the powers on Earth, there was a shuffling and saber rattling that went on-and-on. More meetings and conventions were organized and held in large cities, either publicly or not, concerning the science-and-data about the meteor. There was a good deal of outrage and news media screaming and spin, regarding the incident with the 'Penelope'. It was the first time ever in Earth history that one space-vessel had attacked another. They couldn't keep that sort of thing secret very long, as far as the public. The space-program at that time was very costly, however, it created vast wealth in terms of employment. Science-technology workers, aerospace, design, materials, supplies, fuels, and support industries all did quite well. Lots of jobs, lots of money, many new industries. A good example was the Asian delegation that appeared at the early Vandenberg meeting, that Spring in 2076, early in the unfolding drama, when the announcement was made and the information was released and examined, about the meteor and it's impact on the Mars-program, rolled out like a new product from General Motors. Those businessmen and science-technology merchants were really only interested in selling rocket fuel, but the support industries and jobs and employments, were very well-paying, scattered all over the world, as a new high-tech, expensive, and supposedly 'clean' industry. So it almost didn't matter which side would win, they'd still sell plenty of rocket-fuel. And it really didn't matter that the return-benefit to the general public was disputable, and often disappointing. But it was very important to the support industries and unions and governments, as far as control of the various space-programs, if only in terms of whom they would have to barter with and sell to. So it was impossible to keep the attack on the Penelope a secret, and about the ships that had left Earth for Mars, and also the response of the Western side. Another fine mess.

As the months passed, the outrage, accusations, finger-pointing, hand-wringing, cries of foul and on-going wrangling and news-information shouts and bell-ringing, as well as action in the courts and in the seats of government, halls of power, parliaments and congresses, proceeded apace with the usual lack of grace and manners. Things reached such a pitch that no one really doubted that Earthside hostilities would result, with all the usual forms of warfare, high-and-low, to settle the vengeance. It was an irony to the science-and-research community that had always held for space-exploration to peacefully benefit all, that the first hostilities off-world, would grow from a rather humble and challenging research effort on our neighboring planet Mars. Mars, the god of war, the Roman namesake, seemed to be on payback.

At the US White House in Washington, DC, which by 2079 was occupied by four US presidents at the same time (a four-person council known as the Presidential Seated Council), the four elected leaders naturally stayed on top of the situation as well as they could.

Present were US Presidential Council-Seat Mark Renolds, about age 65-years, known for his freckled face and background as a farmer, and the rest of the group, now including Boline Bouvier, the Black man from the University of Texas, Martha Hazlett, age 45, a widely popular athlete with a law-degree, and a Hispanic man from a large farming family, also with a law-degree and business background, named Martinez Jeses-Gaurrero, age 38-years. There were many decisions and actions to be taken, and population control, responsibilities abroad, concerns with safety-measures for municipalities, and much to deal with for popular opinion and news-leaks, because as usual, the US White-House was relatively powerless. From the White-House and military point-of-view, the base on Mars was an expensive property that deserved to be protected, along with the people who lived there. The actions against the Penelope and other hostilities, and news of the command of Prokov Keeje's arrival of ships on Mars, persuaded the US leadership to take the matter very seriously. And all this on top of the gloom-and-doom associated with the meteor. With a hardened bunker-mentality, the four leaders gather in protected meeting-conference rooms, supposedly informed, and summoned to make choices.

"Well the bad news is that the intelligence community is saying the Eastern space-program is ready and able to launch more ships to the Mars conflict," said Martha Hazlett, from where she sat in the crisis-room, on the grounds of a nearby secret facility. The other leaders were also tossing around options and ideas. Bouvier, the black man, enjoyed puffing on a tobacco pipe, by a fireplace, convenient to an air-filtering fan and de-odorizing ionizer, by a window, as they talked. He was a large man with a sober appearance, befitting his office. "I'm surprised they haven't launched more ships already," he said. "But I guess they don't have the capacity."

"We haven't launched any other ships to Mars yet either," countered Renolds. "Pretty much for the same reason. We didn't have the resources. Berle's ships are within a couple of weeks of Mars. If we could have sent more by now, we would have."

"Well, that may not be the point," said Hazlett. "The battle on Mars will happen anyway. What I'm telling you, what the

CIA and military are telling me, it's fairly clear, the Eastern-bloc is prepared and ready and prepped to launch more ships right now to Mars, from launch facilities around the world."

"Are you suggesting a response?" said Jeses-Gaurrero calmly.

"Well," said Hazlett, "we want to win this conflict and we want to win it decisively. If the Eastern-bloc is able to launch more ships to Mars, it could extend the conflict indefinitely. Even if Commander Berle overcomes the Russian ships on Mars now, with more ships on the way from their side, his forces would simply face another wave. It would go on and on. We send more, they send more. A black hole for money and resources, and dear lives and loved ones, and wasting our time as well."

"I think Winton Berle and his teams are going to kick their ass," said Bouvier. "He has a decisive advantage. They have nowhere to turn to except their ships in orbit around Mars, even to remain alive, for rest and air and food. Berle has a greater number of ships. They have no sustainable supplies. Berle has supplies from Molinari, and the support of the base, which have supplies and equipment. So, the Russians on Mars now will probably lose. He'll kick their ass."

"Agreed, agreed," said Renolds and Jeses-Gaurrero, almost at the same instant as Hazlett, nodding together. It was a comfort to affirm that what Bouvier had said.

"So unless Commander Keeje takes total control of the base before Berle arrives, we should be able to defend the base effectively, or take it back, or just keep in control, and I think we all agree, is that correct?" said Renolds. The other US Presidents on the Counsel Seat re-affirmed, nodding or commenting. It seemed like the battle on Mars was win-able. But you never know. Hazlett, probably the most popular member of the Counsel Seat, moved restlessly around the room, looking through a curtain window, finally finding a seat by an end table and lamp.

"What my sources are telling me is that if we want to end this decisively, for the long-term, that our traditional military can take action now to shut down as many of the Eastern space-launch sites as we can. Shut them down now, in Indonesia, China, India, and others. No more launches," she said.

"That's a heck of an idea," said Renolds. "Looks a lot like a declaration of war to me. Not good."

"Well, yes, it does look that way," Hazlett replied.

"Declared or un-declared, it would certainly be perceived that way," Renolds added.

"Yeah, no doubt about it," said Jeses-Gaurrero. "You're talking about numerous space-facilities, heavily protected in hard-to-reach places, hidden, our forces may not have an easy time. If we just bomb the hell out of them, they're going to bomb the hell out of us right back. We have to look at this very seriously before we attack simply on the assumption they are launching more ships to Mars."

"They've already launched ships to Mars, and they attacked the Penelope," said Bouvier. "They're in the middle of an attack on the people we have responsibility for on Mars, right now. It's already a war, Mr. Gaurrero. You need to accept that."

"A lot of people don't accept it, Mr. Bouvier. The science and educational and academic, the astronauts, they still hope the historic peaceful intentions of the space-program will prevail and resolve peacefully without a lot of frozen corpses floating back across the space-transit corridor. War in space is new, and we have no way to measure the implications. Sure, it's a war, my dark-skinned friend. We still want peace. We want to save the base on Mars and save lives. Attack on Earth launch-sites will only escalate things. We have to prepare for the meteor. Maybe those Eastern space-launch facilities would be valuable in that event. Such as launching rescue ships to the moon and the bases there, or maybe for escape or resources for people in those parts of the world. So, yes, acceptance, but are we really entitled to go on a launch-site facility invasion and shutdown spree? I am not so sure," Jesses-Gaurrero replied. His mind was bright and alert, but he knew his opinion would die quickly.

"We don't have to destroy or obliterate the facilities. We don't have to destroy them or annihilate them. We shut them down in a tactical, controlled way to reduce their ability to launch more ships. It's a strategic move. There need not be a lot of killing or deaths," Bouvier replied sternly, holding to his view.

"They always say that," offered Hazlet smartly. "And then move in with men and guns and helicopters."

"They are 'us'," said Renolds. "We are 'them'." There was a pause in the room as they withdrew from the debate and considered their opinions. Bouvier tapped his pipe on the fireplace. Renolds, the gaunt and tall 65-year old farmer, the more experienced US President of the group, poured himself a hot coffee. The US Presidential Seated Council in 2079 was very similar to today's White House. With four US Presidents, areas of attention could be divided between them, until they came together for votes on difficult or important choices. The shutdown by Special Forces of the Eastern launch sites was only one option, only one proposal, but was favored by the military. The generals and intelligencers felt the Islamic-Russians would soon launch more ships.

Many hours passed, and they wrestled yet more with the issues, there in the private bunker-type conference area near D.C. They had many advisors and information sources, with aides and captains linked directly to military. Bouvier tapped his pipe, Renolds tapped his foot, Hazlet tapped her computer keyboard, and Jeses-Gaurrero troubled to the tune of 'taps' in his fearful dreams of the blood that would be on his hands. It was a long night. They finally settled the terms, on paper, so it could be grasped. Renolds presided over the final vote, with a small computer-log data-recorder used for that purpose, to capture the formality.

"Bouvier?" he said.

"I vote yes. Shut them down," the black educator said.

President Renolds touched a button to register and record the vote. "President Hazlett?"

"Aye," she said. "Yes."

"Gaurrero?"

"I vote yes," said the Hispanic President meekly.

Renolds logged the votes on the machine. "I also regretfully vote to shut them down," he said.

Thus, by what the philosopher would only call an Act of God, the perhaps imperfect chance that asteroid U2753b would strike the Earth in about three years from then, by estimates, nations on Earth would again deal each other the hot burns of war. To their credit, the US leaders would limit the Special Forces Strike Teams to only the launch-sites, with the specific purpose of crippling their space-launch abilities, and that would be that. The formal vote was registered and the wheels began to turn.

"It's all we can do," said Renolds, later. "If we stop their launches, Mars will be safe. So, we disallow any further space-launches on their side, by caveat. I just hope the military makes it quick and clean with a minimum of deaths."

"Amen," said Bouvier.

A space-port or space-rocket launch facility is hard to conceal, but it was well-known there were at least 12 major launch sites for the US, and nine for the Eastern Alliance, and probably more for both sides, hidden or even 'under-construction'. The US military would use air-assaults, ground teams, sneak-attack methods, superior advanced weapons and small-tactics teams of trained soldiers, to make the job easier and more likely to succeed quickly. Move in, overwhelm, take control, and dismantle operations. It was felt that within a few months, no Eastern-bloc space-program ships could launch into space, from anywhere on Earth. The Islamic-Russian space-program would grind to a halt until the Mars Base was secure for the US, who built it.

Bullets, fire, blood.

## CHAPTER 39: 'It Will End the Battle'

Radio communications that were to reach from the
Vandenberg Space-port in California, to the ships in space, and
the Molinari mid-point dock station, were very advanced by
2079. As revealed by Marconi, Edison, and many others in the
late 1800's, radio waves of different types emanate from
different sources, as expanding circles, or directed energy of
electro-magnetic signals or waves. When electricity is generated
at significant levels, such as a standard dynamo, or electric
generators, and high-energy movement, from wire lines and
power plants, they inevitably produce a minor or larger field of
emissions, in the vicinity of the movement of electrons forced
through metal wires. This is more or less true, as a principle of
science, depending on the arrangement of generators and power
sources and wires, and of course much more efficient when
designed and intended to generate radio-waves. In the same way,
a magnet such as a child's toy, will invisibly attract iron filings,
nails, bits of iron-based metal scraps, tacks, staples, etc.,
demonstrating that electro-magnetic influence is one of nature's
most basic forces. In prior generations, electro-magnetic energy
was un-revealed and un-recognized in its potential for many
centuries. The science of the electron, developed by physicists
like Isaac Newton and others, in the 1600's, 1700's and then
applied by science-explorers, late in the 1800's and early 1900's,
became the basis for late Twentieth-Century prosperity,
including everything from light-bulbs, to the television, electric
tooth-brushes, ignition plugs in cars, the radio, the computer,
communications, and seemingly endless applications. Standard
radio, amplitude-modulation radio, frequency modulation radio,
radar, microwave, and things like cell phones; all were developed
to very reliable, near-perfect performance. So in a sense, the

discovery and revelation, and application of electron energy was a liberation of great importance, such that prior generations could hardly have dreamed the ways in which we would learn to use the simple reality of electric energy.

High school students in the late Twentieth-Century watched the launch of probes and robot machines to Mars, to take samples of soil and air, and minor measurements, very similar to the 1960's with the landings on the Moon, and could learn very basic ideas about how science could send radio signals over great distances and even control robot ships and machines on Mars, yet so far away. Science tells us that it takes about eight minutes for the heat of the Sun to reach planet Earth, and give life and joy to the birds and the bees, and that sound-waves move at about 400 miles per hour through the air-mass on Earth, from mouth to ear, and person-to-person, and so on. Perhaps only the Supreme Being moves faster through the Universe than light itself, as light is an energy form, waves or particles, or both, which Einstein and others explored as far relative speed, refraction, amplitude and other qualities. Radio transmissions also have a rate at which they move or initiate to reception from point A to point B, all of these being energy forms, vibrations and types of radiation that nature and the Universe have provided. There are many naturally occurring radio waves. The Earth and Mars are surrounded by magnetic fields, and some stars also emit radio-types of energy. Mars and Earth are variably at a distance of at least 50 million miles, from season to season as they turn in their course. So, for a solid radio-transmission from the Earth to Mars, the delay could be 10 or 20 minutes, or even hours, but seldom less for a radio link to reach Mars, or points between.

Lynn Rogers-Smith, the designated leader of the US Mars program, working out of the Vandenberg, California facility, needed to communicate with the command ships America had sent to Mars to protect the base there. The Commander was of course Winton Berle, and the two of them were long friends. Somewhere out there in his tin-can, at the controls and in full command of the ship's powerful engines, guiding eight ships towards Mars, Lynn remembered him as an affable, jovial, big-hearted man with a large physique, and a grand capacity for adventure and learning, and significant personal strength. All the astronauts were 'the best-of-the-best', they had to be to survive the rigors of space, and also to earn the needed education and training-disciplines. So, Lynn Rogers-Smith was very hopeful and had faith in Berle's command, and his ability to carry forward what needed to be done. But, there was still always the Abyss. He was a good sort, and they liked each other, and that meant a great deal, at least on the more nurturing and 'space-user-friendly' side of the planet-Earth space-program equation. Rogers-Smith, the busty Texas woman with golden hair and freckles on her shoulders and sometimes exposed breast cleavage, had only been in space herself a few times, but knew a lot about it, and was a capable administrator, and an inspiration to all the men and program workers. Perhaps more than anything, she felt passionately devoted to the men and women whose lives depending on her decisions and planning, because with every ship that went into orbit, or into deeper space, there was always a basic danger. She had no taste for war, and no real understanding for military and combat. Her disciplines were research and education, rather the 'hind-most Mother', organizing the use of resources and man-power, for various efforts, mostly now those on Mars. Her job was even often boring, dealing with fuel purchases, assignment rosters, pay-raises or pensions and retirement, design plans for projects 20 years in the future she would never live to see completed, or the technical issues related to the discomfort of some astronauts with a certain space-helmet air-vent.

If Winton Berle was the angel, Lynn Rogers-Smith was his commander, and more than anything the bond between them was real and actual, whether he was a million miles away in a tin can (ship), or sitting across the table in Texas having a beer. It was the type of thing that space-workers like Guy Reisling and Lila Meetek had observed, that the distance between hearts is shorter than the vast expanse of the void between dusty rocks that men might walk upon. So, like before, Rogers-Smith and the people working with her at Vandenberg, cranked up the high-tech gear and machines to connect with Berle's command. There was a constant radio-monitor link anyway, operated 24/7, but this was only a drone-tracker to monitor progress of the ships. In order to speak with Berle, it took a bit more effort, but was fairly routine. Hundreds of years of work and progress, education and discoveries, billions of dollars, thousands of workers and scientists, were brought to task, standing on the shoulders of giants. Codes and commands, signals, text-messages, content types, pre-recorded messages, beacon signals, could reach the vessels much easier. Verbal communication was often the least types used. Many functions were handled this way, though human speech was more familiar comforting. The connection was never lost.

Now working again in the Vandenberg high-tech center, the Western mission-control keeping track of the spectrum of Mars-program activities in space, Lynn Rogers-Smith, together with Branson Porter, the security chief, science-lead Ibrahim Mehudi, along with transport crew commander Okman, and many others, including the Mars-Base Defense Team Planners, and the Military Envoy, General Alberto Gonzales Mortissimo---they huddled together to make the connection 'as one'. Each had their own special interest, and special work, for the success of all. The entire affair was a new adventure for the US military commanders, having never made any real war efforts in outer space, prior to the attack on the Penelope. Thus a career high for the stuffy, stern, big-shouldered General Mortissimo, employed by the White House and Military powers. Who was taking orders from whom was something of a toss-up. In a way, Lila Meetek's half-joking statement had come true, that her boyfriend Guy had started a war in space. But the US was keenly interested in the salvation of the Mars Base in any case, more than just an exercise. For all of them, that night at Vandenberg, they were completely removed from the so-called 'action', and could only track things at the command center on their computers, rather like legends of the Greek gods who would gaze into a crystal pool of mystical vision, to read the fortunes of mere mortals on Earth, and decide their Fate. The momentary goal was to connect with Berle's command and exchange higher orders from the military, with up-dates, especially regarding the US choice to shut down the Eastern launch sites, which would be a relief to Berle and his teams. Winton could continue on his own anyway, and knew his mission and course, and had his own soldiers and commanders for the battle at Tharsis Montes.

It was about 1 a.m., on a dark, starry night under the California sky, with the mist and tule-reed fog moving in from the beaches over the low hills, when they were finally ready for the link to go ahead. The greasy green-gray waters of the Pacific, lashed against the sand and stones of the coastline along Santa Barbara, and the Vandenberg peninsula, near Lompoc and the Santa Ynez Valley, where pop-star Michael Jackson's Wonderland was now a children's museum. The spaceport antenna-array's permanent gaze at the sky, pointed with dishes and motors and engines, was like a sentinel, silent, meaningful to very few people at all. It was strange on the face of the coast there, where the Chumash Indian had rowed their carved-wooden boats among the sea-otters, abalone and clams, hundreds of years before, for this high-tech towering facility to have grown like a glistening magic giant machine-mushroom, there among the creeks and washes and oak-trees, sage-brush, rabbits and coyotes. Like anything else, it was entirely normal, in 2079, and went ahead as an organized and orderly business that attracted very little attention except from those involved. Life around them on the highways and local cities, continued without any real alarm about the affairs of the space-program. Saving the world--- business-as-usual. Even less dramatic or attention grabbing, was that what was happening was 'invisible'---mere radio waves, un- seen. The powers-that-be of course wanted no public attention at all.

"Are we ready?" said Rogers-Smith, working with the communication techies. "Is the call going through now? How long?"

"We're ready now, Mrs. Smith," said the operator. "Connecting now through this monitor-display here. You can use this microphone." There was a pause. Then they all could hear Winton Berle's familiar voice.

"Goooo-oo-ood morning Planet Earth!" he said loudly, through the speakers. "Commander Berle, aboard the space- vessel 'Understandable', nearing the planet Mars, Vandenberg authority US space-ship radio signature ID on file. Coming to you live, via the miracle of modern radio. Can you hear me Vandenberg?"

Smiles around the command-center at Vandenberg. If he had been riding a snow-sled pulled by reindeer, sporting a long white beard and wearing a red snowsuit get-up, it couldn't have been more 'jolly'. The Happy Warrior was in the building.

"Greetings, 'Understandable'. This is Lynn Smith in California. You sound far too jovial, my friend, for travel in the void, dark out there I guess. Serious business at this time. I'm here with the generals from the White House, and the mission-team. How are conditions with you now?"

A pause, static, a tracking beacon sound. "All is well, Lynn. Good to hear your voice," came the response. "Clear."

"The military general Mortissimo in charge of the White House battle campaign and strategy is here with me. You will need to log and confirm receipt of his orders and relay the message to incorporate with your plans."

"Ten-four, Big-Ben," was the response. Mortissimo, then moved to where he could handle the microphone, a hand-held remote that fed to the mains at the tech-center where they worked.

"This is General Alberto Gonzales Mortissimo, US-Army Four-Star. To whom am I speaking, please?"

A long pause. More radio sounds. Then Berle. "A Christmas elf, General. How ya' doing? I got a green suit and bells on, right. Ha-ha. Fleet Commander Winton Berle here, sir."

Mortissimo then relayed the information about the plans to attack and shutdown the Eastern space-program launch sites, and the basic logic there. They wanted the US defensive soldiers and ships to understand, there would be no more Russian-Islamic ships behind them, and that the battle on Mars, if it proceeded, would not be anticipated to last very long. This was critical; all space-travel was tenuous and depended on available resources "out there". A long protracted battle on Mars was very serious, they may never get home at all, even if they won the battle decisively, if only for lack of fuel or other troubles. Like any military effort, supply-lines were the path to success, and they had never been stretched so thin in any battle in history. The information was logged and would be repeated among Berle's command. There were other orders, the military now wanted Berle to go ahead with any capacity to directly shoot down the Russian-Islamic ships in Mars-orbit: a new wrinkle, and a deadly one. The same order was released to the Mars Base itself. This upped the stakes, all those astronauts would die, and their response on Mars, against the base and Berle's teams, would be equally deadly. There was some dispute about it and discussion, if shooting down the orbiting opposition ships was even practical, or would succeed, or wise. The 'doves', who were the science and research people, generally were powerless as far as those choices. Washington had spoken. After a while, details about the position of Berle's nine ships, including the Penelope, and Berle's specific orbital-entry plans, and plan of attack upon arrival at Mars, were relayed by his navigator. There was also information about what was happening on Mars, that very hour. What word had reached Vandenberg gave the impression that there was an early-version of a 'ground-war' or 'surface-combat', underway, but details were sketchy. Enemy soldiers on foot, with weapons, fighting their way towards the gates of the base, with Keeje's ships above, threatening a bomb-attack by air at any time. Berle was eight days from Mars-orbit. The call continued for about an hour, when the 'business' of the call was more-or-less completed.

"One more thing, General Mortissimo, please," said Berle, the radio signal now fluid and compressed, digital-sounding.

"Go ahead, Commander," replied Mortissimo, pleased with himself to be the first US General in charge of a war in space.

"The Russian ships are much like our own. I know these machines. The launch-sites in the East, the Ukraine, and Hindu-Pakistani. The others. There will be data and information on each ship, on-site, probably secret files, protected. Have your men prepared to retrieve those files, at all costs. Get your computer-tech crews to help you recognize what you're looking for. With the right electronic data on their ships, we can end this battle very quickly. If your men go into the launch-sites, that data is even more important than stopping further launches. Your Vandenberg on-site ship-design specialists will know what I mean," said Berle.

Another long pause. Mortissimo seemed un-easy; he really didn't understand what Berle was talking about, or why it would be important. "Confirmed, Commander Berle. I will take this under advice and act accordingly. Thank you for your suggestion," said the General.

"It will end the battle. Do it, at all costs," responded Berle, the static-hum and digital warble making him sound like a Christmas elf indeed.

Another 20-minutes, with greetings and well wishes, and the call was terminated on both sides.

## CHAPTER 40: The Rhinoceros Hut

Over the course of about three days, the Russian-Islamic soldiers who had dropped to the surface of Mars near the base, made an all-out attempt to take control of it by force. Commanded and organized from above, where the ships-in-orbit circled the planet unseen, the soldiers handled the assault in a similar way to a typical modern ground-war taking place on Earth. Wisely, Commander Keeje, the Russian-Eastern Alliance space-program point man in charge of the assault, did not wish to annihilate the base. He knew very well that if he destroyed the base, their mission would have defeated its own purpose. Additionally, the men in the ships-in-orbit there at Mars, very much now wanted to get their hands on the resources inside the base, such as fuel, oxygen recharge, edibles and water, and many needful things---so far from home. And of course if Keeje bombed the base to oblivion (and he had the capacity), there would be no base to take over. Even with bombs and missiles that could fairly easily target and destroy from above, the assault was organized much differently than that. As a military operation on Earth would proceed with foot soldiers, the two sides lined up their men and weapons, and held their positions, as they were able, to fight it out until one side or the other prevailed. Yet, because it was all happening on Mars---quite different, quite the same, and all-too-familiar.

The first round of fighting was composed of mere violent and scattered skirmishes. With the eight descent-pod shuttles, each with about ten men, about 80 to 85 soldiers on the Russian-Islamic side were in place early, walking slowly across the Mars-surface in their walker-suits, and armed, to approach the base. As barren and naked as the surface of the planet Mars actually is, even there at the feet of the Tharsis Montes and the Snikta-Ridge

Volcanic Basin, with its long shale-stone ridges and flat cliff surfaces, larger rocks and stones (a more differentiated landscape than much of Mars), there was no place to hide, very little shelter for the soldiers. The advancing forces were somewhat disadvantaged, having no cover, or very little. So they moved forward in lines with the assistance of some of the electric carts, which provided some cover. As things proceeded the ships-in-orbit dropped even more of the descent pod-shuttles on the glider chutes with the thruster brakes, into the region, to release even more soldiers, and more gear. By the third day, the number of Russian-Islamic soldiers on the surface numbered at least 170-men. The battle was slow and ponderous. The soldiers were in the suits, walking in the lesser gravity, in teams and squads, from attack positions. The defense-soldiers from the base, the US side, had rest, supplies, food, shelter inside the base, energy, batteries, oxygen recharge, within their protected fortress-outpost. The opposition knew the terrain well enough to send down other type of pod-cargos, from the ships-in-orbit, with supplies like this. But this was an enormous task, to parachute them down, and unload them, to supply their soldiers. Over three days, it all seemed much like a high-tech chaos of common warfare, exported by men from Earth, with all the venom and murderous savagery they needed to win what they wanted.

The Mars Defense Task Force Planning Team had made good choices, especially in the placement and construction of the 20 or so 'safety-foxholes', quickly constructed in a circle around the base itself. They had months to build these from scratch. These 'oxygen-igloos', as it worked out, were used successfully by the base defense on an extended basis. If a man was hit or injured, he could be evacuated to the igloos, and back to the base, if he survived. They could be used for cover and to fight from, firing back at the enemy, or to simply enter the igloo's, recharge oxygen tanks, use the communications, etc. They were especially useful at night. They were placed in a series-pattern around the base to repel the invaders. Each one was rigged from available materials, and ranged in size from that of a small portable toilet or phone booth, to the size of a rail car or very large cargo container. With whatever was available, the soil was excavated to a shallow depth, and then build-in the shells, cargo containers, plates, seamless materials, plastic doors, until they had a functional life sustaining, oxygen-sealed environment. They were not very sophisticated. The atmosphere seals were either sealed only from within after the men entered, or in some cases had no seals, and the men could use them fully suited. They were only for the soldiers during the battle. For avoiding enemy bullets and grenades, and for breathing, they worked just fine.

The Mars Base also had more men. In terms of the battle, it was a scene spread out over several square-miles. It had the appearance of an area in the California desert, such as the Mojave or Death Valley, rocky and barren. If you can imagine a series of small campgrounds there in the heat and dust and sand, with men moving back and forth on foot or in small carts, firing weapons at each other, you may get the idea. Separated by a quarter mile, half a mile, here and there or more or less, with lines of men in groups of three, or ten, or twenty, advancing or retreating. All taking place in a breathless atmosphere where the walker-suits were essential, or to face certain death, with the helmets, air-tanks and radios. It was a somehow pathetic vision, at such unfathomable expense and ages of learning, now to kill and steal, there on Mars, away, so far away. In the distant background, the base itself glistened in the Martian Sun, alive with comings-and-goings, and whatever the soldiers needed to do. At an even greater distance, one might see the parachutes billowing, high above, as more of the shuttle-pods came down from ships-in-orbit. They landed to one side or the other, without a great deal of accuracy, carrying supplies.

The bitter truth was that the men on the surface had orders to kill their opponents, like any war. As a fraternity of astronauts, this was painful on both sides. They knew that the men, who had invaded from the Russian-Islamic, were much like they were, trained to work in space, and previously employed peacefully for research and education. They knew that space-travel without any hostilities at all, were dangerous enough, and made for a challenging and adventurous life for any of them. They knew that on either side, the men and women were among the most highly-trained and professional the Earth had to offer: educated, pilots and athletes, with highly specialized knowledge, rare enough to earn the respect of any of them. Maybe not true of the Mars Base teams, but the opponents who had arrived, they were also soldiers-first, and maybe more desperate, more willing to hurt and kill. On the US-side, they had never really anticipated a war on Mars, before the early triggers about the meteor and the Eastern space-program's sneaky plans, almost two years previous. The Mars-teams were novices, very inexperienced at soldiering and killing. Bojji-Than, Juno Amorrossi, Vinces Grant, and other Mars Base leaders, wondered privately whether or not their teams had the heart to kill anyone at all, or would stall and balk, give up, surrender. It just didn't seem likely that the communication workers, food workers, farm workers, life-sustain and launch workers, repair workers, computer workers, research-and-exploration workers, who lived at the base, would suddenly become savage-bloody soldiers, out there so far from home in the mystical and truly rare and wondrous area there on Mars. The fantasy of peaceful stillness and learning-adventures had been suddenly altered by events. Yet, they too had their orders. The command team in the Control Center above the facility, with the maps and scanners and radios and telescopes, controls and computers had made the choice. If they failed to defend themselves, they would be over-run, put in chains, even tortured and killed, and lose the base. Ironically, base-commander Bojji-Than, the Buddhist-Pacifist, had been forced to decide, with plenty of hard-rules and orders from Earth as well. Simpler choices were lost.

Students of war might ask what kinds of weapons are effective on Mars. To fire bullets and end life that way on the surface on Mars, it was not needed to pump a man full of large-caliber shells in a bloody mess. A single puncture or hole in the airtight suits, and within a few minutes the man would be dead from lack of air. It was a sad thing. The suits could not be seriously reinforced with armor or bullet-resisting Kevlar or plates that would repel projectiles, though some attempts at this were made. They simply were not designed that way. The main tactic was to avoid bullets and rips-or-tears in the walker-suits, not to mention their bodies. Even without an Earth-like atmosphere, 'normal' small arms, rifles and pistols, various repeating-fire guns, or grenade-launchers, would work just fine. Standard firearms were easily modified such that an ambient oxygen-rich atmosphere was not necessary for bullets and guns, firing sealed shell-casings with explosive powder.

By the end of the third day, fourteen brave men had been killed. Eight on the Russian-Islamic side, and another six among the Mars Base teams. Bullets, fire, blood, destruction. Were they brave or courageous, were they foolish? It was hard to tell. Some of the men would form groups and move in-line, and begin shooting wildly, at the electric carts, or attack the igloo positions. For the most part, they were inefficient, and didn't get much closer to the base itself. They made little headway. The Islamic-Russian strategy seemed to be to establish front lines that they could hold. As the evening hours came, things got even worse. Surface temperatures dropped to as low as 20-below freezing, very, very cold. The walker-suits were equipped for those temperatures, but not much lower than that. The men suffered greatly on either side, if they found themselves outside on the surface, as the darkness fell, and the warmth of the Sun disappeared, the planet slowly turning. The fortunate ones, on the 'enemy'-side, could retreat to the descent shuttle-pods they had arrived in, and gain warmth, electric recharge, oxygen, and refuel. The pods that they had landed in, served in a similar way to the igloo-foxholes the Snikta-base men had prepared for their side, in that sense. Those men who were unfortunate enough to be outside an hour or two after darkness fell, as the stars appeared, and the Sun swept away, were lucky if the cold was not simply too intense, and they could barely function at all, perhaps finding a rock, or one of the carts, or even covering themselves with sand and dirt, curling up for warmth in their semi-heated suits, waiting in the dark hours, breathing heavy breaths thick with moisture condensing inside the helmets, wondering how the heck they ever signed up for combat on the surface of Mars, something so unusual even at that time in 2079, as to be absurd. It was best to keep moving.

Squad captains in small groups would make their forays against the enemy, firing their weapons, or small-arms rocket launchers, and do as much damage as they could. They couldn't really deal with the harsh-environment, it was all new. The surface terrain was hard to maneuver despite the light gravity. The suits were too fragile and difficult to deal with, yet absolutely essential. Hours would pass between any actual combat. It was a somewhat clownish affair. A rocket would fire at one of the igloos; the men would fire back to defend, again and again, over-and-over.

Toward the very early hours on the morning of the fourth day, one of the oxygen-igloos, to the forward side of the base, which the men had called the Rhinoceros Hut, for some strange reason, was ready to come to life for another day's work, with the expectation of more meaningless battle. It was still mostly dark, but the Martian sunrise would soon appear and the madness of the battle would begin anew. The Rhinoceros Hut was one of several of the 'foxhole-igloos' that were made out of cargo-containers, which were long and square, half-buried in the ground for strength. It had an entrance sealed from within. At that hour, as many as eleven soldiers were crowded inside, wondering what was next. They had done the best they could with the foxholes. There was oxygen-recharge, food, small communications gear, water and food, rest areas, ammunition, medical, and spare suits. But they were very uncomfortable, and the astronauts and space-workers inside that morning, grimly realized that the rigors of space-travel for each of them, had reached a new low.

The Rhinoceros Hut had batteries and gear for various functions, and was about a quarter-mile from the front of the base, at a small dip in the landscape, where the sandy shale-rock skiffed downward towards the broader, sandy plains that seemed endless. The huge mountains protected the entire backside of the base. It may as well have been a large, abandoned trash-dumpster or rusty old rail car somewhere in the hot Mojave Desert back on Earth, or the Sahara. But in this case, it cost many, many millions or billions of dollars to build and maintain, and each of the men inside was suited for surface work on a planet millions of miles from home, airless and alien, which for even one person also cost many millions of dollars, far-advanced from a homeless bum on Earth. But Earth-bums didn't need air-suits, or space ships.

Each one of the defense-igloos had a man in charge, and this was often rotated between the squad captains leading the men out to fight. The man in charge of the Rhinoceros Hut was Peter

Shores, one of the maintenance workers, a base-regular for two years, age 43-years. Like all of them, Peter was strong, tough, smart, well educated, and knew all about his job and his work and the environment and the stakes at hand. And like them all, and common to foot-soldiers on Earth, he didn't know a hell of a lot about soldiering, and leading men around with guns and grenades, and having people shoot at each other. The Martian dawn was coming. The men huddled inside, careful to recharge their packs and clean their equipment, check their suits and gear, check their weapons, eat and drink, and rest. Trading stories and complaints was a favorite past time during the hours as they waited for the next attempt on the enemy, or vice-versa. One man spent at least an hour describing how he and three others had suffered equipment failure when one of the electric carts stopped operating properly, as they engaged with weapons fire, at some ridge a distance away.

"We ran out of batteries," he said. "It hadn't been re-charged, it had been in-use too long. We had plenty of bullets, and so did they. But the damn cart stopped right in the middle of a firefight. Over a ways, Northeast. They had maybe nine or ten guys. We had another seven or eight. The others ended up back at base. We used the cart to sneak attack a position behind these rocks. Our team-commander had us moving in with the cart. We were doin' a raid on 'em. The damn thing died. It made good cover for a while. But, we lost one man. I think it was the new guy, on the life-sustain crew. That kid with the black hair. The fuckers shot him. Stray bullet I guess. There were bullets everywhere. They fired 200 rounds a minute, they got repeaters. They were behind the rocks for half an hour, and then it slowed down. I guess they ran out of something they needed or something, or got tired. All we could do was retreat. We moved back towards the nearest foxhole, firing back at them to guard the retreat, backing away on foot. It was a roust, they blew us away. By the time we were a quarter-mile off, they destroyed the electric cart, I couldn't believe it. Some type of grenade. There was no way for it to burn without an atmosphere, but they toasted it. They must have had chemical weapons. It shouldn't have burned. It melted to the size of a pancake by the time we were far enough away to be safe."

The other men listened to his tale closely, for whatever they could learn about conditions. He went on and on a bit. He didn't realize that the raid-commander in charge of that skirmish was there with them in the Rhinoceros Hut. Often the men didn't know who-was-who, because of the helmets and walker-gear. The only real identification was from nametags on the front of the suits, which were hard to read except up-close, or by radio communications when the men would identify themselves, or in some cases the color of the suits. Some of the men tried to sleep, they grumbled and moaned, made their plans. Then there was a communications alert on one of the receivers linked to a computer, there in the dark hours of the Martian morning, inside the improvised battle-shelter: *beep-beep-beep.*

Shores knew how to activate the call and take the message. Some of the men could hear it, some couldn't (they all had universal radio-links inside their helmets, but the inside of the Rhinoceros Hut was now air-sealed from the outer environment, so many were not helmeted). The signal was from the Command Center at the base, where there was more equipment to track whatever they could about what was going on within a radius of ten miles or so. "Peter Shores, fox-hole-13," he said. "Go ahead."

There was a woman's voice from within the base on the radio. "Hello Peter, this is Carol at dispatch. Heads-up for you. Incoming to your position. Three men on foot from one of the fights, about five minutes away. You need to get your air locks ready to un-seal so they can get inside. There may be injuries."

Movement in and out of the air locks was as simple and uncomplicated as they could possibly make it, for the battle-igloos. Mostly they were sealed only from the inside. They needed air locks just like the base, but for the foxholes, they were very primitive. The men inside would put their helmets back on, and use the suits for air. The entry way, which resembled a thick, hard-plastic clear view door formation, was then un-sealed. The men outside then entered, and the door was re-sealed, airtight, and checked for air-leaks. It took five minutes to re-pressurize, and then the men could remove their helmets.

Carol, at the base dispatch, wanted Shores to have advance time to get ready, so the men on foot could quickly pass inside. He had no idea what kind of injury might be involved; it could have been a problem with one of their suits, or a bad leg from a fall. But the idea that there was an injury was alarming, there on Mars, where almost any medical crisis was a very serious matter for as long as the base had been in-operation.

"Message received, Snikta-base, Shores out, thank you," he said. Then a pause. "Wait a minute. Carol? What kind of injuries? Any info there? Please respond, dispatch."

Another pause. "Copy, Igloo-13," came the dispatcher's voice. "Uh, two of the men had serious moisture and waste evacuation problems inside their suits."

A moment, then all of the men inside the Rhinoceros Hut began to laugh with quite an uproar. Those who had heard what she said understood: the soldiers had shit or pissed in their walker-suit 'rump-kits' and the waste-holder-collectors overflowed or couldn't handle it, or just malfunctioned. It was very serious, actually.

"Uh, copy that, Snikta-base. They over-shit in their suits. Right," said Shores. "Understood."

More laughter. 'Depends For Astronauts' was what they all called the rump-kits, inside the torso leggings. Oddly enough, an extended malfunction of this type could be very serious while on the planet surface.

Shores ended the call quickly. He had the men now get ready to open the air lock. All eleven men were in full oxygen-suit walker-gear and helmets, each man, within about 3-minutes, a bit of a Chinese fire drill. For any space-worker, the oxygen gear was his essential life-support, and he knew how to remove it or get it all back on and working, in a very short time, with a series of swift movements and controls and locks. They didn't complain. They had to do what they needed to, for all of them to survive. Within a few more minutes, there was an angry sound, off at the perimeter of Igloo-13, somewhat distant, guns or small arms explosions. Weapons fire, as if the straggling three men were firing back at the invaders chasing them home. The Mars-atmosphere did somehow carry sound waves, though thinly, yet an odd thing for them all, with the helmets and gear, the sound of weapons, never heard anywhere but on Earth. Then a hustling at the clear-view air lock, the process went quickly, the men entered, the doors were closed again and re-sealed, and then the small space inside was pressurized. A scan of the outer-area by camera showed they were not under attack at the moment. After a few minutes: all clear.

The three men still had weapons in their hands, as their helmets came off. One of them was Chassidy Katola, the 30-year old African-ethnic Health-and-Wellness counselor. The others were surprised to see her as her helmet came off and she gasped for air, seeming exhausted.

"God! These fucking suits! They don't have enough room in the disposal tubes to shit and piss in!! We could fucking die for lack of moisture!!" It was obvious she was furious. There was no laughter now. One of the other men was John Balker, the repair-guy for the base-wall leak integrity daily inspections who had taken Karen Tutturro on her first Mars-walk. The other was a soldier from another team.

"You piss yourself too, Balker?" said Shores. "Heck of a way to die."

"Give me something to eat and a pure oxygen canister for a few minutes, would you?" said Balker. "This whole thing sucks."

So it went. Within another three days, the Mars Base Command Center was alerted that the nine US Mars program deep space vessels were ready to enter planetary orbit, under the command of Senior Astronaut Winton Berle, Fleet Commander.

## CHAPTER 41: Berle's Armada

On the hour it was confirmed that 'Berle's Armada' was ready and now beginning to enter Mars-planetary orbit, the staffers at work in the Mars Command Center sent up a cheer. Shouts and applause, the clatter of desk and workspace items tapped repeatedly on hard surfaces, whistles and bells on-hand, beepers or small horns, all began to sound. Then that joyful noise spread throughout the rest of the base itself, in the halls and doorways, air locks, offices and tech-centers, into the ag-areas, the cafeterias and personal living quarters, and then by radio-confirmation, out onto the planet surface, to the oxygen-igloos, to the men in the Rhinoceros Hut, and into the walker-suit radio-helmet communications systems, into their thoughts and minds, to the soldiers on the electric carts, to the Mars Base lifter-pad launch sites. Then even beyond, by other means, though with a delay, to the Molinari Space Dock, and back to planet Earth, Vandenberg, the US White House, the US media (eventually), and to the public. On the opposition side, the news also arrived. One might have thought it was the Second Coming of God, but instead of home to planet Earth, it was on Mars. Or maybe it was because of the more demanding and hostile circumstance that the men on Mars now found themselves in. In any case, Berle was more than welcome. But like any Cavalry-to-the-Rescue scenario, more hard work was ahead.

"My god," said Juno Amorossi, as it became clear that it was finally true, that the ships under Berle's command, were now ready to enter Mars-orbit. "He made it. Holy Mother."

There were actually only seven ships arriving from Earth at that time. The addition of Guy's cargo-vessel ship, the Penelope, increased the size of 'Berle's Armada' to nine. But two of

Berle's original flight formation, had stayed behind at the Molinari, and were following up behind. So that with two behind (from eight), and adding Guy's ship, the total number of space-ships now speeding into orbit around Mars from the US-Earth side, was seven. The ships that stayed behind at Molinari would arrive later, with the idea of establishing a supply-line backwards to the space dock.

"Connect by radio," Bojji-Than directed. "Prepare a complete situation-info package as a content-parcel transmission, with maps and numbers and figures for the battle on the surface. Send it immediately."

There in the Mars Base Command Center, things had gone on now for nearly the entire month, as a cliff-hangar for them all, a holdout, 'bunker-mentality', or 'last stand'. None of them were familiar with such violence and hostility, and it showed. By this time, the number of those killed on the surface had swollen to 40 deaths, almost even to both sides, with two more dead to the US side than the enemy. Prokov Keeje had become more and more desperate and brutal as his time dwindled, before that awful moment (for him), when the US ships would arrive. His earlier boast that he would walk into the Mars Base himself within only a few days of the start of the battle, was abandoned, leaving a bitter taste of humiliation. So, he pushed harder, and they 'won' a few more. Thus, 18 deaths to the Russian side, and 22 deaths to the US side. Not big numbers. But, one might say, more costly than many of the Earth's war-dead in possibly the entire planetary history, even at the relatively small body count.

Like a passing instant of celestial joy, there was no time then to linger and enjoy the success of Berle's passage. After all, Mars had never experienced so many ships in orbit, even in the 20-year history of the Mars Base. Sadly, they were fighting each other, instead of co-operative education and research. The transport across the Earth-Mars corridor had taken Berle just less than 9 months. But, there was work to do.

"Greetings, Mars Base Snikta," came the initial radio-link from the Understandable, Berle's lead ship. "This is Commander Berle aboard US Mars-program space-vessel Understandable, as I'm sure you understand. Please confirm."

"Welcome to Mars, Commander Berle," came the base radio-dispatch monitor's reply, by radio wave. "Please hold."

The complexities of bringing his ships into proper orbit-position was undertaken from there with a long process, by the flight-path navigators, individual ship's pilots, communications and mappers or radar-scanners, engine crews, and helps from the Mars Base. The Mars-Earth passage, and orbital entry standards were at this point in space-travel history, was at least mostly routine. With all caution and care, they knew they would be safe, and yes, safely returning. Riding to the rescue to save the day was a bit of a mythology they entertained as an encouraging positively. Mature space-workers knew it wouldn't be easy. Immature space-workers basically didn't exist, or stay with the program very long. To be precise though, Berle was a full day ahead of schedule, to his credit, as time being a factor among the dead.

Following ship-to-shore radio formalities and comm-link set-up, from the deck of the Understandable, Berle's voice could soon be heard in the base Command Center. "Ships are setting up stable orbits over a few days here, Colonel Bojii. Your scenario information has been received, my seconds are going over it to get a picture of your little war here."

"We call that the Battle-Plan Quick Start app," joked Vinces Grant, there with the other leaders in the Command tower at the base below Berle's ship, somewhere above floating around like a gigantic leaf on the endless seas of cosmic buoyancy. The radio-link carried his thoughts.

"I guess what we need is a decent-quality Shutdown command-program app for the conflict," Berle replied. "Maybe get a programmer working on that. You know. Peacemaker Two-Point-Oh. Win-win for everybody for that. Colonel Bojji-Than, if you please sir, time is precious here. What's your impression of the conflict right now, and what needs to happen from our side? My orbit-entry will happen just fine, we have some transport supplies, the teams of fighting men and suits, the shuttle-down descent pods, all that. What I really need to know is how you see it, how to end this realistically and soon."

Bojji-Than looked tired. He was almost 65-years old, very fit for his age, Asian in complexion and strong. But they had been on double-shifts since the first sign of the Russian-Islamic arrival. There had been the deaths, each one, another beloved man or woman lost, each one "one too many". And he felt this personally as a painful thing he would carry the rest of his life. There were equipment losses and failures. A small rocket had actually hit one of the outer-area facility-structures, a support-storage for fuel-chemicals, with significant damage. It was a wake up-call. If the same rocket had penetrated to the walls of the base itself, many inside might have suddenly died. And they all knew this. The plan there, as stated before, was to stay calm and use the walker-suits they had inside the base, then re-seal leaks, re-pressurize. But like anything they had to deal with, 'easier said than done'.

"It's a mess, Mister Berle," said Bojji-Than by radio, as their initial conversation continued. The others in the Command-Center and on the ships could listen in. They coveted no real secrecy or privacy at this point. "Mars is a city of errors. The situation is absurd, a total mess, a total waste. We have more than 20 dead, and these were people I cared for, and needed. The invader ships are also in orbit, as you know, we have their positions, mostly. But they change and of course evade our tracking. They could destroy us at any time by bomb or missile from those ships, but so far they won't, they obviously want the base intact. But that could change. They have, oh, at this hour, Vinces, what are the troop numbers on the surface? Please hold, Berle."

There was a pause of about three or five minutes. Grant and others quickly looked at what they thought they knew, there at the base. Then Bojji-Than spoke on re-uptake again to the receiving-link on the Understandable in-flight. The radio garble buzzed and hummed. Staffers at the base, and on the ships, and among the Russian-Islamic teams hoping to intercept their conversation, listened-in anxiously.

"Mother Russia and Friends seem to have about 95 soldiers on the field," Bojji-Than told him. "That's because they are rotating back upstairs to the ships. Our side has about 160 soldiers on the surface. They can only deal with surface conditions for so long. But the fighting is ridiculous, Berle. Stupid. The suits are easily breached for air, so the soldiers' fight from that, with many hours between skirmishes, even days. One team gathers, they prepare their suits with fresh oxygen and for safety, it takes hours before they can tediously head out in an electric cart, or by foot, maybe ten guys at a time, from their positions. They are using the descent-shuttle pods for their surface positions. We're using the igloos. It goes back-and-forth. Our teams will do raids, and then hold their positions to defend a breach at the base itself. But it's slow, it's stupid, the environment is the real enemy, and they know it. Some of the Russians have been lifter-pod evacuated back to their ships; they replenish the teams on the surface, the lifter-pods blast off as usual, and dock to their ships, so the men can rest or replenish supplies, and they also have supply pods dropping down. Lots of them, more than we counted on. We've done an estimate and it now seems they understood their long-term supply needs better than we thought they did before they left Earth, and they are also very stingy with supplies, for obvious reasons. Highly motivated to take the base, you might say."

A long pause, static-and-buzz on the radio-link. "War is stupid, agreed," said Berle to the radio. "Even stupider in space. Okay, Colonel Bojji-Than, I get the idea. Let me ask you this quickly. Do you figure the enemy will at any point choose to just annihilate the base wholesale, just wipe you out, kill everyone, and destroy the facility? Not a pleasant question, I realize. Just the opinion of your staffers here."

The invaders had already had a month to take advantage of the defenseless base, and had not succeeded in assuming control as they wanted. The base had not been taken; no one had been placed in chains, none of the base functional controls were in enemy hands. But there were 22 men and women now killed on the US side. It was serious business. Colonel Keeje now had to face the bitter truth that his bold claim that he would personally enter the base himself within a week of his arrival was now demonstrably un-true. This might presumably simply piss the guy off more, they knew. More than the Russian Commander's pride was at-stake. It began to look as if the Russian strategy was very inadequate to their intentions overall, for whatever reasons. With Berle's armada, and the base teams, and two more ships on the way, things didn't look good for the invaders. The Russians were very elongated for supplies. Entering the base was a survival issue for them now in many ways, or to make peace. The wild-card factor that sustained the drama was the idea that they might choose to bomb or destroy the base and kill all the residents. But all the planners well understood it was easily possible for the Russians. So the hand-of-devastation was still theirs in this sense to manipulate the outcome. "Very perceptive, Commander Berle," said Bojji-Than, the radio-link humming in the Command-Center. "I guess if you want to enjoy the Mars-gravity and a real hot-water bath, you might just go ahead and shuttle down, and sit here with the rest of us asking that same question. Myself, I hate to say it, but it is possible the enemy would make this choice. They have told us so, very early after their arrival. Their terms are complete surrender or die. Their ships could do it. We know they have those weapons. Sad, but true."

A pause. "Then we're going to make that impossible," said Berle. "We're saving the world, Bojji-Than. Maybe two worlds, I guess. There's more going on here, I think you know. Back home. I have orders to tend for the orbital entry, Mars Base. Ending this call. Re-connecting with you..." A long pause, static and buzz. "Re-connecting in ten hours, our position to be delivered as we stabilize. Hold fast, you're in command down there, not me. Greetings to the residents of the US Mars Base and crews from the Understandable and my teams up here. Thank you. Signing off. Berle out."

There was a familiar beep-beep sound indicating a shift in the radio-frequency signature that maintained the link, as Berle's radioman adjusted the hook-up, and the call was concluded. They never really would spend hours on the radio-calls, in-flight, it made no sense, and larger portions of information were delivered in other ways---as 'packages' or content-data streams, such as Guy Reisling's new data-processor gear, in the cargo holds of the wounded Penelope, to re-install and re-establish better radio-links all the way back to Earth.

And there he was, our hero, safe-but-sorry, also at the helm of his ship, also working through the steps to place himself in-line for orbital position with the other vessels, and also listening in on Berle's radio-call to the base. But Guy's thoughts were elsewhere, to a degree. His ship had no fighting men aboard---each of Berle' s ships held 30 men or me, ready-at-arms, and all those weapons. Guy's co-pilot Rob Cowan was back at Molinari, hitching a ride home, due to his medical problems. Guy's cargo-hold had been evacuated of a lot of loose stuff. His entire team on his ship, Tom McGee, Raz Brahman, Arron in cargo, Peter on life-sustain, his other crew for this voyage, what they really wanted to know was exactly how much damage they had done to the Krenika, or the other Russian-Islamic ship (the Saint Peter) when they had been disabled by the EMP blast, there in the depths, and released the space-crap, and believed from their scans that 'something' had hit one of the enemy ships. But they never really found out much about how that all turned out. *Enquiring minds want to know, thought Guy musedly, secretly wishing the worst on his attackers.* Their little adventure with the total power-loss from the EMP had been extreme.

Even deeper in his sub-conscious, Guy's thoughts and his heart were in Idaho, back on Earth, maybe dreaming idly about Rob's voyage, even given his co-pilot's unfortunate psychosis. Maybe the loss of one of his testicles was somehow a good thing? Guy chuckled. Lila would confirm-or-deny that one. He hoped. Idaho, Montana, green hills, cool water-streams, trees with squirrels, fish, food, houses, regular gravity, regular people and friends. And yeah, an approaching giant meteor, true.

*I want to go home, Guy was thinking.*

They all did, every single one of them on both sides. And even in a way back on Earth. They all wanted to go home.

## CHAPTER 42: Quickly Now Milana

*"Viet-Nam never happened, because it's in the past. That's what love says, but we never learn."---Doctor Manny Fitzu, Ph.d., MD, Counselor-Psychologist, US Mars Program, 2077*

Milana, the youngish, mousy-looking Russian girl assigned to the exclusive role of assistant to Commander Rudolph Terchenko, at the KKF/Region-6 launch facility for the Russian-Islamic Hindu-Eastern space facility back on Earth, found herself about that time, walking down one of the hallways there, with her arms full of files and highly selective reading material. Like so many people, Milana was a simple woman, with an appropriate level of understanding and education, and yet really knowing so little as far as what she was involved in, what was really going on, and how her small role would or would not effect the outcome of events taking place. She was attractive, of course, as befit a man like Terchenko, in his leadership capacity. Just under age 30, slender, somewhat athletic, she had dark hair and the pale ceramic-doll complexion typical of the region. Milana was usually found wearing an office-outfit, quasi-military, a buttoned-up gray woolen blouse and pressed slacks, that appealed to the hierarchy, such that there was no mistaking her low position. It was also cold in the area where they had hidden the base, far into the Ukrainian wilderness and woods, where they hoped to be safe from the prying eyes of other nations and powers that wanted to know their secrets and their plans. Ferocious loyalty was one of Milana's first priorities, personally. Even in her minor role, if the leadership felt she could not be trusted, she would quickly be dismissed. It wasn't just the job, even in the impoverished regions there in Russia. She really was loyal. The reality was, instead, that Milana was somewhat innocent and overly impressed by the whole affair, the machinery, techniques and facility at KK-F, the men, soldiers, satellites, space ships, telescopes, and everything that was involved. The personification of all of that, in her eyes, was her immediate superior and sole-purpose in life, the figure of Commander Rudolph Terchenko. Thus, almost a love-slave, and more, his personal assistant might even have laid down her life, to protect his interests, and those of the program. Yet, they were not intimate, and there was no sexual activity between them, though it was sometimes implied from backrubs, winks, and a kiss on the cheek. This only endeared her more so where he was concerned; another man would have taken advantage of her endlessly. Milana moved farther down the hallway, past some of the offices, into a part of the building where she knew she would find her boss. Her soft-heeled shoes made a shuffling noise on the walkway. Maybe Terchenko found his lustful interests satisfied elsewhere, such as the plentiful trade in prostitutes. Maybe he didn't want to fuss with the office politics and personal entanglements. Maybe he was just a decent fellow who found it improper. Maybe he simply had no interest, or had a dysfunctional sexuality, or urological health problem. But for those who witnessed his career (and Milana had only been with him less than two years at that time) there was a fondness between the two of them like a father and daughter, and it was very touching. For all of this, Milana was very thankful, and the long hours of work and other demands didn't bother her much. Even in that year, now late in 2079-80, life in Russia could be very harsh, and she knew this well.

As she passed through the secured metal-and-glass doors, she entered a sub-station secondary command room, or series of rooms, full of computers and rows of desk-seats, radio-systems, and so on. It was immediately apparent to even her that something seemed to be going on. Something maybe more immediate or urgent than the usual monitoring and tracking, or planning and command activity, at the space-launch center. She had been in these areas on the base before and was used to it. Similar to Vandenberg, California, and other US space-program launch-command and tracking stations, there on Earth, the KK-F/Region Six and other Eastern space-program facilities, had command-centers like this, where they organized and kept track of all that was involved in various missions. KK-F was somewhat unique, because it was a 'do-it-all' base for the space-program. Other facilities were only for launches and little more. So, it was a busy place anyway. All Milana was required to do, was to care for the immediate personal needs of Terchenko, who was always under-the-gun, for making day-to-day decisions. Just like Lynn-Rogers Smith, where-the-buck-stops, was on Terchenko's desk, for much of the Eastern program, at least at his level. The Easterners were a lot more secretive, and for many decisions, the process was Byzantine, hidden, mysterious, out-of-view. Milana usually had no idea what was going on, what kinds of decisions were being made, or about the progress of the missions and the ships in space. The Western process wanted to keep their programs transparent to public inquiry and political over-sight, to keep checks-and-balances more in-place, and make better overall judgments. Outside views were even welcomed, and considered fresh, new, or inspired, and innovative. Perhaps for cultural reasons, especially there in Russia, but also in the Islamic regions where they had space-oriented bases, also hidden, there was no 'transparency'. For Milana, this all worked out just fine. She didn't want to know.

She looked around the room she had entered with the materials she had gathered for Terchenko. There were now, for some reason, numerous soldiers, Colonels, Generals and Lieutenants, men with guns, and a lot of unfamiliar buzz, in the room. Some of the men even had automatic rifles ready at their sides. An emergency over-head alert-light (red), was flashing, and there were ringing phones, radios spewing garbled messages with static, beeps and voices and more. When she finally found Terchenko, he was seated at a desk, with men at either side, asking him questions, and she noted that he also had a personal handgun on the desk in front of himself, at-the-ready. "Milana, good girl. Put those down, I don't need them any more," Terchenko said, looking up at her approach.

She placed the books and computer files to one side of the desk, out of the way. The men with Terchenko seemed to be armed-escort, or advisors. "Yes, sir," Milana said. "What's happening, sir?"

"Never mind for now, girl," Terchenko told her. He rose from the desk a few moments later. Then motioned to the two men, and then started walking in the opposite direction from where she had entered, with Milana tagging along. By now, in her ignorance and some fear, she imagined that the meteor they were studying had somehow arrived early to end all of their lives, even the whole world. She hurried along beside them, their footsteps tapped and shuffled. The other men knew their roles, and other people they passed had jobs they tended to, and knew what they were about.

"All right, quickly now Milana. I don't have time for you now, I'm very busy. You must go into my private residence, my room, on the East-side of the base, do you know where I mean?" said Terchenko as they walked.

"Why, yes sir," she replied. "Your apartment, you mean, in the housing across from the parade-quad."

"Yes, that's it," he said. "I will give you the entry-code. I want you to go personally, as quickly as possible. Inside in the bedroom closet, there is a large green leather suitcase. It's locked. It's under a lamp and some towels. It's not very big, it's already packed, it may be a little heavy. I want you to get that suitcase, only that, nothing else. Then go over to the parade-area again. I will send a jeep-transport. Give the driver the code-word, 'monkey', and he will take you a few miles over to a private bunker, by Launch-Pad Number Two, over there. Number One is that way, then over this way is Number Two."

He pointed it out for her, towards the dense hills and green wilderness. There were the tops of towers and tall structures, far off from there. "Meet me there with the suit-case. You need to do this within two hours, you understand. No one will bother you, tell them you are on orders from me, and show them your ID."

"Yes Colonel Terchenko," said Milana. "Please sir, what's happening? Has something changed? I thought we were working on the up-dated star-maps."

They continued walking, now down a short flight of steps to a landing, and then to an exit door. With a latch and a loud sound, it opened to the exterior and the bright day was on them, onto the tarmac and grounds. One of the men with Terchenko and Milana was talking on a cell-phone type communication device, as they walked yet more. "Yes, yes," he said into the cell-phone, with his thick accent. "Yes, here he is. He's right here."

Ignoring Milana completely, they paused. Terchenko took the communications-device. They had left the building onto a grassy area with pathways, and then farther down, larger buildings and towers, machinery, trucks, and the launch-areas farther out. Unlike Vandenberg, the base here was far inland, with high, rocky, snow-covered hills and forests full of pines, far from any ocean. Beyond were vast, flat, naked areas, with only dull, monotonous tundra and over-growth. They had chosen the area to hide the base. To reach the nearest real town or city was many hours and many miles. Across from where they stood, an eighth-mile, Milana could hear a siren begin to alarm.

"Who?" Terchenko said, to the man with the phone.

"Ambassador Black," the man said. Terchenko took the phone, pausing.

"This is Terchenko," he said tersely. A moment as he listened to the call. The siren Milana had noticed continued. Trucks were moving about, and other activity. "Yes, Ambassador." The others could also hear him talking.

"We're securing the facility at this hour. Everything on alert status. Yes sir," Terchenko was saying. "Your information is correct. What we know now, the Western strike team is in-motion to attack, half-a-day at most. A small force, just ground teams, mobile units, soldiers, the Marines I guess, but also apparently air-support. "

There was a pause as the party at the other end of the call took another moment of their time. "Yes, Ambassador Black, that is correct. We think the teams initiated out of Western bases on the Spanish peninsula, and North Africa, Mediterranean. I don't have all the information, sir. No, sir. I do not know that. No, sir. I do not have that information, Ambassador. Yes, sir. There may be other launch facilities, our bases, under similar stress at this time. At various locations, you understand, in Iran, the Indian sub-continent, the Philippines. Air-Command General Ginaloba in Saint Petersberg, he could tell you that, the global scenario. All we know from here, from our sources, the Western is trying to shut down or destroy our ability to launch ships into space. Certainly here at KK-F. This is certain, here, I mean, we are under attack within just a few hours, by reports. I cannot be in ten places at once. It's the Mars Base, sir, of course it is. There's no other explanation. There's nothing here. All we do is launch rockets."

There was a long pause, while Terchenko was listening to the Ambassador, on the other end of the call. Milana was transfixed by what she was hearing. She couldn't help herself from listening. The base was under attack. With all its high walls and security, its hidden and secret systems, and secure communications. The two security men with Milana and Terchenko paused with their guns at their sides, looking about at the scene idly, anxious, nervous. The base was busy, but there was no sign of any hostile attack at all, for the moment. Terchenko finally muttered a few words to end the call, shutting off the phone, handing it back to the man. They started quickly walking again in the same direction, moving past traffic.

"Andrew," Terchenko said to one of the men. "Take her over to the apartments. You, come with me. Quickly now. No time to waste."

For just a moment his eyes met with Milana's, and he looked at her with a smile. "The green suitcase, in my apartment," he said. "The code-word is 'monkey', for the driver, then to the bunker, Flight Pad Two. You have two hours. Hurry girl, hurry."

"Yes sir," she said meekly. And then she and Andrew diverted off in another direction.

"No one fucking declares a war anymore anyway, for Christ's sake," Terchenko muttered to himself, and they went along, too, towards somewhere else. "Why the hell would they??" The alarm siren in the distance continued its noise.

The US White House and military had indeed put their plan in motion. Precision strike teams were executing the same operations at the other known Islamic-Hindi-Russian space-launch sites, to prevent any new ships, launching to the battle at Tharsis Montes. And none others had launched, from the Eastern-program. That way, the conflict on Mars itself wouldn't go on forever, with launch-after-launch from Earth. So while the base on Mars was under-siege, even at that hour, with soldiers and guns and rocket-launchers and so on, and the men in the walker-suits and air-breathing helmets, more than 50-million miles away, the consequence on Earth was very similar combat, against very similar bases, multiplied ten times or more, for that many more launch-sites. Two planets were at war, one so much smaller and distant, with just one small base in the airless mist, and no more than 1,000 men anywhere near the planet at all. The other was now a battle at home, with the threat of Big Bertha, the approaching meteor, Asteroid U2753b, seeming to cause a fuss. Only this and nothing more.

## CHAPTER 43: Keeje Has No Mother

As 'Berle's Brigade' entered Mars orbit, and began to shuttle down soldiers to defend the base there, the complexities of a hundred different scenarios and situations happening on the planet Mars, and the battle at Tharsis Montes, were truly beyond the ability or capacity of anyone there, or any of the individuals in command, or under command, to understand or comprehend. One-by-one, Berle's ships entered pre-planned orbital trajectories. Working with teams below at the Mars Base with their tracking tools and communications, they organized the orbits of the ships into a systematic and arranged set of standard pathways. At the same time, seen or un-seen, tracked or not tracked, the Russian-Islamic ships were also circling the planet. Mars is smaller than Earth, and considerably so. Yet, still a vast area, so all of these large inter-planetary ships could easily orbit at the same time, without bumping into each other. On the planet surface, around the base, the soldiers already in combat continued the skirmishes and assaults and raids. The men in the oxygen-igloo foxholes, the men in the base itself, the preparation of descent pods, the electric carts, the protection of gates and entry-ways and small roads around the base, and structures. Small missiles, lifter-pods that could go from the surface back into orbit, to dock with the ships there. A lot of busy-bees, a lot of high-tech labor and training. Today's fan of space-travel and space exploration, might be able to picture all this, as if the NASA shuttles of today, like the 'Atlantis', along with the Russian ships, were in orbit around the Earth, intent on attacking a very small area in the Saraha desert of North-Africa, or in Death Valley, or Mojave, if there were a base in these regions. It all took several days. Berle's ships stabilized their positions. His plan was to immediately send down soldiers, and waste no time.

The US had a lot more experience on Mars than the enemy. It was their base, after all. The US sent over four to eight ships per year, now 20-years of this, very similar to Guy Reisling's cargo-ship. The same techniques were used. There really was no other way to get men down to the surface. The large so-called Mother Ships released the descent pods, they entered the thin atmosphere, and used the thruster rocket-brakes, and parachute-gliders, much like small air-craft, targeting their landing, which was a rough-slide, or dirt-glide roll, until the descent-craft stopped, in a cloud of Mars-dust, and could stabilize, and release the soldiers from inside.

Not including Guy's ship, 'Berle's Brigade' at this point had six ships, full of men and gear. There were between 20 to 30 armed-men ready for combat, equipped with the surface-walker suits, on each one of those ships, for a total of about 145 men in the first wave, ready to join the fray. Guy's ship was for cargo-only, and his men would only be in the combat as needed, and the two other ships in Berle's Brigade, were left behind, from the Molinari space dock. Those two ships would be another 45 days, or less, before they would arrive. It was all a lot of work, and as seen with the Russian-Islamic ships earlier, Berle's teams dropped down to the planet, the pod-pilots, the ships in orbit released the pods safely, the soldiers and men ready and prepared along with the shuttle-pod captains, the navigation to the area near the base, and planning for the lifter-pod ascent-rockets, to bring men back into orbit to dock again with Berle's ships. It was a big job. Even with the navigation-planners, computers, telescopes, radio and radar, math-plotters, mapping, high-tech life-sustain, schedule and hourly monitors, the philosophy was only "if anything could go wrong, it would go wrong". Berle's teams had planned ahead, and knew what they were doing. They had many months in space to work it out ahead of time, during the crossing. Now it was all put into practice, for better or worse.

At least four of the new-arrivals reported problems with their life-sustain suits that made surface-work impossible, needing replacement suits. One descent-pods experienced a very, very harsh landing by accident, with several non-life threatening injuries. There were minor communications failures, and confusion or mix-ups. More than once during the Cavalry-Rescue style avalanche or Berle's teams, Commander Bojji-Than of the Mars Base, and Berle himself and his seconds and thirds in command, organizing the assault-rescue, came to disagreement about what was going on, and what was the best to defend the base. In other words, who was in command? This was sometimes disputed during it all, as things un-folded, but the needed unity was always there. The US side only wanted to pushback the enemy, to secure the base from any possible overwhelming take-over. The base would be secured, even if they had to kill the enemy-soldiers in their tracks, something many of the 'doves' on the US side did not wish to see at all. Perfect bloody-hellish harmony in their purpose, dread and drudgery in their hearts.

Winton Berle was not a military man, and had no real military experience. He was a space-man, a lot more experienced with trips to the moon, only an astronaut or space-pilot, working on the exploration of cosmic objects. He knew the ships, the engines, the suits, the technology, piloting ships and descent-landers, people-transport or cargo. He didn't know much about battle and warfare, few of the astronauts did. His teams included military generals and strategists, battle-planners, and they had completely planned ahead for the battle at Tharsis Montes by this time.

Complex hardly describes all that was going on. On the surface, the attacks and skirmishes and raids continued, hour-after-hour, day-after-day, as before, with the teams moving back-and-forth, firing small arms, small rockets and grenades, then retreating to the life-sustain stations or, for the enemy, to the descent-pods that had brought them down. There were only a few significant victories for either side. It was a slow, inefficient dance, in the nearly airless environment, and the heat-and-cold. At one point, about 40 days into the struggle, (from the very first arrival of the invaders), one of the Mars Base rocket launch towers half a mile from the larger base, connected by roadways, that was used to send ships into orbit, was seriously damaged by an all-out assault by the enemy in greater-than-usual numbers of foot-soldiers. The launch pad was damaged, and other equipment and vehicles, and deaths on both sides. The Russian-Islamic moved in by night, under cover of dark-and-cold, with about 50 men, an unusually large group; most of the skirmishes and assaults had less then ten men ion either side. That particular launch pad was defended, but they did not expect the attack, and at first the small arms fire and small-rockets, grenades, etc overwhelmed the US-side men there. The launch pad was then commandeered, over-taken, for a time, as they fought, and was under enemy-control. There were eight US deaths, as the US Mars soldiers battled to take it back. The launch pad was seriously damaged as the enemy retreated; choosing to destroy it rather than surrendering it back, there on the sands and in the airless, breathless habitat, with the so-thin rays of the Sun enveloping all. It was sad to see the men die on either side, much like a ferocious Marine combat battle of blood and fire, but fighting over a Disney-land hotel or University campus, or a Las Vegas sports arena, instead of a valued military goal, it made no sense. They gave their lives bravely, defending their home on Mars. So far from home, so far from family and friends and the cool green hills of Earth, they were among the first Earthmen to die on another world, in this way. There was a mystery and a sadness to it, that they all felt, as if a change or shift in their consciousness had been invented for wandering souls, so very far from home, to find their way back to the eternal grace of the

Creator's love, that is, to die. If they did find a way.

Away to one part of the Martian sky, beneath the silent void, the space-vessel 'Tolstoy' seemed to float on nothingness, like a great fat metal bride to meet her lover, jealous and raging war, the natural formations of a sandy reddish world below. The Tolstoy was in-orbit now, at high-speed, but it seemed not to move at all. The ships resembled very large ocean-cargo ships from back home, but like huge Jumbo-Jets, rounded in shape, like large angular high-tech potato forms, useless with their 'wings', engines to the back, topped by control rooms or pilot's helm, and with doors and cargo-release gates, and communications-arrays, solar-collectors. Each was larger than the largest commuter jumbo-jet that had ever flown the skies of Earth. The Tolstoy's main engines were silent, but smaller engines maintained the orbit and maneuvering rockets. Running lights flickered. The indigo void beyond, full of God's own stars, were it so. Prokov Keeje worked from the ship's pilot deck. He also would move himself personally from one ship to another, among his vessels, by means of a small temporary docking transport that could go from ship-to-ship, still in weightless orbit. He portrayed himself now as a bit of a Darth Vader-type, in his command role. He knew no other way than his cruelty. He was a trained military man and pilot. They would win control of the Mars Base, or probably die. But there was no black cape or black mask. It was only perhaps his attitude that set him apart, as his teams plotted and schemed to snatch victory from the jaws of defeat.

The Tolstoy's command-deck was as tense and highly charged as any Star Trek TV show Klingon battle-ship scene. Keeje had his sources about the battle, the men and ships, and so on. He worked directly with two military commanders, Colonel Aslan Brattles, and a General they all called 'Little Stalin', both sent from Earth specifically to win control of the Mars Base to the Eastern space-program side. It was Brattles first voyage into space, and he didn't take to it well, seeming to sweat and heave and puff a lot, even without any stress. Common to many wars, the battle-commanders now felt the dread of their task, and sometimes cowardly despair, as they surveyed what was going on: they seemed to be losing. They were out-numbered, far from their supply-lines, and the idea of even returning home safely seemed very challenging indeed. They couldn't destroy the base just to take control of it, which would be a major loss, rather than a gain, but they might have to. Of course they had a plan ready to put into action. The idea was to use their ship's ability to project Electro-Magnetic Pulse emissions, of the same kind that had disabled the Penelope, and target key strongholds on the US side. It was Colonel Brattle's idea. He was a short and stout Caucasian man.

"Our greatest victory so far was against the cargo ship during the crossing," Brattles told Keeje. "The EMP overwhelmed their ship from a distance of less than two miles. They were completely disabled for weeks. We can do the same here, now, on Mars, against their ships, and against the surface targets."

"I agree," said Keeje. "It's a good plan."

The other commander, the Little Stalin, had been working with the technicians on how the EMP projectors could be employed. Only four of the Russian-Islamic ships had the EMP devices of sufficient energy levels. These were devices developed on Earth, that generated a huge blast of electro-magnetic radio-waves, trillions of watts in voltages, an extremely powerful short blast of energy, which if properly directed, could disrupt all electronics wherever the energy would pass through. "Our victory is poetry," said the General, scratching at his shaggy beard, in his Eastern-tongued accent. "We have the

EMP's in place and ready. We move on their ships and facilities on the ground. Target carefully, one-by-one. Not difficult. The EMP's recharge in a short time, half a day. This war is now poetry. Our victory will be poetry. We can win this battle, in this way. You agree, gentlemen??"

Commander Keeje and the other man (Brattles) confirmed, as did the other workers and technicians. It seemed like it just might work. The victory against the Penelope was a significant success, in the use of the Krenika's EMP generator. The Krenika, and her Captain-pilot (Zolotny), were included in the present formation in orbit around Mars, on the 'invader's' side. The General was right. The Penelope had nearly been lost. If they could apply the same in the current conflict, it could turn the tide in their favor. Zolotny savored the idea at another shot at Guy's ship, privately or with his friends. The minds at work schemed gleefully, spinning new ideas by the hour, there on the flight-deck of the Tolstoy. An EMP could be directed at the Mars Base, the US soldiers, disabling communications and the walker suits, EMP's could be used against the lifter-pod ascent-rockets, and the downward shuttles, and against the US ships in orbit. The Russian ships had to move into position close enough for the EMP's to be effective, for each target. The seduction of the idea was enough, and they began to put it into action, in complete secrecy. All ship-to-ship communication was now deeply encoded for security to keep the plan hidden from the Americans, and to maintain the surprise attack advantage.

Within a few moments, a radio-link alert began to sound on the Tolstoy's monitoring station computers there on the flight deck. The communications man was a youngish Islamic astronaut named Mohhamed-Ishtar. *Beep-beep-beep.* The technician activated the connection with his technology, the radio went active. Keeje and the others paused, waiting. Mohammed-

Ishtar identified the incoming message, confirming twice.

"Commander," he said. "It is the Americans. The US leader of the arriving ships is requesting to speak with you. Colonel Winton Berle is his name, authorized of California-Vandenberg. An experienced astronaut, he is in charge of the US ships in orbit, sir."

Now Keeje's eyes darkened and he seemed to grin wickedly to himself. His nemesis, his opposite, whoever he was, he must soon die. It did not make for pleasant conversation to know this. He took a moment. The null-gravity environment on the Tolstoy and aboard all the ships in orbit lent itself to rooms full of very busy men who necessarily would float about in mid-air like balloons, looking at times somewhat foolish, sideways or upside down, pulling along by hand-rails from point-to-point.

"Tell him I refuse to speak with him for five hours," said Keeje. "I am too busy. Schedule his call for five hours from this moment. Book his call ahead like he was only a personal call from my mother, which can wait, when I am working, you know."

The other men on the ship's flight deck laughed at his joke. They always laughed at his jokes. Mohammed-Ishtar returned to his radio-station to relay Commander Keeje's message back to the US radio-dispatch, also laughing to himself.

"Keeje has no mother," the radio-man said to himself under-his-breath. Then the call was relayed.

## CHAPTER 44: Fire, Bullets, Blood

*"Permit me," said Mars. He raised his right hand, and the red sword floated up and across the mat, dipping momentarily as it did at the edge as if bowing, and moved to his hand. Mars gravely sheathed it. "A remarkable man," the Samurai said, exchanging bows with Mars. Then Mars turned and walked out of the dojo. From 'With A Tangled Skein', by Piers Anthony, 1985*

Berle liked to play solitaire with a standard deck of playing cards, whenever he had spare time, or needed to wait, during his space-voyages. Which was often enough, voyages through the Abyss were very long, with almost nothing happening, the ship's engines hummed and moved the mass-weight of the vessel forward at a standardized pace, the science and technology eliminated un-anticipated events or dangers successfully (or else they would surely die), and despite advances in ship-design and engine thrusters, the journey to Mars still at least almost a year or more than a year, even when the planets were in a favorable arrangement. Weightlessness was not unbearable, and could be pleasant. But it did make a card-game like solitaire, an unusual pleasantry. The paper-cards, ornamental with kings and queens and spades, the numbered cards, and so on, would not rest to a table-surface as they would on Earth, or Mars, for that matter. So he could either use small adhesive stickies on the backs, and also the fronts, because in solitaire the game is played with cards face-down until the player turns to see what Fate has dealt him; or, he could work the game with the cards actually suspended in the cabin air, one-by-one, or in small groups, like butterflies hanging in an invisible tree. This didn't work very well, but was hours of fun. Weightlessness also could produce nausea, and Berle mostly felt the same about war, killing, and battles among brothers over the trinkets of planetary properties like moon-bases, and the enormous waste of resources, time and people he knew and loved. He loathed it, but there he was.

He and Mars Base had decided to try to reach out again to the invaders, and Prokov Keeje specifically. So, they set up the attempt, the same as Bojji-Than had done before. They knew the Russian-Islamic vessels were 'out there somewhere'; in orbit the same as Berle's Brigade. But Mars is a large place. They had to wait, scan and track the skies of Mars, spot and identify what seemed to be the Tolstoy, or one of the other opposition ships, and then use standard radio-channels and frequencies, and broadcast their 'CQ'-signal ('hailing-frequencies'). A radio-transmission "hey you", they hoped Keeje would pick up and respond to. The technology was not so advanced, and maybe never would be, that such a thing was a clockwork-simplicity, easy, or flawless. It could take hours just to pick up the signal they wanted. So, Winton Berle, like a lovable cop passing out second chances to naughty international space-men who may wish to repent of bloody attempts to take control of other people's off-planet bases and facilities (killing a few friends along the way, blowing up expensive items) Berle was patient. When Keeje's radio-man returned the first volley message-relay, that they would wait five hours, refusing the call, making them wait, or if ten hours, or 20-hours, it was no big deal to Berle's pride in the matter, or the jest, "like a call from his mother". Berle was not naïve or gullible. He knew what he was dealing with. So the deck of playing cards kept him busy, his thoughts empty for a while, resting, wait-and-see. The 10-of-Clubs, the Two-of-Diamonds, the 10-of-Diamonds, a single Ace (Clubs again), and so on, gently bobbed in the cool cabin air on the flight deck of his ship, the 'Understandable', in a circle about three or four feet around where he sat and waited, held firmly to his captain's-chair by the magnetic strips. No good for solitaire, that hand.

His men, there with him, were pensive and anxious. The Understandable was now the command-ship for whatever victory they could pull out of the whole mess. Berle's second pilot, the African man named Ben Jazreel, and his radio-operator, Raza Brahman, along with two battle-planners, also waited. And waited, and waited. More than 30 regular surface-soldiers, who had made the journey along with them, had already disembarked to the surface, ready and armed, in their surface-suits, to fight Keeje's forces back and defend the base. The chess-board and the players, that Berle was trying to keep track of, was far more complicated and dynamic, than he was able to keep in his thoughts at any one time. It was also a multi-level game, with the soldiers fighting on the surface, as well as the orbiting ships, and the base itself. Typical of space-travel in 2079-80, Berle and his chieftains really couldn't see any of what they were dealing with, from the helm or flight deck cabin of his ship. They worked on computer-screens, maps, radar, tracking data, lists of monitored environmental information, lists of people and positions. But it was like working from within a tomb or far-off tin can. The ship buzzed and hummed with blinking inner lights and the constant drumming-buzz of the engines. In most of Berle's career, the same technology was simply for keeping track of educational research and exploration they were doing in space. There were few if any instances of his work and the technology and ships, when it was in his hand to bring hurt or harm to people, he wouldn't have even thought of it.

So, they waited, and things unfolded however they were going to. About three hours had passed with no other signal from the Russian-Islamic ships or the Commander, the notorious Prokov Keeje. Then, Raza, the radio-operator, picked up a relay from the surface of the planet. It was of course the Mars Base.

The link between Berle's command ship and other US-side ships-in-orbit, to the Mars Base, was stable and reliable. Raza Brahman picked up the signal on his desktop array of tech-gear, the Understandable's communications-center. He initiated the sequence to receive the call. Mars Base Commander Bojji-Than, and his staff, had some idea what Berle was dealing with. The battles and firefights, the raids and skirmishes, continued without much real effect, aside from the take-over of the launch tower (launch pad), and that struggle, with the deaths. But now, as Brahman learned, something else was about to overcome them all, and Mars Base needed Berle to be aware.

Berle didn't take the call personally, and while the connection was on-going, during Raza's work, back-and-forth to the base, he was content to believe it as an information-advisory, or an update. It was standard-procedure for data and planetary conditions, environmental information, to be shared more-or-less 24/7. Conditions were to be informed to the ship-pilots and others constantly for safety's sake. Brahman continued with the information coming in, on his computers and radio-gear, for about 20-minutes. When he was finally satisfied, he ended the call with their sign-off, and paused to collect his information. Like a good space-worker, he would make a report to his captain, so the flow of information would move towards better decision-making as circumstances were changing.

"All right, Raza, all right," said Berle, who had witnessed the call, and patiently waited. "What did they send over now? Don't worry, you won't hurt my feelings. Lay it on me. I can take it."

Raza pulled himself across the flight deck with the handrails and straps they used in the weightless environment. "Sand-storm," he said clearly. "They've got a seasonal sand-storm building up on the West. Environmental and Mars-weather report confirms. Big sand-storm."

Berle huffed and crunched his shoulders together with his smile disappearing. The news brought a dark look to his features. He took a deep breath. Sand storms on Mars were legendary and well known. As thin as the Mars-atmosphere actually was, almost none, the sandstorms on Mars could be vast, lasting months at a time. Created by the heating and cooling of the changing seasons, the storms built up over long periods, until the ancient dust and sand of the Martian surface would loft into the thin air, as huge clouds of reddish, thick, dust and sand. The storms would cover hundreds of square miles, and included electrical activity as well. The atmosphere on Mars is not very thick, or dense at all, so the storms were not like those back home. But they could move very fast, a hundred miles an hour or so, and be very violent and destructive. Prior to the establishment of Mars Base, there was nothing here to destroy. And for many ages, the Martian winds would blow; the storms would build up and then settle down, into the distant past, century after century, endless into the former. Current photographs of Mars show the rippled dunes, or flat areas, or rocky places and mountains, where the sand acted as an abrasive agent wearing down whatever rocks and mountains or hills remained, like a gigantic jelly-bowl of moving sand-paper air, slowly wearing down the features of the planet. Explorers like Vinces Grant, in charge of exploration and surface research, were often delighted that whatever rocks and gems and minerals the planet might have hidden away, were sometimes exposed by the wearing-away of the elements. And there were gems and minerals they had found, some of them quite unusual. Rock hunting on Mars was a favorite. But when the sand-storms came, it was a joke among the residents that early science-fiction writers like Jules Verne or Edgar Rice Burroughs, imagined that Mars had vast canals of water. It was a sad mistake. All of the water they used on Mars was brought in from Earth, stored in tanks, used for hydroponics farming indoors, and to sustain their lives. It was all recycled.

There was almost no atmospheric moisture. Some could be collected in small amounts from dew-point collectors. The planet had polar caps, supposedly of water, or possibly 'dry-ice', but they had not been explored or exploited. The obstacles were too great. Indeed, almost every drop of potable water on Mars was precious. The recyclers, and atmospheric dew-point collectors, were meticulously maintained. Sand storm appeared about once or twice a year on the planet, and could be predicted to some extent. When they did manifest, the residents at the base found it was truly hellish, fraught with danger, even without a battle or war going on. The base itself was sealed and airtight, for life-sustain, but with a sandstorm, the external functions were compromised. Antennas would fall over or be damaged, telescopes would be ruined or toppled, entry-ways and gates, less-fortified portals, the smaller launch-sites, any movement in-and-out of the base, the off-site tanks, storage, buildings, all fragile enough that the wind-storms would do some kind of damage, inevitably. When the storms ended, they would start repairs, fix the damage, but it could take months. Yet many of the functions were essential for their survival, there on another world, in an area 'colonized' by America, only a few miles square.

Berle, his planners and navigators, base-management, the mappers and navigators, they were all well informed about just how angry the Martian environment could be. Berle's teams knew, Guy Reisling knew (about the sand-storms), and so did Commander Keeje, and all the Russian-Islamic pilots. Now, with more than 200 soldiers in the walker-suits on the surface at any given time, moving in-and-out of the foxholes, or the 'enemy' shuttle pods, fighting with small arms, the specter of a sandstorm presented a considerable difficulty. The man-in-charge, Winton

Berle, took the news with a grim resolve. He took the rest of the report from the radio-operator Raza, and took the hard-copy data-sheets to look them over later.

"All right," Berle said, now in conference with the radio-operator and his co-pilot, and the other men, on the flight deck of the Understandable. "How big is it, how far off is it from the base and the battlefield, and how long is it going to last? Do we have that information?"

The navigator for planetary orbit working with Berle, a man named Peter, was charged with all the details of anything to do with the ship's navigations into orbit. His work was important to position the ships in orbit, tracking the positions of the Martian moons, to place the orbit paths over the base-region, also radio-linkages with communications blackouts to the base or their other ships when out-of-range, all ship's positions at all times, and this sort of thing. So it was natural for this operator to review the maps and data and try to understand the course of the sandstorm.

"The front of the sand-storm might be two days off from the immediate region of the base and the men on the surface," he said. "It's not a particularly big or significant storm, maybe 200-miles across at the front, and then bigger or smaller, changes. But it could grow and it could be prolonged. So at the very least, within 42-hours or so, the base will be right in the middle of it, for at least, oh, at least another three days after that. With any luck, it will pass over in about a week, from previous experience. So, we're lucky, in a sense. Some of the sandstorms will last a month, and be quite larger. This one is early in the season. The normal seasonal storms won't start up again for another four months."

"All right," Berle said, again, taking it in, pausing, with a small cough. "What effect will it have on the men on the surface, in the foxhole-igloos. How will the suits hold up? Do we have any experience on the use of the suits during the sand-storms?"

The analyst, Peter, continued to speculate. "The winds are only going to be maybe 50-miles per hour tops. You have to remember, the Martian atmosphere is very thin. So, to be honest, on the surface, a man in a standard life-suit would be pelted with the grains and dust. They won't last long. His suit could be compromised with sand in the joints and small vents and the electronics. "

The flight deck of the Understandable hummed and buzzed and beeped around them, with the ship in motion. It was now three years beyond the original announcement of the threat of the danger and drama concerning the approaching meteor they were calling Big Baby Bertha. It was ominous for them all. The journeys were always difficult. Now the stakes were higher, and it all seemed more important, urgent, and critical to the purposes of the American public and government, and the Earth's world government. So all the space-workers were ready to do their best, even to lay down their lives. They also understood, that to sustain their own lives and 'win', may call for the application of every means at their disposal.

"I can't do that to these people," Berle said. "This is madness. Get that fucker on the radio! Keeje! Dammit, this war is over, and soon!"

The deck of playing cards now flew away from where he had sat all this time, taking it all in. They spread out across the flight deck, in the zero gravity, like limpid petals of some honky-tonk flower-tree, kings, queens and aces. Peter, Raza, Ben, and the others, watched Berle hoist himself away from the main-area, then down a hatch-portal, near tears in his anger. They knew their man, a leader and research-educator, and a pilot. But not a warrior. And they had never quite seen him quite like this; he was livid, bitterly determined. Planet surface walking, in a deadly sandstorm, with weapons and guns and rockets, a living hell.

"Give me an hour," said Raza. "I'll boot the call."

## CHAPTER 45: Please Comply

The US Mars program in that year of 2079 had major operations centers, either on their own or as part of larger programs, at facilities other than Vandenberg, California. Similar centers in Texas, Florida, Utah, Colorado, and Alaska off the mainland US, were fully functional. Lynn Rogers-Smith was the chief administrator for all operations. But her main contribution was overall leadership, not necessarily as sole-proprietor of her own kingdom, doing whatever she wanted. About that time, she was summoned to the space-program center in Washington, D.C., during the crisis. With her went her best advisors and seconds, Science-Lead Ibrahim Mehudi, Security-Lead Branson Porter, Transport Chief Okman, who had replaced Berle for overall ship's-systems while Berle was in-transit (Okman had much of this kind of knowledge), and the others, each with their own support team or small staff. Mars-Base Defense Task Force Generals and 'real' military men joined them. They ended up together in a Virginia US military-reserve, where the tech and gear they needed, and security, was readily on-hand. There to meet them was Margaret Hazlett and Borgalt, one-half of the current-year Seated-Council of the US Presidential Office, along with their considerable entourage. All this togetherness was arranged by default, to bring forward for this group all that was known to be in-play regarding three main topics of immediate concern: 1) the approaching asteroid U2753b, 2) the conflict on Mars, and 3) the US military strike-force assaults on Eastern space-program launch sites, to end and prohibit all current Russian-Islamic space-launches. For the Mars-program, only the war on Mars, and the impact of the strike-force teams against the Eastern launches, were on the table. Rogers-Smith was not interested in the asteroid, aside from wondering about the fate of the world like everyone else, her own safety-fears, family, future, etc. It was nerve-wracking for them all, and of course she was up-dated, but she was in no way in a leadership role concerning the meteor. The Mars Base and Mars-program were her venues. Concerning the Eastern launch-sites, and the US strike-force teams, her specific interest was what Berle had so strongly suggested (and more than once), that somehow they could find or locate, and acquire, the remote radio-transmission signal codes, from the Russian or other sites, in their offices or computer files or lock-safes, that would possibly control or activate and manipulate important enemy ship functions. These signal codes were for safety. All the ships had them, for rescue-type arrangements for the men in space. They could open docking-ports, and things like this, even if the men inside were dead or incapacitated, or suffocating, in emergencies. Would it work? She had no idea. For the White House, and other space-program leaders, it was all about the meteor, and the rest. Saving the world, yes, and it began to seem possible they would succeed.

The Virginia base was typically secluded, safe, protected, well equipped. Offices, halls and doors, reinforced windows, computers, radios, monitors, soldiers, etc. Not very fancy. The sandstorm they had encountered on Mars was now well into its first 24-hours, meaning so much to them on that barren world. Mehudi seemed as if he had a flu-virus. Porter was alert and mostly angry, and somewhat incompetent (some of them felt). Hazlett and Borgalt were dressed in a casual way, handling things in a command-style as best they could. During this period in the US, there were four Commanders-in-Chief, not just one, and two were present. They had all now been dealing with the crisis for two years or longer, almost three years. After a while, Mehudi was able to take the center-of-attention. He was not out-paced by the other scientists, but well respected.

"I agree with the science on the asteroid," he was saying. "Of course the details are all very esoteric. The tracking is extensive, exhausting. We now have a date for a possible impact, as some of you know. April 24, 2081, eighteen months and 24 days from now. For two years, tracking and trajectory seemed to place the meteor perilously close at that time, even a direct-hit. This object is the size of the island of Manhattan or even larger. A hit would be extremely destructive. Some of the science gives us hope. It might be a near miss. Various survival efforts and means to change the course of the object, or destroy it, are underway or in-place. For now, with your attention, I'd like to add my own opinion, as Science-Lead for Vandenberg, if I may."

Those present knew Mehudi's opinion, as he had not been silent. Some agreed, some did not. He was thought to be too liberal, a 'dove', which didn't fit the scenario, so he was mostly dismissed. But he was also a man on the inside of the space-program, well known, so his opinion mattered. President Hazlet didn't like him. They had been in dispute on other issues before. And he lacked for physical presence, no athlete like herself, and many of the pilots and astronauts.

"Much of what I'm telling you today is seen here, on this computer of mine," Mehudi said. He was referring to a rather simple computer set-up he had arranged at the meeting-place, part of his personal work-area. "Maybe you can view this if you look closer, I apologize, the display is too small. To be brief, I think Bertha is going to miss, or we can coax it into a miss."

"All right, please Doctor Mehudi," said Hazlet before he could go further. "We've heard this before. Can you prove your point or give us some certainty? Even with the most optimistic view, the government still must prepare. I know you realize this."

"It's true," said Borgalt, the black man, also a seated US

President, one of five men of African-descent elected to the US President's office since Barack Obama in 2008. By 2079, the US had changed its leadership organization such that four US presidents were 'in-office' at the same time (three men and a woman at this time). "We have no choice," he went on a bit. "We will launch to stop the meteor, and make plans for impact, even if you tell us everything will be fine, two long years from now. We can't take the chance, even on very-well researched false hopes."

Mehudi paused. "Well, that's fine," he said. "Your job, not mine. Blow up all the meteors you want. You will do as you must. Let me just go over the idea quickly. The meteor is in motion, from the direction of the far side of the Sun. It is a rogue, not in-pattern with other meteors. The path-projections and trajectory models are very accurate; we use the same methods for all flights and re-entry. The speed, mass-weight-density, and the Earth's normal orbit, are all known. But my office did a different study. We looked at the roll, the way the object is turning on itself. There's no doubt the meteor is spinning slowly in space. The standard calculation includes this, but mostly looks only at the overall physics of the mass-weight-density and speed-velocity. If you include the spin, it's a different story. Other projects say a hit on April 24, 2081. I'm giving it a near miss, within less than 10,000-miles from Earth, the distance from New York to Australia, maybe. The difference here could save us all. Add to this the idea that instead of destroying it, we simply nudge it to a slightly different path, like a billiard ball on a pool table. That's the nutshell version. I can back up what I'm telling you here. I thought you might like to know. Thank you."

There was a pause. President Hazlet huffed a bit to herself. Her view would later dominate; of course, she was far more powerful in her position than Mehudi. The other reason was clear all along, that being an overwhelming public fear. Mehudi humbly began to pack up his tech-gear and so on, after a few minutes. Some there at the meeting looked at his computer models and maps before he left, but it seemed to only mock. Nothing so simple as a near miss would save them. Only perhaps an option, but not a real hope. Fear-based decision-making trumped any calming influence from Mehudi's findings. The sessions there at the Virginia base moved ahead. It was a grinder. Hours of talk, details, choices, power-moves, minor or major.

Next up, however, was a matter of intense interest: the radio-link with Berle's command ship, and Prokov Keeje in the 'Tolstoy', would now include a distance-link to their Virginia location. Amazingly, the set-up was reliable and stable. Smith, the Presidents, the military generals and others, would listen in or contribute. Smith recalled how things had gone with the Penelope. Not very encouraging. Berle would leadoff. No Russian-Islamic Earth-side authorities were on the call-link, for whatever reason. It would be recorded and archived. Smith didn't expect much. They felt they might talk him (Keeje) down and end things. That would be the preferred outcome. The enemy wanted the Mars Base badly, they were also afraid. It meant survival.

The communications-gear and satellite-links, antennas, source-monitors, data-compressors, the Virginia-base radio-operator, all worked fine and were ready. They gathered around the kiosk like some kind of elite power-trip space-man computer camp-out. The signal-ID went through, with distance-receiver confirmation. Within a few minutes, they could hear Berle's voice, garbled by the tech.

"This is Commander Winton Berle of the US Mars space-ship 'Understandable'. You have our ID and frequency-signature, Tolstoy. Please respond."

Static, a pause. "I am Colonel Prokov Keeje of the Tolstoy, in orbit around the neutral territory of the planet Mars, under authority of the people of Russia and the Ukrainian-Hindustan Jakarta space-program alliance. You are a piece-of-shit, Berle. I have 20 men dead by your hand, or more. What do you want? And please inform me, who else is on this radio-link? Do not bother to deceive me, sir. I have no time for this."

A long pause. Again the transmission was delayed. A long ways even for a radio wave, the voices compressed, like muddy-waters. Each speaker or statement was preceded by a gap of sometimes minutes-long, though mostly shorter.

"Colonel Keeje, you are vigorously advised to stand-down your forces attacking the Mars Base. I repeat, your ships and soldiers will immediately withdraw all hostile actions. The US and global Earth governments have long-standing treaties and agreements with Russia on space-exploration. This base here is a US property of high-importance and high-costs. Your actions and your Earth-side authority are out-law and you know it! Your men and mine as well died for a meaningless and illegal cause because you commanded them into a pointless aggression against a peaceful research station! We are in self-defense to save our own lives, obviously. I'm giving you a chance to survive, and go home in safety. Do you copy? Please respond Tolstoy. Berle out, holding link open."

More static-buzz, 'dead-air'. A moment. "Message received, Berle," Keeje's voice came back at them. "How else do you plan to insult me, or is there anything else?"

Another pause. It seemed almost for a long enough moment, they thought the connection might have been lost. Rogers-Smith saw her opportunity.

"Gentlemen," she said, as the radio-operator gave her the nod on her use of the gear, and a hand-mic. "This is Lynn Smith, Vandenberg Mars-Command, with the White House Counsel members, in Virginia, USA, on Earth. Can you hear me?"

A long pause. The radioman there at Virginia-base confirmed by his computer-screen, tracking the frequencies and in coming signals, with beeps and noises. "Berle's ship has signaled a wink, 'yes', back to us. They have the call, we're still good," he said. Another few moments. "Wait," he added, "now Tolstoy. You're good, Ms. Smith. They can hear you, even if they won't admit it."

Small laughter at his joke there in Virginia. "All right," said Smith to herself, bracing for what she had to say. "Repeating. This is Mars-Command with the US leaders from the White House. I believe you have my signal. Colonel Prokov Keeje, greetings from the US President's office in this dark hour. Commander Berle has the full support and authority of US powers and resources. You must please comply. US forces have recently shut down Eastern space-program launches and orbital launch sites, and there will be no more ships to assist you, or to help you and you men get home. No one will rescue you or your men in Russian ships from Earth. Your supplies and fuel will run out, and you will have no way home. Confirm with your sources on Earth, if you wish, please go ahead. This is your last chance, for lack of a better term. Stand-down all hostile actions and weapons at-the-hour, and US Mars space-workers at the site, will assist your rescue and voyage home. We want no more killings of these wonderful men and women. The Mars Base is protected by the full-might of US and Western space-programs and military. Also please confirm to your sources on Earth: the urgent panic concerning the meteor next year, or two years from now, is basically over. We have reliable science that this meteor may not hit the Earth at all. Ever. Things have changed, Colonel Keeje. You can come home now. End transmission, copy-please. Mars-Command, Virginia-base, Earth, out. Holding this link open from here."

Then she stopped, as if resting from a track-meet sprint. "Good job, Smith," President Hazlet said dryly.

"Hear-hear," added Borgalt, the 'other' President.

"Did it send?" Smith asked the radioman there on-site. He checked his monitors. Static-buzz, pause. "It appears to have sent, Ms. Smith, yes," he told her. Now the others with them, Mehudi, Porter, some of the military men and Air-Space Officers, the two Presidents, relieved the moment with a short applause and muffled cheers for Smith. She seemed almost to blush, her Texas pink-and-freckles passing beads of sweat. Would it work? They had no idea. But she had stated the case well, true-to-form.

Berle's ship signaled now, then by voice. "Thank you, Virginia." Then another wait. They could hear next when the Tolstoy spoke, it had a different sound, different energy-use or signal-driver. So they knew it was Keeje.

"We will consider your claims, United States. The Tolstoy will terminate all communications to your radio-link at this time. Fuck you, Berle. Keeje out. End transmission."

Another long pause. At least it was clear. Keeje had bombs. Keeje had ships. Keeje had men. Keeje was desperate. If he was really that crazy, he could kill all the residents on Mars at the base. But he would never get home to Earth if he did. His mission was like sending a man into a dark cave full of snakes, with no way out to back out. But now they could choose to live.

The radio crackled again with sound. "Okay, Earth-base. Berle out, ending transmission. See updates on regular links. Santa Claus and company needs some hot chocolate. Signing off, Commander Winton Berle, aboard the 'Understandable', above planet Mars, US Mars space-vessel ID Uk42B."

Then silence. They all felt something had budged, just a bit. Within a few days, Ibrahim Mehudi suffered a serious cardiac arrest while swimming in the heated-pool of a local hotel-resort, and was hospitalized. A 'dove' was down, but the 'dove' would never really die.

## CHAPTER 46: A Small Green Suitcase

The campaign to shut down Eastern space-program launch sites, and space-launch capacity, had been planned and organized following the decision at the Presidential US White House. It seemed 'simple', to the Generals and strategists, but of course it was not. The West had plenty of military muscle, but these were 'strike-team' maneuvers, not extended operations, with a very singular goal: each known space-launch site by which the Eastern alliance could launch ships into space, would be 'shut down'. In other words, Western military would take control by force, and then disable, etc. Of course the East would resist, and so on. The war over the fate of the Mars Base had now taken the complexion of a full-size deadly conflict at home. It was war here now, too, for better or worse.

The Mars-capable space-programs of the world in 2077-82 were several and varied. The West was dominated by the US, which had re-invented their space-program following about year 2020, when private companies had proven that orbiting 'space-resorts' for tourists or wealthy adventurers, were possible, safe and profitable. The US had a number of global partners, too, and the space-industry utilized all the same resources anyway, regardless of which side you worked for (the fuel, computers, tech, training, education, ships and machines, etc.) For the East, given a disparity of political stability, 'will', and revenue, the space-program organizations were more fluid. Russia dominated, with a long history of successful flights and launches. The Islamic nations formed a very strong unity with the Russians later, including Iran-Iraq, Saudi, Indonesia, sub-continent India, and to a certain extent China, Japan, and African. The Russian sub-states in the Ukraine and Balkan regions were very useful, isolated and 'under-the-radar' on the global scene, ideal for secret bases. South-America, Australia and similar Western-oriented powers aligned according to their interests. In general, the space-programs were mostly beneficial as high-paying jobs. No one, or only a few dreamers, seriously believed in any true off-world riches, colonization, far-off travel, or new worlds. Yet, it remained very true, that near-Earth local solar-system exploration was very exciting, because Earth had now finally ventured successfully into her own local planetary arena, with many hopes and dreams for all kinds of possible benefits and discoveries.

So, for men like Bojji-Than, Ibrahim Mehudi, and other 'doves', and women like Lynn Rogers-Smith and even Lila Meetek, this new war was a terrible defeat. The vision in space-exploration had been purely educational and research, something they all took great pride in, and which kept them 'safe'. But it wasn't hard to understand. With the predicted meteor only eighteen months away, there was a panic. The Mars Base meant survival, resources off world to help survivors on Earth, and much more. As noted, some of the Eastern-program science-philosophers even speculated that 'true alien' visitors from far-off inhabited worlds, would 'meet' them on Mars, if the Earth were devastated, based on very obscure and highly-questionable so-called 'contacts' (the so-called Edinburgh Society). But this was exclusively an Eastern fantasy, also motivated more to demonstrate the extremes of fear involved. So, if war and killing ever really made sense, they could all at least try to comprehend the 'why'.

The Eastern-program launch-sites included nine or ten significant facilities, scattered around the globe. Ukraine's KK/F Region-6 was only one, and was a command-center for many of the others. Three more were in the Russian Motherland and Balkans. Sites in Indonesia, sub-continent India, Iran, deep

Africa, South-Asia (Korea and Viet-Nam area peninsula), and a highly secret and unique launch site in the South-Arctic ice-mass. The West knew all of these could launch ships towards Mars. Of course the East had already spent its early capacity to launch ships with Prokov Keeje's flotilla of five heavily armed and supplied vessels, more than a year previously. But more launches were planned, to support Keeje, and then eventually to completely occupy the Mars Base for the East.

"No more enemy launches to Mars," was the mantra at the White House and US Western military. Moon-launches, and Earth-orbit launches were also on the table. It was a historic time: Earth war and conflict, based on space-launch and space-travel capacity, with the ostensible logic of salvation in the face of an approaching destructive cosmic object, the meteor.

So it finally began. Air Force would lead, with heavy bombing of sites, followed by ground-forces. US Western military organized strikes from deep-ocean ships, and with heavy equipment and ground-gear at strategic staging locations. Soldiers, planners, vehicles, and weapons: the Generals and politicians were quite pleased, and yes, blood would be the price. And yes, the East would respond, with the critical avoidance of a nuclear-bomb conflict banished and unlikely due to nuclear-weapons restrictions in 2045-55, when almost all nuclear weapons were eliminated, destroyed, dismantled, finally, at long last harmless. But, "you never know". On the ground at places like KK/F Region-6, the scene was fairly typical, sudden terror and hell from above, they had come, the jet-aircraft attack ships with their missiles and bombs, and then the soldiers, the tanks, more fighting and weapons, more ground-bombs, etc.

Milana, the mouse-like female personal assistant and devoted servant for Rudolph Terchenko, Commander at KK/F, found herself escorted by the driver of small transport with another soldier, across the tarmac and grounds of the deep-woods Ukrainian space-shops and launch-sites, clutching the green suitcase her master needed from his personal dwelling-apartment. "All hell had broken loose", one might have said, they were under attack. By that time, a series of jet-aircraft strikes were unleashing the poison-flames and burning chemicals of their hot bombs and computer-targeted missiles. Shrieking jet engines whined in her ears overhead, first one, then another later, then more. They screamed, then *boom-boom-boom!* The facility gates and outer-security, the hangers and fuel-storage, communications buildings and electrical-power structures, the command-buildings and computer-worker areas, and the towering launch pads including some dormant ships. *Boom-boom-boom!* Smoke and flame, great-balls-of-fire, like the burning stars in the far-off cosmos (but smaller). They assault may have been compared to a military attack on the US space-command center in Houston, Texas, or the one in Florida.

Milana was dazed and in shock. The transport driver wheeled their vehicle ahead, towards the safe-bunker where Terchenko and other leaders were taking refuge. Men were running, sirens were also screaming, guns were firing, other missiles from the base-forces and small-rockets, trucks, loads of gear and soldiers from the base, and more. Like slow-motion, the dance started, then paused and even seemed to stop, yet continued, then started up again. *Boom-boom-boom!* Young Milana gripped the green suitcase. It hardly mattered what was in it, or what it was for, she would never even know. She truly loved Terchenko, in her way. Whatever it was all about, her role was very minor now. She wanted to be safe with him, he represented power and control. The transport vehicle electric engine hummed and ran efficiently, the driver knew the way. The cold Ukrainian breeze tossed Milana's dark hair into her face. Away far beyond the walls of the base, in the woods and forest trees, among the rocky hills and snow, small animals and creatures, goats and rabbits, large and small local birds, moles and badgers or wolverines burrowing in the warm Earth, insects like beetles, ants and roaches, lizards, the whole Earth was somehow aware, the beasts ran and scattered, shaking with the blasts, "the humans are at it again". On the way were the soldiering US ground-forces, their black boots in the snow and dirt, they're grinding wheels and cleat-track treads, on the small roads, or no roads at all, engines and machines. KK/F had been taken by almost complete surprise, though arriving slowly.

"Up ahead now. Exit quickly with Boris when we stop. He will take you underground. Good girl," said the transport driver to Milana, in his thick dialect. She nodded. It was all going to be okay. Somehow.

Similar scenarios played-out at other Eastern space-alliance launch-bases all over the world within the period of a few weeks.

The Mother-God of the indigo Death-Abyss between two worlds drew a breath into an airless infinity. Somehow it was satisfying that Mars, only really a half-sized dirt-globe planet spinning empty and free, ageless, had somehow lived up to its legend, despite every opposite intention for those involved in learning about Mars and her secrets, bringing war. Or was it only a coincidence? Mars was called by the name of the old-Earth Greek-Roman 'god', but it meant nothing except high-sounding conversation for late-Earth university psychiatrists. Others felt that even minor gravity-pull from Mars, Jupiter or Venus and beyond, could affect events on Earth, like the Earth's moon pulled on the ocean tides. God yawned, *"Not again."*

The Penelope, Guy Reisling's transport-cargo space ship, was also now in orbit around Mars. She was 'better now', following the EMP disaster with the Krenika, the mocking voice of Captain Zolotny (then), the two weeks of complete power-loss and near-death critical life-sustain failure, and the crew's brave efforts to restore their passage. It never left Guy's thoughts, how close they had come to death, he felt responsible. He had also lost Rob Cowan, his co-pilot and also his friend, and he felt bad about that too. Psychosis in space was an occupational hazard, and Rob's gene-trigger antibiotics would later be reviewed by program medical doctors for any future use in space. But it was too late, of course. Perhaps Rob would recover, of course he would. Perhaps he had even already found a transport on its way back to Earth from the Molinari space dock, although Guy didn't know of one. In off-hours and sweet sleep, Guy dreamed of Lila, her long legs and hot thighs and lusting, heaving tits full of life's gift, inside, warm-and-squishy, kissing him, so dear, so very dear. Her love sustained him, it was true.

"What's going on down there with the sand-storm, Tom? Anything?" Guy asked his navigator, Tom McGee. Again at the helm or flight deck, the ship was in stable-orbit position with Berle's other forces. The crew of the Penelope was thankful they had only cargo to deliver, not bombs, and not men with guns and weapons, or weapons for themselves to fight with. The communications-tech gear that first-time Mars-voyager Karen Tutturro needed was all they had to deal with. It was still the case, that the Mars Base radio-link to Earth was compromised, un-stable, and unable to transfer large-size data streams successfully, or to maintain long-term contact, 24/7, as was desirable. Guy's cargo was supposed to solve that issue. Only a sandstorm or two and a war might stop him, aside from the usual dangers of space flight. *All in a day's work,* he thought. *Keep a grip, Guy old-boy. It's just a job.*

McGee kept a perpetual grin. Like most of the space-men, there was an odd sense of roller-coaster bravado, a sort of 'true grit' mentality, like a wild ride they would survive if they kept their strength and didn't panic. The rest of the temporary flight-deck team on the Penelope, (Arron the cargo-man, who worked from a wheel-chair and was technically without the use of his legs, along with the life-sustain men, the propulsion men, the communications men) were now on alert, as the ship's orbit and cargo-disembark held their focus. During deep-space passage, things could get very dull and monotonous. Now in-orbit, there was work to do.

"Sandstorm, uh, let's see, sandstorm, sandstorm, oh yeah, here it is, let's see," said McGee. His computer-mapping and surface scans were at least equal to Berle's and the others, along with data and info from the base. "Into the third day, passing over the base region somewhat to the South, not a direct hit. Wind-speed on the surface about 70 or 80 miles-per-hour, pretty serious, considering they got fucking atmosphere anyway. No way we can get your shipment down there in that. We can't even consider a shuttle down with the containers. I'd say, looking at this, you need another 17 or 20 hours for a safe descent. You try to go down in that, no way, Guy. No way. They'll crash."

Peter, the youngish, dark-haired radio-tech and communications monitor, joined their discussion, from his kiosk work-area. "Your radio-gal specialist or whatever has been leaving you messages, Guy. She wants her equipment right away. They need to talk with mama back home the way they want to do it before the Russians blow them to bits, I guess. Maybe just to say good-bye," he said, then laughed at his own joke.

The restored Penelope hummed and buzzed; the flight deck seemed like a Christmas-morning light show of computer-screens and displays, green or red alerts, main-buss controls, monitors and measurements. The crew had the inner running lights dimmed for 'sleep hours', so the cabin-area was semi-dark, rather than the usual bright-neon light. The men floated about as usual, still weightless. Like the others, their bodies had been absent from normal gravity for more than a year. This was overcome for medical purposes by exercise and centrifugal-spin rooms that allowed times for normal-weight workouts, and also diet and other means. The flight deck alone was the size of a rather large McDonald's or Burger-King hamburger-joint on a city-street on Earth, or maybe a gas-station convenience store (without the snacks). On the other decks, other crewmembers, now eight men total, kept watch on the main-engines, life-sustain, power-systems, external ship integrity, pathway-navigation, communications, and many other systems. The cargo itself was lifeless and still, waiting only, like a baby to be born, there in Arron's cargo hold area. They all still wanted to know if the heavy aluminum window bar they had released into space with other ship's trash and junk, towards the passing Krenika and other Russian ships, had done any damage or actually hit the hull of that space-ship (the Saint Peter was the one that actually took the hit), during the EMP blast attack, now months behind them. It was only right they should know. The Penelope, mother's womb, was overall quite large.

"That would be Karen," Guy answered. "Okay, sand-storm. So, okay, who the fuck is winning down there anyway, Tom? Our side, I hope?"

"We're just cargo, Captain Kirk. Berle isn't telling me any hourly on the battle and all that. Probably our side, though, if I'm any judge. We have more men, more ships, more resources and supplies. My opinion for right now, is all," said Tom. "I can get you more specific data if you want."

"And the other two ships from Molinari," added Arron, who seemed to be merely hanging around on the flight deck, at that time, not his usual post. He was working on the descent-shuttle plans and navigation for his load, actually. Arron was a large-jawed man with a huge forehead and deep-looking features, somewhat bearish. He had lost the use of his legs as a younger man to a virus, but was a fully active-duty astronaut. Weightlessness suited his disabling infirmity. During the voyage he had grown a huge, fuzzy beard. "Ten days and they'll join us. That will change things, too."

They paused in their talk for a few moments. One could view from Tom's monitors, of a more-or-less real-time view of the surface of the planet, in real-time. It was a primitive telescope-based video image, not very sophisticated, just a reddish-green image, with grid-lines and numbered hatch-marks indicating kilometers and various positions. But it was clear enough, the large mass of sandy wind, like a cloud or storm system, which was exactly what it really was, was concealing much of the usual view.

"I can't imagine what it's like for the guys on the surface in the walker-suits right about now," said Guy.

"We call that, 'hell'," said Tom.

"No shit," said Arron, in agreement. They all took a moment to consider their fate and how they could help their side, and the 'men in hell', far below. Ending the conflict somehow seemed to be an easy first-up idea along those lines.

## CHAPTER 47: Guy On Mars

Spirituality among the space-workers in the time of this crisis was quite broad and 'buffet-style'. During the period of the conflict over the Mars Base, and the battle at the overwhelmingly gigantic Tharsis Montes mountains, which was consuming so much of their time and money, resources and lives, successful views included a multi-level mix of all the Earthbound historic religious paths and disciplines, concerning soul and the endless human voyage. Foremost among those discussions was science-truth and Mother Nature, perhaps because work in space called for very focused and practical minds with most or all such issues 'settled'. Residents on Mars Base and at the Molinari space dock, the pilots and so on, were not restricted or prohibited from various forms. The traditions were many and varied, and the US space-program leaders were committed to what had been most successful in the past---an open-and-honest acceptance and tolerance for all kinds of views. Whatever strengthened and comforted the space-workers was thought to be helpful and useful, but there were obvious guidelines and exclusions. Fanaticism, vulgar and hateful ideas and teachings, subterfuge, destructive views, were rejected as unacceptable because of the need for unity among the workers to get the job done. The majority of the US space-workers had spiritual paths they had grown up with, so it was a matter of policy, for the good of all concerned, and it seemed to work just fine, that positive-successful spirituality was accepted. The Buddhist view, such as that of Mars Base Commander Bojji-Than (and many others), was prevalent and common.

The wisdom here was that despite expansion in knowledge of the wonders of Nature and the curiosity of the human mind, their was no Supreme Being that could be contacted directly or dealt with one-on-one, or known as an omnipotent Person, with characteristics that would explain so many things. So in a very 'real' way, there simply was no 'God' in the space program, although all the wonders of the Universe lay before them in new and exhilarating ways. As usual, a conundrum, a riddle.

Much like Guy Reisling's co-pilot Rob Cowan's adventures on their recent passage with psychosis, in that season, a very real danger for the space-travelers, especially with long-distances to Mars, was the development of delusions and intense dreaming. This had many components that the doctors and counselors and space-program organizers were familiar with. The endless darkness of the Abyss, the absence of gravity, the simple dangers of an airless, ice-cold or intensely hot externality, the confined spaces and lack of expanding horizons and things to see in a natural setting, with no up, no down, the disorientation of directional qualities, the sense of vertigo, the removal of workers from lands and grasses and lands and 'cool-green-hills' of Earth, the movement of the ships at high-speed and yet seeming motionless, these all made up the psychology of space-laborers such that with the wrong combination of personal circumstances, psychosis could manifest in some of the astronauts, and it was a known factor that was screened and cautioned in staff matters with each launch or new worker. Religion didn't help, but on the other hand was not necessarily harmful. Part of what the conflict on Mars at that time had demonstrated, and the great distance-travel of opposing teams, was that deeply religious Islamic and Eastern space-workers in alliance with the Russian and Hindu, had their own difficulty adjusting their beliefs to the space-programs. Health, overall vigor and wellness, and the standard good-looks and athletic sex-appeal of the astronauts all worked together, so that only the best-of-the-best were really going to be able to handle the challenges involved. These were very educated people. And yet, here they all were, with all these viewpoints coming together, at war, fighting and killing, and using all these costly resources, to control or defend the facility built at Snikta-Ridge Volcanic Basin on Mars. Maybe it was understandable, given fears of the approaching meteor back home, and the levels of predicted damage to the home world, so well researched, even apocalyptic.

By allegory, from a more loving or dreamlike point-of-view that some shared, if a Supreme Being really existed in His high-level of intelligence and consciousness of love, billions of miles away from Earth and the Milky Way galaxy, the awareness of those at war on Mars, their minds and opinions, might have been impressed with a sense of love and compassion and sharing, and the reasonable qualities of a choice to cease fighting and killing, and make peace. Highly impractical from a creature point-of-view. Rare indeed in the history of Earth warfare, and now off-planet. Was Mars really any closer to the Big Guy? Did it matter? Whatever strengthened them or sustained them, their strength and their sustenance was their own, and their beliefs, and these applied to the ends and means organized by the powers in charge of the affairs at-hand. "Only meant to be", perhaps the soldiers on the planet surface may have dreamed, as the conflict continued, with the dead even quite so few, being no less painful. One thing that could be said about the world's first war in space was that the casualties were quite limited. The roll call counted down, and none of the names of the dead were greater or lesser or higher or lower, than those of the survivors and leaders. By this time, the US side had a total of 42 deaths, and the invaders had a total of 57, or possibly a few more.

*"I am spirit, I am man, I am space-flight,"* were Guy's *thoughts.* Like a feather in some kind of cosmic wind, or a leaf that fell from a very, very large tree, like a strange dream that had no beginning and seemed to have no end, Guy Reisling, the

transport pilot and commander of the Penelope, found himself drifting down-down-down. Like a star, or bit of magical meteoric gemstone, weightless a long time before entering the Mars-atmosphere, drifting, floating, down and lower still, to the planet surface. He had aged more than a year during this time of the conflict over the base, and was in the prime of his life. He knew what he was about, what was intended on his part for the success of the base-defense and the overall mission. So here he was, drifting, floating-floating, and drifting. Floating in a tin can.

There had been a change of plans. The Penelope was intended to deliver the communications-tech gear needed by radio-specialist Karen Tutturro, which had left Earth aboard the ship, and been arranged by orders, long before the current conflict began. There was a lot of 'stuff' to get down to the surface. It was all in long-distance crates, high-tech boxes and power-supplies and devices, carefully packed and protected. These would replace or repair the high-powered data processors on Mars already. Better-designed, more-functional parts-and-pieces, all very high-tech, to correct the errors made years ago by the base-designers, the decaying and deteriorating protective seals and external casings that were inadequate to the temperatures and sand conditions, and thus had broken down or mal-functioned. Tutturro was also a newbie on Mars, or, from Guy's perspective, a fellow Mars-novice, anyway (he had not spent much time at all down on the planet surface, on typical missions in his five years of space-travel).

"We need ten-degrees left-roll in 2-minutes, Guy," said his shuttle-descent pod helper, Peter Swain, the dark-haired young astronaut assigned to work with him at this point.

The technology and radio-problems weren't very interesting to Guy. It was not a standard situation. As the Captain of the Penelope, he would not have usually piloted the descent shuttle-pod to the surface. But things had changed.

"Mark ship's timer to 120-seconds, star-board flight-thrusters two and four blast at full-charge in two minutes, 10-seconds should hit us ten-degrees left," said Guy.

"Confirmed for thrusters two and four at ten-seconds, full-thrust in two minutes, got it," answered Peter.

When the Penelope was disabled halfway out from Earth, and then later docked at the Molinari, the man who would have piloted the cargo shuttle-pod descent, was injured, apparently a significant neck-bone pain. It wasn't obvious, even to the doctors at Molinari, but the designated shuttle-descent pilot had a problem with his neck, it hurt, and was a hindrance to his performance. For whatever reason, he had significant pain and difficulty, now, months later, in turning his head from side-to-side, visual acuity, distracted from attention. Maybe it had been some of the jolts and rock-and-roll style ship's movement, during the delay caused by the Krenika, or maybe he hurt himself while working in the tight spaces and crawl-holes as they tried to repair the Penelope's electrical and life-support with such urgency. It might even have been the high-level of EMP energy burst that had somehow bothered his nervous system. But it was a legitimate complaint as far as Guy was concerned, after talking with him. Piloting the shuttles down to the surface was hard enough. Now, at this moment in Mars history, they had just cleared a dangerous sandstorm in the landing area. There was also the on-going armed-conflict, men with weapons, guns, small grenades and small rockets, who at any point might decide to shoot down a small opposition transport descent-pod shuttle. To make matters worse, the Mars Base could no longer dependably provide the so-called 'harbor pilots', experienced orbital helper pilots who would guide down the descent-pods or vessels. With the war, or conflict, apparently Bojji-Than, the Mars Base commander, could not provide this safety-service for incoming transports like Guy's. No surprise there.

*Sand storm, guns and bombs, an off-duty pilot dismissed with a mysterious neck-injury, and no harbor-pilot assistance,* Guy thought to himself. *Be a good space-man and do it yourself. This entire mission has been screwed from Day-One. And here I thought I was getting my wings back!*

120-seconds passed almost as a thought; Peter Swain fired the left-side two and four thrusters, ten seconds at full. Floating and drifting, rolling and tumbling through his own life, with a job to do, Guy, and the others back on the main ship. The descent was like those on Earth, a fast burn, slowing into the thin atmosphere from high-speed. The pod was released from the orbiting ship, then orienting to a proper descent approach. The other man with him, Peter, was a younger, novice astronaut. Together they would handle the parachutes and deployment lines, the thrusters, braking into the planets thin surrounding vapors and layers of gaseous, wispy environment. The shuttle pods for the Penelope were much like the others with Berle's armada, and the Russians, about the size of a rail car, circular or ovoid. Friction and speed caused heat; the landings were more like controlled collisions with the surface. Almost all of what was needed for the radio-system repairs had been loaded into just the one shuttle-pod.

"Boot up the rock-n-roll! We're going down!" Guy said with a strange joy in his voice. Peter was able to quickly start-play a classic rock-n-roll tune on an internal sound-system, in this case a Steely Dan song from the late 1970's, called 'Kid Charlemagne'. Somehow appropriate.

Below, the battle was now about 70-days long. The sandstorm had passed in about five days, dwindling in potency to a memory. The men fighting on the sands and small flats and rills or among slanted rock-walls, creeping towards the shiny-steel face-fronts of the base, and the walls and stones of the Tharsis Montes, they became One with their essential life-support walker-suits, their weapons and ammunition, the helmet radios and view-sets, and the priceless oxygen supplies, water-supplies, and batteries. Taken together, on both sides, the sandstorm took five lives, not even associated with violent battle or attacks. Five of the oxygen-igloo positions had fallen, the US soldiers either killed or rousted to retreat. Other parts and pieces of the outer-grounds and facilities were under enemy occupation, giving the Russians a great victory, because they could now regroup and restore power and supplies, or just rest. At least five or six of the foxhole igloo positions for the West, were now useless by virtue of their locations in too advanced placements, in other words, the enemy had moved forward far enough past where they had been placed, that they weren't much use, like fighting a tide that had long passed. There had been three assaults on tow of the main entryway air lock sealed gateways into the base, and also some damage. Small air-leaks had resulted, and the panic inside was almost palpable, even a very small air-leak to the outside could eventually kill everyone inside the base, for lack of air. The overall scene was about ten square-miles, littered with the descent pods, leaning over on their sides, or even half burned and blackened, surrounded by surface-men desperate for the comforts and supplies inside, And then the small groups of soldiers, ten here, 20 there, two or three, and the still-quiet bodies of the fallen, dead where they had been shot or burned, the oxygen tanks and batteries quickly recycled by the living. Some areas had been seriously burned by repeated small chemical bombs and grenades. The electric carts and small transports were used with increasing effectiveness, like rolling raids of Michael Jackson moon-walkers, 20 or 30 Neil Armstrong's and Buzz Aldrin's, each one armed-and-dangerous. For the individual men, it was a nightmare. The suits were not made to handle that kind of stress, they could be very uncomfortable, and although they provided air and cool or protection from the hot-cold surface temperatures, they were no match for even a single bullet. Moisture collected inside the suits in even a short time, sand and grit from the surface would ruin the suits in the joints and boot-hemmings, or gloves, the electrics would fail or falter, and the men would have to eliminate their own body-fluids through the catheter-based internal 'rump kits'. Yet they could not escape their suited-up inner-worlds for even a moment, outside on the surface, or death would be instant. And all the time, they needed to carry out orders arriving by radio links to their helmet sound-systems, as the enemy was shooting back and trying to kill, or destroy base property, or run-and-hide.

From the Command Center at the high-point platform above the base's main structural center, Bojji-Than and his staff-authority viewed the scene and counted the cost. Day-after-day, it had become a vigil and waiting game. The actual fighting was very, very slow, not fast at all, the suits and atmosphere and heat-cold conditions were too demanding. Things seemed to grind on-and-on, and then the reports of the dead. Bojji-Than was watching his small research kingdom and educational outpost of science and explorations become a desolate waste of mindless death and ruin.

"My god," he said. "This is utter insanity. My god, my god."

"Or to put it another way," said Vinces Grant, "My Favorite Martian took a shit that stinks more rotten than Bill Bixby's cancerous left lung."

Another one of Vinces's less-than-well-received jokes failed to cheer the gloom.

"Who is Bill Bixby?" asked Juno Amorrossi, there in the control room where they could view the monitors and computer data-trackers, as things proceeded.

"Oh, a TV star from the mid-Century in the 1900's, in the US," said Grant.

"Not a TV show, not a movie," quipped one of the radio-alert systems dispatchers, in response, dropping a disc recording with recent daily reports onto Juno's desk.

"Say it often," he replied, yet darkly.

By this time, the descent pod piloted by Guy Reisling and the novice Mars-astronaut named Peter Swian, had safely achieved landing status and upright-stability, about two miles from the base, in a somewhat un-contested area where the fighting had not been very intense, because it was not strategic. The situation-monitors could view the tiny electronic speck of energy indicating the pod's radio-beacon signal, and also up-dates from the Penelope, from within the Mars Base control towers and rooms.

"That's the cargo guy," said Grant, reviewing the tracking with Bojji-Than and the others.

"I hope he brought some answers," said Bojji-Than. "I truly do."

## CHAPTER 48: Think Positive, Boss

Guy Reisling, himself something of a Galaxy-Baby, at least to his so-distant lover Lila Meetek, managed to place the cargo-descent shuttle pod almost two miles off-course, when it finally touched down near the Mars Base. It might have been considered a near miss, or a serious error, to land the pod so far from the base. But as any aircraft pilot will tell you, any landing you can walk away from is a success. Two miles on the surface of Mars was a great distance. The pilots typically landed the shuttles much closer, as near as a quarter-mile, for obvious reasons. So, it was a little humiliating, for the cargo pilot who had steered the Penelope more than 40-million miles from Earth to Mars, to screw up bringing his stock of high-tech radio-gear down to the surface. *Maybe that's why we have spaceport shuttle-descent specialists (the space-harbor guides),* Guy thought to himself. *These things are poorly designed.*

Bringing the descent pods down to the surface was a very specific skill. In any case, by the time the large, ovoid, somewhat circular-looking air-craft had come down through the thin atmosphere from the mother-ship in orbit, and the fixed-wing paraglider 'chutes' had deployed, with the thruster brake-rockets to slow and guide, and then glided at a long angle to the low, flat, sandy surface, with it ancient red rocks, both Guy and his co-pilot Peter Swain, were very dismayed to find they were so far from the base.

"Holy shit, Guy," said Peter, a youngish novice astronaut with dark black hair and boyish features. "It will take us two days just to reach the base."

"We blew it royally," Guy replied. Now they were un-buckling their safety harnesses, shutting down the power-systems, and starting the process of securing the pod with external extended hydraulic 'feet' that would push it into a suitable angle or attitude against the surface, because they would often roll or tilt over. The landings were rough. The pod was about the size of a rail car, now rested on the sands of Mars. The two men could feel the familiar pull of planetary gravity, for the first time in just more than a year: less strong than that of Earth, but so much more than the weightlessness of space. The pods usually skidded to a stop along the surface. Mars was amazingly silent, still and quiet, even with all that appeared to be going on with the battle and fighting, the ships, and so on. The surface was virtually void of any real sounds or noises, seeming endlessly empty of any ambient noise at all. The pods had to be stabilized and prepped to unload; otherwise it was very difficult to get in and out. This took some time, and had to be done properly by method. The para-glider fixed wing 'chutes' were now discarded on the sands behind, empty of the wind or atmosphere, serviceable enough as they fell to help the critical descent, thin though the 'air' was. The sands and plains of Mars were very flat, quite suitable for this type of landing. By the time they finally skidded to a stop, for some quarter-mile of landing-runway, with a cloud of dust and sound behind them, the silky plastic wing-forms and cables, gently dropped behind them, like bizarre flower-petals from the sky. And then there they were, perhaps literally in the middle of nowhere, with the corpuscular, greenish-red Martian sky and landscape all around, silent and still. It was something only a space-traveler in 2077 could appreciate, the sense of complete self-dependence and total reliance on knowledge and skill to survive even for a few moments more, and with a very specific and demanding job to complete.

Man was now the alien on Mars, and somehow yet fit right in, as if in the desert regions of the Sahara or California's Death Valley or Mojave, hot, or freezing cold, empty, vast, waterless, with the high-tech pods, the base-structures, the ships launch pads, the men in walker-suits, and now the foxhole-igloos, and manifestations of the battle and struggle, and the 40 years of progress there on the landscape, occupying no more than 10-square miles of this tiny, barren planet, the fourth in the orbit around Earth's sun. Much like wars had always been, the military objective at first review seemed utterly pointless and without any value whatever, certainly unequal to the deaths and immense expenditure of resources. But none of this really mattered. They had a job to do, they had to survive, and they were far from home, with only themselves to depend on. The other immediate task was to prepare the cargo. This was again the high-tech communications gear and replacement parts, needed by radio-specialist Karen Tutturro, on her very first voyage to Mars. As rough-and-ready as the landings could be, the cargo, though in safety containers, could be damaged or dislodged, so it was standard to inspect the cargo for functionality and readiness to move onto the airless surface and to the base itself. Still working inside, they used a checklist quickly for this review, and also estimated the procedure and method needed to reach the base. The Mars-gravity was now causing both of them physical stress. They had long been weightless. Heavy breathing and huffing, sweats, even a small labor was at first strenuous, their hearts racing with the 'change'.

"What do we have from the Mars Base tracking, Pete?" Guy said, as they worked.

"Last contact was three hours ago, signal indicated they had our position and descent," Pete said.

"Gee, is that all?" Guy joked. "I thought we'd get a marching band or parade to welcome us."

Pete laughed. He enjoyed Guy's jovial attitude and positive sarcasm. It was something they all worked at, as the labors of space-travel proceeded. "I doubt it," Pete replied. "Not with all that's going on. They have our position and they know the cargo on-board and applications they want. With the Russian-Islamic fighters, and a weeklong sandstorm, what a mess. We're small potatoes. Our next scheduled contact is in twenty minutes, according to the radio-log. I'm on it."

"No one is going to be happy that we put this thing down two miles out, that's for sure," Guy said. They continued working. The cargo was in large, hard-plastic containers, sealed with foam and metal bindings. Each container was labeled for contents, and indeed they had been jostled around quite a bit during the landing. There were five main containers, each about the size of a Dempsy-Dumpster garbage bin on Earth, such as found behind some food-market back home. There were three smaller containers, no larger than a traditional street-corner mailbox or an ATM machine. As the 'Penelope' had been in orbit around Mars with the other ships, Guy's teams had communicated directly with the base, and channeled Tutturro's information, as far as the specific items she needed. Her job was still to repair the high-tech data-compression processors for the Mars radio system, the external antennas, and signal-computers, moving large data-streaming files at a high-speed efficiency back-and-forth to the receivers on Earth. The equipment she needed really was very important. There were redundant units that remained high above, still in the cargo bays of the Penelope, carefully guarded by Arron, their cargo chief. So it was no small matter as the battle at Tharsis Montes continued, for Guy and his second Peter, to move the containers into the Mars Base as quickly as possible. It wasn't just so the Mars-residents could call home and say hello to their families. And it wasn't just so the Earth transmitters could send entertainment, news-updates, music-files, or current hit movies, for the Martians to enjoy. These were important too, in a 'normal' year, here on Mars. Without his realizing it, Guy's efforts on the radio equipment, might even play a critical role in the outcome of the battle to control the base. Winton Berle, the Commander of the US Defense Armada, and planners back on Earth, understood why this was the case.

But until it all came together, there was no use counting chickens yet un-hatched.

Twenty-minutes passed, and the signal came across from the base. There was the familiar alert-buzzer within the cabin of the descent shuttle pod where Guy and Peter waited: *beep-beep-beep*. "Should I answer it, Guy? What if they want something?" said Peter jokingly. "I have a serious sleep-debt going on here, personally. This gravity is killing me."

"You can rest in suspended animation on the way home, Pete. Go ahead and pick up the call," said Guy.

It was another hour before they were in their own walker-suits, and had exited onto the Martian surface. The in-coming communication was no surprise. The Mars Base radio dispatch was from the base control-center, where Base Commander Bojji-Than, security man Juno Amorrossi, launch-manager Matt Currison Von-Templar, and explorer-researcher Vinces Grant, and the others, were trying to manage the crisis. A team with two electric carts was being sent out to meet them. They would all make their way back to the base together, dealing with whatever Russian-Islamic fighters who might try to stop them, as best they could. Certainly, the invaders, or Eastern takeover fighters and leaders, had spotted Guy's shuttle. As empty and lifeless as Mars was, any ship or vessel was fairly easy to track with radar and other means. So of course the US teams sent out to pick up Guy and Peter and their cargo, would be heavily armed.

The best that could be said for the Mars Base defense forces was that they were 'hanging on'. The Russian-Islamics with their ships and men, their own descent shuttles, their weapons and equipment, were not giving up by any means. The peculiar strategy of the battle-scenario had been invented more by Nature and fate, such that the Eastern space-travel teams could not really successfully surrender or end the battle. The simple reason was that they were all of them more than 40-million miles from home. Even on a personal level, the Russian battalion commander Prokov Keeje, and his generals and battle planners, and the soldiers, plotting to use their EMP projectors as weapons, were urgently motivated to gain the resources, protection, and sustainability of the base itself. They may have no longer cared at all about their original mission and their commanders back on Earth, such as Rudolph Terchenko, there in the dark woods of the Ukraine, at the KK-F/Region Six base, and the goals of the Eastern-Alliance space-program, or the fate of the Earth-residents, with the dangerous approach of the meteor, where it all began. Like most wars and battlefields, with the bloody hells and destruction, and killings, it hardly mattered, at-the-scene. By the time Guy's cargo reached the Mars-surface, the enemy had not given up at all, and it appeared they would not do so under any circumstances this side of death.

From the base Command Center, the progress review of the battle looked pretty pathetic on both sides. Numerous deaths, slightly more to the Russians; major destruction of some of the oxygen-igloo defense posts; at least one Mars Base launch-pad worth billions of dollars, seriously damaged; systems-breach on out-laying fuel tanks, and some undefended roads and entry-ways, and external infra-structure, with the enemy able to establish secure positions; near-breach of base oxygen seals, and at least one of the larger defense positions; an assault of horrific venom at a base entry-way airlock, used for agricultural and maintenance carts and machines going in and out of the base, over 15-years or so. The invaders seemed to feel that this air lock gate more vulnerable, and indeed it was. If the airlock had been breached, those within knew they could die for lack of air, the sudden rush of internal pressure equalizing to the outside, so the survival suits for those inside the base were prepared and readied. With heroic effort and deaths, the airlock gate-entry was secured for the US, and the enemy pushed back to a powerless retreat. They had lost many electric carts, other transports, and many walker suits. All of the losses were irreplaceable, except over many years. If there were to be repairs at all, new equipment, gear, suits, machines, devices, computers, electrical computers, seals and airlocks or gateways, and so on, would have to arrive from Earth at tremendous expense and effort. The sandstorm had also done damage. It wasn't the biggest they had ever suffered. But the base and all the equipment, and the men outside, were pelted with flying Martian dust and sand lifted airborne by high-speed winds, for about a week. It was amazing, with as little atmosphere as there was. Meanwhile, the battles and skirmishes, the surface-soldiers on both sides, continued the slow dance of attack and counter-attack, rocket-fire, gunfire, damage, death: war, the namesake of Mars. With the fragile walker-suits they had to use, the lack of surface air they could actually breathe, the soldiers could only move very tediously from battle to battle, or new positions.

In the skies above Mars, unseen from below, the ships in orbit also orchestrated attack and counter-attack. This too was 'different'. Berle, now with nine ships, was also 'holding on'. There was so little he could do about what was happening below, he was truly not in control of the battle at all, there was just no way. The threat of the invader ships applying the same EMP projections of tremendous electrical-radio wave energy in short-duration, which had disabled the Penelope months prior in deep-space transit, was not un-considered. The ships in their orbital paths were greatly distant from each other. The EMP's were only destructive at fairly close range. So they also danced, there beyond the reach of the surface gravity, circling the planet.

Guy Reisling and his friend Peter Swain now stood on the Martian surface, and had a chance to look around. Of course, it was a stunning experience, so rare. Bright and clear in an odd light, no visible sign of any war or death nearby, as if pristine, totally natural, though hostile to man's survival, by no particular cause or intent of anything or anyone. Flat, wide open, then far beyond, towering far higher than they had imagined, the humongous Tharsis Montes, red, stony, rocky, and higher than high, reaching upwards like giant gods.

"They say those mountains are five times higher than Mount Everest in the Himalayas back home," said Peter. They could communicate through the standard walker-suit inter-radio sets within their helmets. Once on the surface in the suits, they could not take them off again until they somehow found an oxygen environment.

"Guess they're right," said Guy. For once, he had little to say. Hours had passed by now. Within another short time period, they could see the electric carts, far off, like the strangest kind of distant ants, or bizarre taxi cabs, trailing Mars dust, moving at about 15-miles per hour.

"There," said Peter. "That's them. They made it."

He pointed, and Guy could see it too.

"Unless it's the enemy," Guy quipped coldly.

"Try to think positive, boss," said Peter.

## CHAPTER 49: 'Help Her!'

So as Guy, the space-transport pilot, with his Second, Peter Swain, no more than a youthful and enthusiastic third time Mars transport-ship specialist, recruited by caveat to work the descent-shuttle pod to the Martian surface along with his boss, given the physical problems of the regular descent-shuttle pilot (a neck injury)---as these two finally welcomed the crews from the Snikta base, arriving in two of the electric surface crawler-transports, large enough to move out again in a couple of hours with the valued cargo (communications gear), and then leaving the descent pod abandoned there, in the Martian dirt and cold, maneuvering back towards the two-miles distant US Mars Base, wary of course for the hostile forces of the 'invaders' (in this case, Invaders From Earth, the Russian-Islamic Ukrainian Hindu space-program assault teams, intending yet to overpower and take control of the US base, with no sign of letting up or surrender, at this point some two months or so into the struggle), as all this was going on, the limitless sweep of the eternal indigo stillness between the two worlds, seemed somehow to bend back towards Earth, the home of them all, leaning like a mind, a knowingness of deep quiet, back and back, until the same awareness arrived gently there in the dark woods and snowy wilderness, among the hills and mossy oak-trees, and the frosty breath of marshes, at the KK/F Region-Six space-launch command center.

Captain Reisling found himself riding in style in the Mars electric-cart surface-transport. All the men were in the air-sealed surface-suits. Six men in all, four with weapons. The Mars landscape could not have been more bleak, rocks and dust, sand, small rising shales, far beyond the vast giant mountains of sheer impenetrable reddish-brown or silvery stone, higher than they could even view from the surface. If there was a breeze or thin ripple of faintest atmosphere, they could not feel it; if there was warmth from the distant Sun, they could not enjoy it. If there was a sound, they could not hear it. The suits were not comfortable, but better than to die instantly without one. The little cart hauling Captain Reisling and his load hummed with its dynamo, the wheels bounced across the dirt, top-speed, 15-miles per hour. "Mission accomplished," Guy thought to himself. "I wish I was making love to Lila again in zero-gravity. That was fun." For both Guy and Peter, having set foot on Mars now only half a day's time behind them, the adjustment to the lesser gravity was still hard on their bones and muscles and their hearts pumping life-sustaining oxygenated blood, and they could tell the months in deep space had left them weak. Weak, when they most needed strength. A thin wisp of dusty sand and ancient Martian dirt rose up behind the electric cart.

How similar were his thoughts to those of Commander Rudolph Terchenko, the KK/F Region-6 leader and head-honcho in charge of central strategy for the Eastern Space Alliance in that year. Maybe it was true, that all men of adventurous lives and powerful position or demanding circumstance, were almost entirely motivated by the lovemaking they could find along the way, and the women they adored. As cold and dark as the woods of the upper-Balkans and mountainous hidden Russian regions were, Terchenko, even in the hours of disaster now for his base-command, dreamed idly of the touch and delightful romance of his devoted assistant, young, mousey Milana, the 30 year old Russian student. "My porcelain doll, my pink-knees and nipples, my cold treat, my secret," he thought. Yes, there had been romance and lovemaking. He the older, the master and Man-in-Charge, with so many wanting from him the right choices and decisions to save the world, and steal away a home for them on another---men at his command and billions of dollars in wealth to control and send ships and machines to the far reaches of space, that is, Mars. And she, somewhat the simpleton, though intelligent to be sure---smart enough to serve him unwavering, wise enough to tend his personality or moods, simple in her ways to understand she must always 'stay out of the way', and keep their secrets, including their romance. Not that any of the commanders and generals and space-technicians or Russian-Islamic astronauts were shy at all about their women. Indeed, it was for them a boast, for the most part. Space-program women were athletic and highly trained and very liberal in affections or sexual ways, for health reasons, of course. Terchenko disliked the Islamic-Hindustan types; they were more demanding in their cultural disciplines. But Milana, well, there had been hours of peace and enjoyment, chatting foolishly and drinking warm brandy, hot-steamy bathhouses in privacy, and more. He hated to admit that he loved her. It diminished his power and presence among the others, when she once again became the obedient servant, fetching him books, connecting phone-calls, researching matters by computer, gathering information from classified sources when it was too tedious for him to attend to personally, getting him meals or drinks, rubbing his back and neck. Sex, or romance, between them apart from all those mighty days of the crisis, was a similar task. But Terchenko was never really heartless.

Near or far, whether on Mars, or in deep space, or in the launch commands of the Russian wilderness in their secret fortresses on Earth, it was for this, the loving moments, the tender hours of embrace, the kisses and forever-loves, that they did all their adventuring and deeds. No man ever forgets a lover. And no distance or national preference or religion, no war, no gods or planetary travels, would truly separate.

"I wish I was back on Molinari making love to Lila in zero gravity. What a kick!" thought Guy Reisling, there on Mars.

"To hell with all this war and space shit", thought Terchenko. "I want to be with her now back in the hotel, in the hot-pools, as naked as she was born and me as well!!"

Within a period of about 40 days, the US and the West had managed to cripple the Eastern space-alliance launch capacity, just as they had planned. It was an important decision. By disabling the Eastern ability to launch more ships to Mars, the Mars Base and the battle on Mars, would at least have some kind of end-point or exit from conflict. Otherwise, whatever eventually happened with the meteor (now disputed and in-doubt for an actual collision with Earth, more thought to become a very thrilling near miss), the Russian-Islamic would simply launch more ships, and more again, year after year, until either the base on Mars was won to their control, or the base was destroyed, or they ran out of ships. So the West was quite pleased with the well planned, shock-and-awe, multiple strike-force, air and ground hits on the Eastern space-program bases. And it worked, at least for the short term. It did however start a 'small war' back home, hemisphere vs. hemisphere.

Ten or twelve launch sites were targeted. Each was able to send manned ships into space, to the moon, and to Mars. These were the prize of their space-programs, in places like Indonesia, India sub-continent, Ukraine and Russian sub-states, far North, one in the sub-Artic ice-mass, and Hindustan states, all secret and hidden, for their part, due to military concerns. In the West, the launch sites were more public, like Vandenberg, though not entirely. All such launch facilities, however, were vastly expensive and costly in terms of simple wealth: space ships were not cheap, and neither the fuel, man-power, training, launch pads, support technologies, command centers, education and technology, etc. The space-programs created jobs, an entire industry, but they were also a power unto themselves. Much like earlier at Region-Six, the Western forces massed and assembled without warning, and then moved in by air, and later by ground-troops. After five weeks, it was over. Almost all of the Eastern bases were seriously wounded for any new launches. Of course they defended themselves, and of course lives were lost, it was unavoidable. The main strategy for the West was simply to blow-to-bits the important machinery that made the launches possible, as this was done by air attacks and bombs. So the towering launch pads were soon aflame, or fallen like tin towers of majestic technology, burnt and dead, blackened with soot and ashes, the chemical bombs burning hotter than the launch-rocket engines that would otherwise have lifted new ships into orbit. Obliterating these types of structures and more (the command centers, support structures and infra-structure, vehicles and housing, computer-centers and communications)---it didn't take long. Jet aircraft attack ships flew back and forth over the longer runs like hellish birds of glistening metallic hue, releasing their burdens, like eggs of death. Below, on the ground, in the dusty dry hills of the burning deserts, or the cold dark woods to the North, on roads and hidden paths, US teams of soldiers approached their gates, and there was yet more fighting. The East was furious in their public discourse, the accusations flew back-to-back, at the World Council, and other global leadership groups. And of course each side would next drag out the specter of the atomic bomb, as a last resort, escalating the matter to a critical juncture. The media, the global press, and the public, all joined in the fray and frenzy, there were protests and even riots. Things looked bleak, and only the uneducated or uninformed could truly forget that a huge asteroid, calculated for speed and trajectory by the finest minds on the planet, was now less than a year away, the first of that kind in perhaps millions of years. Once again the end had come, or soon was upon them all.

Terchenko and other command-center big shots had found refuge in one of the safety-bunkers at KK/F-Region Six. Milana, the Russian schoolgirl with the porcelain complexion and mousey brown hair, found herself among them, and was very glad of it. Her fate without Rudolph was unsure at best, probably to die on some frozen street-corner in Saint Petersberg selling cheap vodka to passing soldiers. The assault on the base had been horrific. Many dead, and much of the base non-functional. There in the safety of the protected underground bunker, Terchenko and his bodyguards and other commanders, and a few soldiers, could only wait. They didn't really know, there had been all the usual balls-of-fire and sonic booms of thunderous explosions, above, unseen. Gunfire, some follow-up on jet-aircraft tracking, and a certain amount of defense, and buildings on fire. It happened very fast at KK/F-Region Six, and it wasn't truly over. Terchenko and his gang (mostly the core-leadership at the base, a few soldiers and body-guards) had scooted quickly to safety, as leaders will do. The centralized government forces hadn't even arrived, in terms of standard warfare, although there was some back up, promises to defend-retaliate.

Terchenko was again on his high-powered cell-phone. "We are lost, the base is destroyed, and there is no power. Yes! We are in the bunker! Consult with Mumbai, they have details. Of course I know about the other bases. Yes, yes, sir, of course, yes, you are right of course. Three days. No real information. Yes---"

Suddenly there was an unfamiliar sound, as the entry-way to their room (a sort of large conference area, with gear and ante-rooms where they could sleep or rest), was under assault. A loud banging---bam!-bam!-bam!

"Commander! Attack on us now! Quickly!" One of the bodyguards realized what was happening. He grabbed his rifle, a repeating high-caliber type. Two other soldiers there with them also leaped to action, and grabbed weapons.

Gunfire, rapidly spewing metal against the doorway. Shouts and commands from without. "You inside! Prepare to surrender! US Marines! You will not be harmed! Lay down your weapons!"

More pounding against the door---bam! bam! bam! They seemed to also have a crowbar, or winch, some kind of machine.

It was a thickly protected door with reinforced steel, designed to resist burning bombs and other high-powered threats. Terchenko's phone-call now was meaningless; he threw the phone down and grabbed a pistol. The door to the room flew open, and then it was 'here come the Marines." The KK/F base-soldiers immediately began to fire their weapons, and for about five minutes, the room was filled with the terror of gunfire. The Marines backed out to positions at the door, as the soldiers inside tried to kill them. Smoke, flame, rapid gunfire--

"Oh my god!!" Milana cried out, as a stray bullet caught her in the neck. She gurgled wretchedly, grabbing her slender throat, falling over. One of the other men inside was hit as well. Then Milana lay on the cold floor, bleeding, choking, gurgling, and staring at the ceiling. Soon the gunplay quieted down. The room was filled with gun-smoke, papers and file-cabinets now toppled, books flying around, computers ruined by bullets.

"Enough! Enough! Stop!" cried Terchenko in his booming voice. "Damn you Americans! You killed her! Damn you to hell!!"

A strange silence.

"Help her!" Terchenko screamed.

Within the hour they were all taken as prisoners, and Milana passed into the Next World. Rudolph Terchenko could not find tears, grim and saddened, bitter. Those same Marines had specific instructions regarding the computer safety-codes that all the space-programs used with their ships (these could remotely open hatches and airlocks in the event the men inside their ships somehow lost consciousness). It was a standard safety-measure. The codes were a computer-generated signal, specific to each ship or group of ships, and operated from a high-powered transmitter. Winton Berle, the aging astronaut running his war back on Mars, fully understood the value of those codes, and their use. So this group of Marines, and those working the US assaults on other Eastern space-launch bases, had their orders. As the 'clean-up' at KK/F-Region Six was completed at last, the computer-unit and codes, along with a relay device and operating instructions, for ships launched from that base, were found in Terchenko's green suitcase, which Milana had retrieved for him, as the assault had started three days ago.

Now, if Guy's communication tech-gear cargo was successfully installed at the Mars Base, and the coded-signal safety shutdown data could be uploaded to Mars from such as Vandenberg, Berle, and others, felt they could conclude the war on Mars quickly, and without numerous deaths.

CHAPTER 50: Surface Attack

Meanwhile, back on Mars, things proceeded as one might expect. No one can see the future, or know events before they happen. But there in the year 2080-81, Mars-Earth transport space-pilot Guy Reisling, may have dreamed or desired, or wanted or wished, for clairvoyant powers of his mental faculties, that would have allowed him to predict events. *Things would be so much easier if I was God, instead of merely human,* Guy thought to himself. Sadly perhaps, because Guy was a fine man, now in his mid-years, now just over 40 years old or so, healthy and good-looking, full of joy and enthusiasm, and also quite strong and serious and also very skilled. But almost anyone might have had difficulty with their task, there on Mars. And like all men, facing the dangers of Mother Nature, even extreme dangers, such as the horrors of the vacuum and void of space, and the desolate and inhospitable climate of Mars, Guy, his co-pilots and space-workers, and also the men operating the Mars Base, were really very frail, and liable to die or be injured or become incapacitated, at almost any time, while working there.

They thought it work and labor, but it was much more a struggle for survival, and a conquest for dominance, and a conflict with the invading Russian-Islamic warriors and their ships and machines. The previous labor and work effort had been education, science and research, to the benefit of all of them, and the US Western space programs, and indeed the entire Earth population in many ways. But suddenly, or even over five or ten years, that labor had become war, struggle, and killing and destroying. So, yeah, a lot of work for Guy's cargo team to transport the communications-technology gear, in the crates loaded into the tediously slow electric carts, moving now over the flat sandy shale and sloping rills and small dips or risings, some two miles, from the spot where Guy had erroneously landed the descent pod, from the orbiting Penelope, their home for more than a year in their motion from the planet Earth across airless and freezing nothingness, to visit Mars.

The sandstorm had now passed and it was now a relatively pleasant day on Mars, as the Martian world turned beneath the local stellar object, and it's thin, barely warming rays, at such distance. All the men in the two electric carts, who had come to pick up Guy and his Second-Pilot, Peter, were wearing the essential surface-walker suits. The packed and ID'd crates were in each of the two carts, but even riding in the carts, the men would be deprived of protection from heat and cold, and breath, so the suits were needed, even while riding along, including of course Guy and Peter. They weren't singing cheerful songs or happily humming tunes with lyrics about their victories, and it was a somber and serious hour there among them. The battle for control of the base had been going on far too long, and they knew that if things didn't reach some kind of a conclusion soon, that the entire affair could easily collapse, with many deaths and losses. So they had their weapons as they moved forward, intent on their somewhat minor assignment, to get the communications gear two miles or so from the now lifeless descent pod, into the safety of the Mars Base. In other circumstances, it wouldn't have been considered difficult. Any work in space or on other planets, as the leadership had learned, was always dangerous and called for very specialized knowledge and skill. They had all done this kind of work before.

Guy's Second, Peter, was the least experienced among them, there in the carts with the cargo gear on the way to the base, like bizarre allegories of the Oz story, entering a high-tech Emerald City. But, now, the sadness and sorrow of cruel enemies to protect themselves from, and even fight back or kill, to save their own lives, so they could someday return to enjoy the cool green hills of Earth, if the colossal meteor-strike that God seemed to have planned, did not destroy the Earth after all. They reminded themselves from time-to-time exactly what seemed to be really going on and what their motivations were, or were supposed to be. Cheerful enthusiasm, stress and depression, was always important with the space-workers. It was depressing, however, without a doubt. Because even after talking it out hour-after-hour, and analyzing data and information, all that any of them could say with any certainty, was that they were fighting and working and struggling and dying, such that if and when a very large meteor significantly altered the livability of the home-world with catastrophic-Apocalyptic levels of destruction, that the base on Mars would persist in some sort of US-available functionality as a resource for whatever survivors remained. In their thoughts without a doubt, the souls of the billions of lives back on Earth, men, women and children, who wanted to be happy, cried out to be rescued from certain death, and a Midnight End-of-the-World scenario. US control and preservation of the Mars Base would in no way save those lives and loves on Earth, from a gigantic natural disaster, were it to take place. So, it was depressing, and made the labor tedious and hardship seeming to be quite futile.

The visiting meteor-strike, Asteroid U2753b, had only grown in their minds and group imagination, there floating through space from the direction of the Sun, closer and closer and closer, with all their telescopes and radar and tracking devices, still telling them it was set in its path to intersect directly with the planet Earth, even in a very short time, now, as time is measured by space-travelers. Was a direct hit really inevitable, as the astronomers, mathematicians, and astro-navigators, had assured, with their plots and charts? Had the science-teams and space-workers and military back on Earth, found any way at all to seriously deflect the meteor, about half the size of Texas, from its path? What kind of preparations had been made for the survival of a few, or many, or millions, or one or two, or cities and lands? What kind of panic was going on back home, the men there in Mars wondered, with their families and loved ones, still supposedly safe in real houses with real air and real sunshine and real food, and not required to wear space-suits, just to take a walk around? Were there riots on the streets of large cities back on Earth? They didn't know. Had war and conflict broken out on a major scale? They didn't know. Had the Eastern forces and the Western forces even resorted to the nuclear bomb, in anticipation of the meteor-strike, for whatever reason? They didn't know that either. All these questions and the painful realities of global or inter-continental and inter-planetary politics, had answers and corresponding facts and truths that someone, somewhere, surely knew or kept track of. But for Guy Reisling and his Second-Pilot, and the soldiers riding along with him in the two electric carts on Mars, leaving behind little tracks in the ancient sand, humming with their electric engines, dust trailing behind, they really had no idea. They couldn't keep track of all that information, and still do their specific tasks and jobs.

The two electric carts rolled along at about 15-miles per hour, top speed, with strong wheels and dynamo-battery motors, designed for just that purpose, turning in the sand. Somehow it dawned on Guy and the others, that even if they successfully delivered and installed the communications gear and could communicate more successfully with the planet Earth, at a higher-level of data-transfer, all they might be able to learn from whatever information was shared, was news of more disaster, more ruin, and even complete or absolute failure. *Yeah, sure,* Guy thought to himself. *I wish I was God. I read those books. I guess if I was God I'd save the world.*

By now a good way into the second mile towards the Tharsis Montes Snikta base, they had thus far not encountered any hostility or enemy troops, or soldiers in the walker-suits. They passed more than one of the hastily prepared oxygen-igloo foxholes, like the Rhinoceros Hut, that the US Mars Base had built all around the exterior of the base, for the fighting men to take refuge in during the battles ahead, at that time. Along the way there was even a certain amount of line-of-sight friendly contact with their own side. Men from the outposts around the base would spot them and wave or fire weapons, from a distance. There was no road here, but only a route or pathway that they knew, and by now they could see the base, a mile away or so, an impressive array of structures and buildings and air lock gates and small roads, communications towers to one side, and far off the launch pads, one of them now burnt and nearly destroyed by a fierce battle. The humongous Tharsis Montes Mountains had not moved. Or, might have been said to rest there in perpetual motion, like the planet itself, they there as mountains five times higher than the highest mountains on Earth. The view was hard to describe, the rocky, stony cliffs rising and rising, so tall, so high, and far distant, then still further off, only above and upward, endless to gaze upon. From almost anywhere in the region, the Tharsis Montes behind the area where the base was built was more than just tall. It was like a home in Tibet, that a visiting foreigner who had never been to the Himalayas, thought, or was heard to comment, saying it was 'in the hills'. Almost any description or words were an understatement. So it was not true at all there was nothing to see, or that Mars was quite so dull, there, anyway, or only rocks and sand. On the other hand, the Tharsis Montes had no life or vegetation on them at all, more like a gigantic wall, than anything else, than 'scenic'. Most all of them could only ignore what they saw then, and dismissed from gazing endlessly to dream of Mars, while they went about their lives and work at the base, a year or six months or more, at a time, or assigned schedule, and then home by ships like Guy's. They were not sightseeing, but may have wished they were, on such a grand voyage.

They were now about 30-minutes from the gate they intended to enter into the base. One of the Mars Base regulars, in the front cart, now seemed agitated and alert, pointing off to the left. "Look at that one," he said. "That's one of ours. Those men are in deep shit."

Now they could see more clearly what he meant as the two electric carts moved ahead. They were all connected by the intercom radio sets inside their helmets. Each could hear and monitor what the members of the small group were saying, it was a simple matter to program the radio-units in the suits to do this for any group, especially the men working outside the base or in the battles. There were six men in all, four from the base with weapons, and Guy, along with the descent co-pilot, the young man named Peter Swain. A short middle-distance off was another one of the US Mars oxygen-igloo defense position foxholes. It was burning, apparently from the chemical-weapons used by the invading forces intent on taking over the base. Like the other foxhole contraptions they had devised, this one was dug halfway into the sandy soil, created from spare materials from the base, like large crates, storage units, etc. The route into the base with Guy's cargo had already taken them past one or two of these defense positions. The same 'route' was used for the wounded, fresh soldiers and goods or weapons, ammunition, supplies, etc. This igloo-foxhole they were viewing was a solid football field distant from them, or twice that, and they could see the enemy had completely defeated the position, perhaps even recently. The tiny structure was burning with dark-chemical smoke drifting away on the almost non-existent Martian atmosphere. Bodies of three or four US Mars defense soldiers were scattered around or nearby, in some cases at quite a distance from the lost position. The dead were still in their Mars-surface walker-suits and equipment. Like many wars, the victorious would cannibalize whatever useful gear and machines, devices or weapons, they could take from the wounded or dead. The digital-sound of Guy's voice to the other men, now pitched painfully in their ears.

"Damn," he said. "That's bad. Where the hell are the others? If it's our guys around, great. But, I don't want trouble with this. All we need to so is get these boxes into the base."

The man who was the temporary group captain, waved his gloved hand in the air. The little cart, and the one following behind by a dozen yards, puttered along slowly. They were only now a few minutes from safety. "Go ahead on the dedicated channel to base-dispatch. Find out what's going on and let them know we'll need the airlock gate operational within fifteen minutes. Find out which gate they want us to use," he said, to his partner from the base, another Mars-regular.

"You Mars-guys got this stuff worked out pretty good," Guy joked.

"It just depends," said the team-leader. "I knew some of those guys." Now they could hear the muted sound of the other man using his dedicated channel to reach the Mars Base dispatch monitor for instructions. The surface-war, now over a period of about 70 days or so since the ships arrived, took place very slowly. The men in the air-sealed suits could not move quickly, the carts were very slow, conditions were hazardous anyway, and there was no other way to 'fight' than to ponderously and tediously move groups of men from position to position, and try to establish strongholds or small victories over the 'enemy', and try to push back or eliminate the opposing forces.

In Martian skies above, various ships (eight commanded by Winton Berle from the US, and five commanded by Prokov Keeje), maintained their orbital paths, and could release supplies and more soldiers, to descend to the surface, or even bring some men back up into orbit, which was more difficult. It was a great overall viewpoint for the leaders, so high above it all, and they attempted to secure battle plans of their side. But they were really very ineffective. They were not dropping bombs or missiles, although some bombs were attempted against the men on the surface, by Keeje's command. The ships in orbit were too high above to function like jet-aircraft bombers or attack aircraft back on Earth. They weren't designed that way, and the orbits were as much as a hundred miles above or more, moving 14,000-miles per hour, and the surface targets were quite small. Keeje still did not have the option of destroying the Mars Base, which only meant mission-failure for him, and no refuge or resources in the event they won the battle completely as they wished. 40-million miles from home; the only exit-option for the Russians and Hindu-Ukrainian ally now at Mars, was victory. So the battle in the skies was equally inept, awkward, a wasted effort, and no real progress on either side. The orbiting ships were thousands of miles apart around the planet Mars. The best they could do was communications, tracking and navigating positions from their lofty points-of-view. Many of the ships on either team were often lost from radar screens and mapping for days or hours at a time. Keeje still wanted to defeat Berle's orbiting ships, by using the EMP, or Electro-Magnetic Pulse device, that had disabled Guy's ship so many months before in deep space, nearly ending the lives of all his crew. He might just do it, too.

Even from their command-positions, the battles on the surface looked somewhat like a loathsome group of hostile snails on a sandy, desolate chess-board, which was hard to manage, initiate effective action, or really follow what was going on, or 'who was winning'.

The same was not true down on the surface where Guy and the other men in the small carts, now found that entering the base with their cargo would be challenged by the enemy. Without much warning at all, as the second-officer with Guy's group was trying to reach the base-radio dispatch, there was a sudden flurry of chemical rocket-grenade fire at their transport. First one rocket, then two more, then three to five, came flying from across the sands, directly at the carts. These were small rocket-fire bombs with chemical explosives, probably the same sort that had destroyed the oxygen-igloo outpost foxhole they had seen earlier, with the bodies around. Naturally, the men in the carts began to panic. The radio-chatter from man-to-man was soon a din of shouts, orders, commands and curses.

"That's it, that's it! They've got us! We're under attack! Positions to cover! Positions to cover!" the temporary lead Captain cried out.

"Dammit, where are they? I can't see them! Where? Which way?"

Two of the small missiles had exploded impressively in front of the carts, doing no damage beyond the terror in their thoughts. Two others missed, but flew past them rather near, burning speed with a noiseless sense of high-energy rocketry, and poor aim. Both the lead cart driver and the second cart driver, had to turn quickly and change direction, to avoid enemy-fire and also the burning-smoking chemical puddles of fire and phosphorus, now in front of them.

"Weapons! Weapons ready! Prepare to return fire by line-of-sight!" said the leader. The men huffed and labored inside their walker-suits, with condensing body-sweat and vapor inside their sealed private worlds, the inner-helmet radios chattering with electronic noises and beeps. Each now found a spot around the stalled carts, to save his own life or the others if they could, and the mission.

"Why don't we have any of those fucking rockets?" one of the men said.

About then they could see the opposing squad, there on the sands off to one side in the direction they were moving. Apparently having overtaken the small igloo defense-position they had seen previously, it was a strategy unknown to Guys' team, for the enemy to more-or-less blockade the route into the base, from that small and bloody victory. There were ten men or more, ahead of them, and one large electric cart, this one more like a wagon or bus, brought down earlier by the invaders in their own descent pods from the orbiting ships, and then assembled for use. This was a large number of men for the battlefield to accommodate, usually the groups were moving as squads of three or four or five men at most.

"Because we're peace-loving science-researchers, that's why," another man on Guy's transport joked sourly.

Things were now much more serious there on the surface near the base. The invaders had made successful assaults on one of the launch pads, and almost entered and destroyed one of the airlock gates (at the agriculture deck). Vicious and motivated they were overpowering some of the worn-out and exhausted igloo-foxhole positions. They had more men in the immediate vicinity of the base than before. It had taken them those few hard weeks to establish their positions and strongholds, and they were not giving up. The small arms fire that they were used to, now began in earnest between the men with Guy and his cargo, back-and-forth, and the Russian-Islamic-Hindustan surface forces who trying to stop them from reaching the safety of the Mars Base, with its high walls and other defenses. The weapons were mostly high-caliber automatic rifles with standard cartridge-shell bullets designed to operate on Mars. It was only slightly amusing to some of them that the traditional weapons used on Earth, also functioned on Mars. The guns were altered to function with no oxygen atmosphere needed, to fly to their targets, killing and ending life with as much (or little) excitement as any common liquor store hold-up on Earth.

Guy's cargo-carts and their six men in all were now at full stop. The men took cover. The weapons fired--*pow-pow-pow-pow! Ratta-tatta-tatta--!!* Even by use of their radio-helmets to actually hear anything, the vibrations and noise were frightening. And then more: *pow-pow-pow-pow! Ratta-tatta-tatta--!!*

Some of the enemy bullets and shell-ordinance hit the carts, with thuds and pings and bits of plastic or metal flying off in concentric-circles of debris.

"Radio support back-up! What the hell is with the base!!?? Are they inside there too!!?? What's with dispatch??" the team-leader was saying to his second, the man who was trying to connect to the base by radio.

Now more chemical rockets, and a near miss explosion. Smoke and flame---*whoosh!!!* Two men from Guy's team took their lives at-risk to stand in-view of enemy fire to volley repeated sprays of automatic bullet-fire at the ten men they could still see, only now perhaps an Earthside retail mall parking lot's distance away, guarded for cover by their own electric 'bus' (cart). Apparently the 'bus' held the heavier rocket weapons they were using, very effective. *Pow-pow-pow-pow! Ratta-tatta-tatta--!!*

In this way they could drive them back, at least for a few minutes at a time. Despite himself, and his life as a cargo-pilot, Guy found he was entirely willing to join in with a weapon of his own from the cart. It was a bizarre moment in his life, to be sure.

"Base security says back-up soldiers on the way from another position. ETA five minutes. They want us to hold out. The radio-gear from Earth is critical," the radioman told them now, the six-man cargo transport having now become a life-or-death struggle. They all could listen on their radio-helmets above the sound of their own heaving breath.

"Heck guys, I really didn't mean to be critical or anything," said Guy now, on his own helmet-mic, to the group. "You know---judge not."

Laughter all around then more weapons fire, on both sides. *Pow-pow-pow-pow! Ratta-tatta-tatta--!!*

## CHAPTER 51: My Friend Who Died On Mars

Inside the Mars Base Command Center, the only real comfort or confidence they had, was that the base had not been breached by the invading soldiers or orbiters---yet. The idea was un-thinkable, and all they thought about, as far as preparations and the surface fighting beyond the more secure base walls. All else seemed true chaos. Nothing but orderly simplicity and carefully organized daily routine had sustained life at the Mars Base for more than 15 years. It was the first of its kind, and normally a systematic research and educational facility, albeit high-tech, and even on unforgiving Mars, more-or-less reliably livable, safe, enjoyable, and productive. Now, perhaps because of its value as an off-world resource for Earth-in-Crisis (again), so much of that was lost, or in jeopardy, at-risk, in need of immediate repair, failing functionality, and of course liable to brutal hostility. For base Commander Bojji-Than, the circumstance might be compared to forcefully crushing a beautiful flower, or burning a rare book, or pouring a rare bottle of wine into a toilet. Why? He knew why, but he truly did not. The resources could be shared more easily than they could be taken by forces traveling so very far through space, at such great expense and endangerment to human life, and loss of technological resources that had taken many, many years to create, and learn to apply. There were similar bases on the Earth's Moon, also apparently in disputed control or world authority, as part of the panic concerning the meteor. The success of the Mars Base had come from much of what was learned by creating livable bases on the Moon. Bojji-Than was also depressed. It seemed that no matter what advances Mankind would make in history, society, justice, human rights, or technology and science, that the dark-side of the human heart

427

would so often see advantage again, and strike like a man-serpent. But he had a job to do, and his life was also on the line.

"How can they get inside safely? What can we do right now to get those men past the enemy?" Bojji-Than began his inquiry, with the staff on-hand in the base Command-Center. This now included his best, and though fatigued and full of doubt about the ultimate outcome, determined to survive, save their lives, and save the base, which was their home. Vinces Grant, the expedition-research planner, had become essential to the entire affair, by virtue of his familiarity with the local Mars-mapping systems they used, and his ability to pin-point the surface-scene outside, using telescopes, radar, satellites and tracking. Under other circumstances he would only have applied his skills to sending out teams to collect rock-samples or look for weather-system changes, or search out various features of the terrain, or plan missions to areas where they thought they might find water or ice, or explore meteor-craters, or plan out landing sites. Now those same tracking systems and maps were a view of the battle-field, at the moment, the fate of Guy Reisling's shuttle-cargo lander and his team, and his shipment of communications gear.

The base commander, along with Vinces, Karen Tutturro, Juno Amorrossi, and Matt Currison, gathered around an array of computer screens and plotting-machines, by which they could more-or-less 'see' outside the base. "This is them here," Vinces said with some authority. "On this they look like dots. These dots are the two electric carts from the pilot's descent pod, way over this way, two miles out."

"He really screwed up," said Currison. "Any other pilot would have landed within a quarter-mile, which is standard."

"He's not a descent-pilot. That's the pilot of the inter-planet cruiser itself, Reisling, and one man with him. The descent pilot had some problem, so he did it himself," said Bojji-Than.

"Well, he probably wishes now he hadn't," Grant added dryly. "Six dots, six men, your two newbies from the cargo ship in orbit, and four of ours. And over here, hardly 300 yards away, looks like ten or twelve of the bad guys, and at least one large electric cart, more like a bus from the mass-weight indicator. Doubtless full of weapons. So, two to one odds, man-to-man, twelve to six, not good. They only have about, well, a quarter-mile to the base, they would use Gate-Four, probably, the cargo-gate. But our enemies here have blockaded the way, which seems to be the plan. And of course we have our other soldiers on the surface, in trouble too. It's a mess."

There was a pause. The Command Center was a buzz-and-hum of technicians, and communications gear. The base-radio communications inter-link was now a constant background of voices. "This is Mars Base dispatch, please identify and go ahead."

Now hearing the voice of the man assigned to lead the team that was supposed to bring Guy and his cargo safely back into the base. He was talking to them from within his own surface-walker suit helmet-radio link. He had now been on the surface with only the suit to sustain his life for more than 15-hours. The link was fuzzy and full of blips and buzzing, and they could also hear other noise, like the roar of the guns and sudden loud shocks and jolts, like he was bumping around while speaking to them.

"Mars Base, this is your cargo team leader," the man was saying. "Uh, Houston, we have a problem. We'll never get your gear inside the airlock gate. They have us pinned down, we're way out-numbered, and they have rocket-grenades. We may not make it. Please advise."

The base radio-dispatch operator for that shift was a man this time. His voice wavered. "Hang on, cargo-team leader. We'll get you inside. A team is responding, looking at your situation right this minute. Hang on any way you can. We're with you."

"Don't waste any time, Mars Base. This ain't looking too good here," he replied from a quarter-mile off, there in the ancient dust and sand, huddled with the others by the carts.

The command-team on-hand could hear the conversation on the base-monitoring radio-system.

"They're just going to kill them, Bojji-Than," said Juno, the once-bored security man. "It won't be long. They'll just burn the hell out of them and take the cargo, then hold the position for anyone else trying to get inside."

"Commander," said Karen, the Mars-virgin communications-specialist. "That cargo may not seem important, but it could change the outcome of the battle. The base needs a rock-solid link back to Earth. Things may have changed, or they may have new information. Not to mention any orders, or useful applications. There's talk about safety-codes for ship's system shutdown from enemy launch-sites back home. We need those."

"I'm aware of that, Ms. Tutturro," Bojji-Than said, rather coldly. "Right now I'm more concerned with saving lives. But of course you're right."

Within only a few minutes, an action was initiated.

Outside the base, things hadn't changed much for Guy and his cargo-team. The two electric carts that had picked up him and his Second, two miles back where they left their descent-shuttle pod with its now useless para-glider chutes resting in the sand, were now stopped dead, forming a 'V'-shape, with all six men using them for cover. Just a ways off, were the 12 Russian-Islamic surface-fighters, and a larger electric 'wagon'. They were far enough apart that the only fighting was in fits-and-stops, with one side making an attack, then withdrawing, then repeating, for either team. As long as the Mars Base cargo-team vigilantly fired their own weapons at the enemy-position each time they tried to advance, they could more-or-less hold out. The Russian-Islamic men didn't want to die either, and even a single bullet or grenade would do short work on the fragile air-sealed suits. Additionally in their favor, the deadly chemical-burning rocket-grenades would not reach the Mars-team carts without the other side moving forward almost half the distance, because they had a limited firing range. So, if they advanced, with as little cover as the flat and featureless Martian surface offered, the Mars-team quickly drove them back with their repeating rifles. Only a short way off, the Mars Base seemed like the Enchanted City of Oz; both Guy, and his Second, and any of the invading forces, had been in deep-space transit mode far too long. Reaching the base and finding safety inside was a shared dream, now bringing them to fight-to-the-death, as if a hot-water bath and normal breathing-air and decent food and getting the heck out of the sweaty, smelly, uncomfortable surface-suits, into normal outfits, was worth all the trouble after all.

"Cargo team-leader, this is Mars Base monitoring, are you there?" came the radio-signal.

"I have your signal, base. Go ahead. Give us some good news out here," said the cargo-team leader. It had been about ten minutes since he had asked for help.

"The Gate-Four airlock is ready, and defended with a team of 10 armed men. Another 15 soldiers are heading out from Gate Four in three carts to your position for back up, get you inside ASAP. ETA for your rescue-team is maybe 20-minutes, they had to suit up. Can you hold on?"

Suddenly a huge explosion seemed to fill the atmosphere near the Mars-team carts with white-hot fire. A large chemical explosive had found its mark. *BOOOOOOMMMM!!*

The men scattered and hunkered down or rolled onto the ground. The carts shook and rattled in their rockers and small rubberized-metallic wheels. Shouts and curses. "Damn! Where'd that come from??"

"Man down! Man down!"

"Return fire or there'll be another!!"

The chemical-phosphorus fire wisped through the thin surrounding 'air' near the carts in only seconds, and then vanished, as if consuming itself, and nothing else to burn. Two of the men at the Mars-team carts found their strength, and rose up from behind the carts, un-harmed, with their rifles blazing, at enemies un-seen. *Pow-pow-pow-pow-!!!--ratta-tatta-ratta-tat!!* Long, fierce moments passing.

Now Guy realized in an instant's personal agony of guilt, that the 'man down' was his Second-Pilot, Peter Swain, who had helped him land the cargo descent-pod. The youngish-novice space-voyager had landed hard on his back in the sand with the explosion. His walker-suit seemed to have a rip or tear, just above the right ankle, like a burn, or as if a glob of hot chemicals had landed on it. Guy quickly was with him. In his emergency gear pack was what amounted to little more than a high-tech roll of duct-tape. Yes, duct-tape.

"I got air-loss," Peter was saying, his breathing suddenly labored, the intercom radio link carrying his fearful speech. "Help me, okay??!"

Guy was now urgently wrapping the suit-breach with the high-tech tape, just above the boot. If it could be sealed off quickly enough, the walker-suit's air-tanks and pressure-system pumps would re-fill Peter's only life-sustaining oxygen, maybe long enough for help to arrive, and save his life.

"Damn it, Pete! You're a space-man! You don't NEED to breathe, kid! Didn't they teach you that in training? Don't be a wimp!" Guy was saying, on the helmet-link.

More gunfire, now from the other side, back at them. A hundred rounds in a minute both ways.

"Another hit like that and we're done!" said one of the men.

Peter began to realize, as men will, that his life was over, but Guy wouldn't give up. "No air, no air," he whimpered, then choking. "God---"

"It's re-sealing, then it'll pump up!" said Guy. He looked over Pete's equipment, as the young astronaut lay there in the sand, moaning, and dazed. Then as he inspected the suit, he recognized what amounted to Peter's death-certificate. The fall had also damaged his helmet, with a small seal-breach at the collar, and no airtime left for duct-tape. The man was literally bleeding his air into airlessness, moment by precious moment. A tiny, fragile smile seemed to pierce through to Guy's heart with Peter's last words to his boss' ears, by their radio.

"Don't worry, boss. It's just another way home," he said.

"You and your fucking jokes, Peter. You and your---"

*BOOOOOOOooooooooOOOoooMMM!!!!*

A second chemical-grenade now hit, this one with lesser accuracy. A long hellish moment, stunned silence.

"Cargo-team leader, this is Mars Base, please respond."

The three electric carts sent from the base were now just beyond the enemy-position, and had begun an attack from the base-side, as the radio-monitor tried to re-connect. Fifteen Mars Base soldiers joined in the fight, and slowed down the invaders considerably, even at that moment. At the same time, over a period of some 20-minutes previously, they were all surprised to see a streaming whitish trail of thruster-rocket brake exhaust, high above, seeming out from there by several miles, and a small silver pod-shuttle then releasing its colored para-glider chutes, which opened into the fragile, blue-green-reddish Martian sky at mid-day, or late afternoon. Somehow, one of Winton Berle's ships-in-oribit had dropped a descent-shuttle pod right into the middle of the fight, there a quarter-mile from air lock Gate Four, full of 30 US Mars program soldiers. Although it took another 40-minutes for the shuttle to land, skid to a stop, and unload, the battle was over for the opposition. 50 US Mars Base defenders soon now put a cold-dead stop to the temporary blockade, with all their rocket-bombs and rifles still. Five men had died on the US side, including Guy's crewman Peter Swain, by the time it was over.

In another two hours, Guy's shipment of communications gear, along with himself, the original four men sent to pick him up, and the lifeless body of the crewman Peter, entered the Mars Base safely at the Gate-Four airlock.

By now Guy found himself becoming bitter about things. After admission at Gate-Four, Guy and the other men on the cargo-retrieval team could finally remove their air-suits safely, as the air lock was sealed behind them, and the carts and cargo crates were also brought within, and the gate itself was manned with armed-soldiers. The entire base entryways (air locks) were now guarded at all times, as things proceeded outside, a dismal dance of pointless destruction. There was a period of mourning for poor Peter, and then Guy endured a brief medical exam himself. Within another few hours, he realized pleasantly that he was finally here on Mars, a real planet much like Earth, with real-gravity holding him down, real food and drink, and among friends to share with. It was something of a transitional consciousness-shift for him.

"Welcome home," one of the intake workers, said to him, a busty woman assigned to his meals and medical care for at least 24-hours. Guy was now able to rest, and take a meal with other Mars-workers. The base itself was an amazing new experience, but of course he had no time for enjoyment. 40-million miles, a near-death deep-space ship-to-ship attack (the first of its kind), a desperate in-transit life-support repair job, docking at Molinari and a few hours of joy with his lover Lila, flight-formation and orbital entry to Mars with Berle's Brigade, orbital-descent with his cargo onto Mars, a short surface-trek by electric cart in the walker-suits that nearly ended in a disaster that would have made the entire past year a total waste, the sudden battle-death of a non-military crewman he was personally responsible for, and a Mars-surface gun-battle with invading troops from the Eastern Earthside space-program, astronauts like himself who were also 'lost-in-space', and now here he was. The meal was a tasty clam chowder with hot coffee and bread.

"Yeah, thanks," said Guy. "Welcome home."

## CHAPTER 52: Look at the Radar, Kick

Lynn Rogers-Smith, program manager for the US Mars Western and North-American General space-facilities, found herself in a mess as torrents of heavy rain that delayed her personal transport auto, on the drive from the Los Angeles region, North, towards the Vandenberg US Air-Force Space Port, about 120 miles North of Santa Barbara. Many of the leadership-class, definitely her league and co-equals (though few realized it), enjoyed driving their own vehicles when they could, on open highway trips. A limousine, a bus or helicopter, or private plane, just wasn't the same. To be alone, and let one's mind settle-in about things, a long automobile drive by highway, was private, perfectly safe, empowering, semi-secret, and far more enjoyable. The pressures of her work, and the world situation with 'that damn meteor', melted away for a while, at least as much as she allowed herself. And it was important. The death of Ibrahim Mehudi, a personal friend and confidant, and a true-blue ally when it came to the politics of all the science and research, in contrast to the powerful military interests, it was a severe blow. While they were in Washington, D.C., arguing policy and decisions regarding attacks against the Eastern space-program launch sites, by this time a far done-deal, and mostly successful, though at high-cost to the US reputation as a peaceful nation, and also in lives; at the four-star hotel where Ibrahim was staying with his lovely wife, he suffered a massive heart attack while swimming, a normal exercise routine for him. Mehudi was not so old, thin and vigorous for his age, but with cardiac-health, it was 'just one of those things". And now he was gone, and all he had to offer. It was of course a lonely feeling. Yet, strangely, all his knowledge, his considered opinion and understanding, it was all 'still there', being as much truthful science and inquiry that might never really change, even with the loss of a learned man. She drove on sadly through the California rain.

Her thoughts were drifting and scattered, the coastline and Pacific waters were deep blue, tipped by mild crests. In the rainstorm, there were circles of the water that seem to be speckled with more of the falling torrent than elsewhere, like pans or bowls the rain had found, as wide as a mile, and then beyond, the color of the water seemed to change, more flat or plain-normal on the surface-tension waves and swells, and a person on the highway could see the cloud layer, and slanted streaks of rain, above the bowls or pans in the open sea. Her hydrogen fuel-cell electric automobile sped along through the hills, anonymous.

By now, Lynn Rogers-Smith knew about the fall of KK/F Region-Six, and her opponent, Rudolph Terchenko. So did the Seated-Council at the US White House, and many in the space-program and military. Otherwise, it was a secret. Few could keep track of what was true, or not true, regarding the crisis with the meteor, or the fallout with the space-program. Disinformation was more common than actual useful facts. Most people realized this. Rogers could only wonder what Terchenko was really like. He seemed a gentleman on the radio-link to the Penelope cargo-ship, now almost a year in the past. A gentleman, that is, until they were betrayed, and the Krenika attacked, or was ordered to attack, and the radio-link was suddenly dead. Was he tall? A big man? Or short and stocky, like many Russians? She recalled his rich, deep Russian accent. The Little Texas Cowgirl deep inside Lynn's subconscious was laughing. As if it mattered. She smiled. She obviously could not be more pleased the son-of-a-bitch went down.

But now they had the Eastern space-program remote-entry codes and computer applications and sequence-initiation devices they had wanted so badly, or, at least, the ones used at KK/F

Region-Six. The same computer-stuff Winton Berle had spoken of. And he ought to know. Again, she ran the notion through her own understanding so she could comprehend. Even with Earth's deep-ocean submarines, (at which the use of, the Russians excelled), if they sank, or lost power, or had a crash, the men inside, and the costly material inside, records and machines or data, or even things like uranium-plutonium weapons, were they installed, would be lost forever, with no rescue possible. So, the ship designers wisely created this method, to open the hatches and doors externally, so rescue workers could open those entryways by remote control and get inside. It was the same with the space ships, a safety measure that made sense to everyone, certainly the astronauts. Over many years, the same approach and technology could accomplish other things, beyond just 'popping the hatch'. Some were designed to re-start life-support systems, or re-start engines or power-systems. All well and good, but more paranoid military men feared to recognize that if an enemy came into possession of such remote-safety technology, as they had stolen when the Marines shut down KK/F Region-Six; well, it was obvious, they could cause trouble, anyway. So, they were kept hidden away under lock-and-key. And much of the space-program technology was secret, it had to be.

By the time Lynn arrived at Vandenberg, everything was ready. Orders from the White House had cleared the procedure, and Winton Berle, the US space-fleet Commander, still in orbit around Mars and running his war there, was online. Rogers-Smith was hustled into one of seven Command-Centers, and found the group was impatiently monitoring the radio-link to Berle's ship, the 'Understandable', with empty chatter, back-and-forth to Mars, just to keep the line open. This had been going on a while.

"Here she is," said the University astro-space specialist Willy Atta-Bowman, who had introduced the science and facts about the meteor, at the past meeting-conference, in the Spring of 2077. *What is he doing here? Rogers-Smith thought to herself. Oh, of course. He's replacing Mehudi, that must be it.*

She was escorted to another one of the radio-user stations. "Use this," an operator said. He handed her a headset. Within a few minutes, she could speak to Berle directly.

"I don't want to corrupt this line," Berle said. "You understand. I already know what you know. Let's not discuss that. I don't speak Russian or Arabic, so, you get my meaning."

"I know a bit of Russian," Rogers said. "But yes. I suppose your other sources on Mars can work out the details on all that. It's not important."

"The radio-gear from Earth finally reached the base," Berle said. "It should be ready any time. I can communicate with the base any time for updates, then back to you from my ships, rather than the base itself, for now. When they give the word, you can send over about a gazillion lines of code in hours, flawlessly."

A pause, static and buzz-hum on the radio line. After a few minutes: "Tell me about conditions and the course of the conflict, Commander," Rogers-Smith said. "We have plenty of time. Please go ahead."

Mars was fat and looked dead in the dark sky and stars beyond, far, far away, like a red tennis ball that had been left in the mud too long, and the mud had dried out to white. From a distance, no one would have known all that was going on. The ships in orbit, the battles and men on the surface, the men and women inside the base, the struggle to survive and also the struggle to dominate, and then there was Mother Nature. Even inhabited worlds look mostly dead or lifeless from far off, or by telescope.

Karen Tutturro, the Mars-virgin communications-specialist assigned to make the voyage, prior to the start of all the fighting, now found herself again in a life-sustain Mars walker-suit, outside the base, on the surface. Only this time, she was accompanied by ten armed soldiers, also in life-suits. After Guy Reisling's electronic-tech gear arrived safely inside the base, at the cost of several lives, Karen was quickly told to put her task into effect. She had been planning the work, doing systems analysis, looking at the schematics and diagrams of the radio-systems, for months. The problem had been isolated, and she was confident she understood the job completely: external signal-processors had been installed years ago, with the same kinds of insulating 'rubber hoses' and seals, that were used inside the base. And because of the exotic Mars atmosphere and environment, this was an error; they had corroded and failed, and caused a certain amount of system failure. Un-shielded UV rays, high levels of ambient radiation due to the lack of filtering clouds, and other factors, were to blame. The base had a lot of external long-distance radio-gear, antenna, towers, generators, processors, links to the inside-system, transformers, etc.

"What if they attack?" Karen said to her partner and assistant, the Command-Center life-support lead, Charley Barron, who was also hand-holder-in-chief for the fix-it excursion. There would be several trips out to the radio-arrays and outbuildings before they were finished. Barron was a likely candidate to work with Karen, and seemed able to reassure her, even in all the chaos. He was also good with the Mars surface safety standards, and the radio-tech.

"Well, same as the others, I guess, Karen," he told her. "Don't worry. They are very slow out here. We have ten men.

Let's just concentrate on your job, get this over-with."

They had taken two electric carts out, through the Organic Materials airlock gate, next over from the entryway where the invaders had almost successfully entered the base, about ten days ago. That air lock was under heavy guard, and all the others, now. The assault had been horrifying, most of the regulars within the base were certain of the worst. And the worst was unspeakably horrible to consider. Sudden air-pressure loss throughout the entire facility. Friends and co-workers gasping for air, grabbing air-tanks and suits. Russian-Islamic soldiers in walker-suits, armed with heavy-weapons, stepping over unconscious suffocating bodies in the halls. Then the real fall of Snikta-Ridge base, the real take-over, and whatever the enemy had planned for those who lived and worked here. Now ten days past, but not in their thoughts. Two electric carts, with ten men, Karen, and Charley Barron, and one other tech-specialist with tools needed, rocked and crinkled roughly over the tiny rocks in the sand, just outside the base-walls, nearing the radio-array buildings where she would be working for hours. No attack, no small rockets overhead, no repeating rifle bullets trailing fire-smoke streamers.

Much work now, hours passing. The five orbiting Russian-Islamic ships had now been at Mars for 53 days, give-or-take orbit-entry time for five ships. Commander Berle's total of nine ships (including the 'Penelope', now under his command), had been in stable orbit for half that time, with the two ships lagging behind for supplies from the Molinari space dock now on-site. Enemy dead numbered 80 men, as near as they could figure; the US side had lost 72 men. The Mars Base had suffered serious damage and extensive failures, and also valued gear and essential equipment. Who was winning? It was very clear, now, the enemy could not hold out much longer. Their supplies and energy-systems needed to be restored, and a trip home, back to Earth, for them, looked very un-likely for success, part of the reason they were so desperate. The invading fleet-commander, Prokov Keeje, was also extremely pissed off, having anticipated a much easier total victory, prior to Berle's arrival.

Encoded text-transmissions were the most secure for Mars Base Commander Bojji-Than, to relay information about the radio-gear repair installations, to Berle, in the "Understandable'. Voice-communications might be more easily intercepted by Keeje's teams. Berle had a fairly straightforward plan. Once the Mars Base radios were back-to-prime, Earth-sources would move over the computer-applications programs and special codes, taken by the Marines from Terchenko's green-suitcase, at the KK-F/Region Six combat, where the pretty-pink porcelain youthful beauty Milana had been regrettably shot to death. The transfer would not take more than a few hours, if Karen Tutturro did her job well. Then, working with Berle's teams, and the Mars Base teams, the final stroke they had planned would be a systemic invasive remote shut-down of all the enemy ship's docking port airlocks, and inner safety-power systems, and anything else they could control remotely. Done properly, Berle was fairly certain he could disable all of Keeje's ships. All this was set-up as their plan by text-code messages between Berle and his tech-teams, and the Mars Base. With all five invading space-ships floating powerless and without navigation, unable to support the ground-troops they had released onto the surface, any of the ground-forces on Mars from their side would have no support, no central communication, no systems restoration or oxygen-energy reload, no water re-supply, not even food-source. And no one to tell them how to fight the war to take over the Mars Base. For all purposes, the US Mars forces could consider it a slam-dunk win, at that point. Then a mop-up operation, maybe negotiations or surrender on the Russian side, and then peace.

"What now, Santa Claus?" said Ben Jazreel, the African-ethnic second pilot, on the command-deck of the 'Understandable'. Still in concealed orbit, the ship hummed and buzzed. The command deck was getting a lived-in look, to be sure. After all, they had made the long, long Mars crossing, and were still in null-gravity. It was a standing joke among them, that 'Kick Berle's feet never touch the ground'. Everyone needed a rest, without a doubt. But the ships were steady, maintained daily and restored on schedule as in deep space.

"We wait," Berle said. "Hate to say it. All we can do. Word from the Mars Base is she'll be ready in another few hours. The physical installations have been done, they're running tests, and re-configuring the settings. A transmit is ready on their cue, from Earth. That content will take another few hours to upload, then the teams at the base need to fuck around with it and make sure it has retained integrity, in other words, make sure it still works like it's supposed to. It's not just data. It's a program, an application. The code has to be exact. So, that's ten hours for all that to get done. At that point, I'll give the command, and we'll go ahead with what we had mapped out on the enemy. The key for us, here, and our other ships, is to track and monitor the enemy, so when we're ready, the shutdown codes can be applied as quickly and efficiently as possible. It can all still screw up at any point along the way, of course."

"You know best," his co-pilot said. Ben was a big-jawed man, very dark of his skin, an astronaut many loved, a good friend, as they all were. They were quiet a short while, not talking, just waiting as Berle had said. Ben reviewed the local plotting grids, screens they used with very simple radar-scopes projecting outward around the 'Understandable', in a circle for 500 miles, looking for space-junk or space-rocks, etc. Standard.

At some point, however, Ben seemed alarmed.

"Uh, hey Kick," he said, alerting Berle, who was apparently sleeping in his captain's chair, strapped down comfortably so he wouldn't float away. "Wake up, Kick. Look at this."

Berle was quickly alert, shaking off the light slumber. "What? What you got?"

"Look at the radar," Ben said. So he did. It was indeed a rude awakening then, for Commander Winton Berle.

"What? Tell me. I can't interpret this right now. What is it?"

Suddenly ship's automatic alarms were sounding loudly--- *beep-beep-beep!!*

Lights were flashing. "Right there, on the radar! It's one of theirs! That blinking dot! It's not ours!! Five miles out to the North meridian. One of Keeje's, Kick. Tracking us speed-to-speed."

Berle paused, then rubbed a large hand heavily across his face. "God-dammit. Five miles? He's gonna' EMP us. That fucker. He's gonna' EMP us in about two minutes, just like they did the Penelope. Holy fuck. I knew it."

*His co-pilot was stunned, but only for a moment then he announced their sudden crisis to the entire ship's crew, below or on various decks.* "All systems for energy and life-support to back-up minimum immediately. Secondary support and oxygen to ready-status for all-hands. Save your command files on the navigational!! This is not a drill, people. We are under attack. Please comply at your posts. All hands, please confirm to command deck immediately."

Additional alarms all over the ship, initiated from the command helm: *beep-beep-beep!* The command-deck's inter-ship radio-systems started alive with chatter from the different systems-stations: engines, life-support, navigation and controls, monitoring and communications, cargo and staff, etc. Ben's commands were now announced ship-wide on the 'Understandable's' inter-com, they all could hear him, loud-and-clear. Berle studied the radar quickly. The navigator and radioman on the command deck, also now were suddenly alert.

"EMP? Electro-Pulse? A zillion volts? Goddam I hate this shit!" complained Raza, the radio-operator. "Our systems will go down!"

"Well, we know what happened to the Penelope, the transport ship. Raza, we only have a few minutes here. Rip off an emergency broadcast, one to the base, one to our other ships, while we still have power," Berle ordered him.

"Yes sir, I'm on it," Raza replied, his hands and mind fast and furious at his controls to meet the need. *Beep-beep-beep!!*

Command voices heard: "Main engine shut-down sequence initiating. Emergency-power systems online. Opening internal locks and gates for free passage without power. Suit-room and life-support gear ready to automatic. Orbital tracking stabilizing to thrusters-only, if we're lucky. All tracking and monitoring dropping out, we're going dead on fuel-storage three and four. Silent running, silent running!"

There was a sudden surge in their computers and lights and basic functions, like power had dropped out suddenly. The command deck grew suddenly dark. "There it is, on the radar, Kick, see?" said Ben, the co-pilot, indicating their remaining monitor for external environment tracking. "Huge pulse. Huge. Mother!"

Berle could hardly see it at all, but he trusted Ben. The co-pilot had detected an energy wave, which are not a visible form anyway.

"Here it comes! Brace for impact! Hang on! Save your lives, men! We can do this if we…" Berle was shouting loudly into the ship's inter-com. The whole crew could hear it all.

Then the Understandable was almost instantly awash in an electro-magnetic energy wave equal to many billions of voltage-watts of raw radio-wave electric-energy signal, similar to what happened to the Penelope. The pure power of it now predictably fried all their main electronics, just as Keeje had hoped. Berle's ship went dark.

## CHAPTER 53: Clam Chowder and Wine

*"I don't care if we win or not, or if we take the Mars Base, or even if we even get back home to Earth. I don't care anymore. I just want to hurt them,"*

*----Russian-Islamic Mars Program Commander Prokov Keeje, at the helm of the 'Tolstoy' in orbit around Mars, 2081.*

The 'Understandable', a more superior space-vessel than the Earth had seen in all the general history of aviation and rocketry, had taken the same sort of high-energy radio-wave blast that had crippled Guy Reisling's cargo ship in deep-space transit months ago. The Electro-Magnetic Pulse theoretical weapon device had now been proven twice, in the course of the world's first actual space-war. Internal electric-dynamos within the 'Tolstoy' revved up for hours to charge large-output Tesla-coil type output battery cells, by which an increasing-amplified charge was doubled, and doubled again, and exponentially increased, then released all at once to a certain type of antenna-array outside the ship, mounted to exterior points on the belly of the Tolstoy. Using the same sort of science by which a typical auto combustion-engine 'coil' would amplify a low-charge for just an instant, to a high enough level to spark across a typical 'spark plug', the energy level raced higher and higher, until the modulation-wave was 'dumped' suddenly as radio-wattage, simple magnetic waves, for a brief, short pulse. Billions of voltage-watts passed through the form and shape of Winton Berle's ship in a matter of seconds, frying his electronics and computers in the process. Better perhaps than an explosive missile, the EMP's were easier to target against the enemy, because the signal expanded like a circle, outward, while a missile-bomb could only perform as a very narrowly targeted 'death-arrow'. The only drawback was that the Tolstoy needed to be close enough to her prey for the EMP to maximize effectiveness, and could only use this from a few miles away.

Maybe it was merciful, maybe it wasn't, but Keeje could feel his personal sense of appropriate vengeance swelling in his teeth like a hideous gladness. *Now they will pay, he thought. Their lights are dead, their air-supply has stopped working, they have no way to open doors, or find their oxygen suits! The ship is drifting and may easily spin out-of-control into the thin Mars atmosphere, and then burn up on the way down to a horrid and final crash, if it even reaches the surface at all before incinerating. The have no computer-function to help them re-start essential functions. Berle's men can only survive without help for a few days at most, then they will choke on their own exhaled carbon-dioxide, breathless, heaving, or with any luck puking their worthless guts out as they die. No radio to reach out to their so-called friends, either. This is my hour of destiny, Berle. Good-bye my enemy!!!*

And it was so, much as the Tolstoy's battle-commanders had planned, even for a week or more, tediously tracking the enemy ships from a great distance, setting up the cumbersome and delicate EMP device and systems, and then waiting for the right hour to move in slowly to a targeting distance against Berle. From the Tolstoy's command-deck, a telescope-camera provided a picture of what could be seen of Berle's ship---the Understandable was some three or four miles distant, floating lifeless with no external lights or engine activity, on the Martian horizon, about 100 miles above the surface. Berle's ship had tilted a bit nose-downwards, but only a few degrees. A space-pilot would have known however that for an orbiting spacecraft, it was a serious irregularity. They would only orbit properly at a planed flight more-or-less horizontal to the planet skyline. Keeje's staff could not be more pleased, just then.

"That did it," his second pilot offered. "I think without a doubt the signal wiped out their electronics. That ship is dead, and soon they will be dying too. Good job, Commander."

"Thank you," Keeje said coldly.

A pause. There were ship's controls to trim and tend to, business-as-usual for the helm of their ship, very similar to Berle's ship, except they had power, ship's command and life-sustain. The communications-radio man was alerted to an incoming transmission, with a loud tone. He worked his equipment. He turned towards the Commander.

"Captain," he said. "It is the Saint Peter, 200 miles south of our orbit. They have some sort of intercepted radio-link to the Mars Base. They want to connect you. What shall I tell them, sir?"

"Ask them how?" said Keeje, now curious. "I have another card to play. This may be to our advantage."

The radioman worked on the connection, they could hear him speaking to the other ship, the Saint Peter, (one of the Russian-Islamic Eastern Alliance ships), on his headset, in native Russian. Other communications were in other languages, Hindi, or Arabic, at other times. It seemed to be taking almost an hour, the messages and links going back-and-forth.

Finally, the radio-operator was able to offer a simple solution. "The radio on the Saint Peter linked to an emergency communications from the Mars Base as they attempted to reach their other ship, the one we just disabled—"

"He calls it the 'Understandable'," the second pilot chimed in.

"Perhaps their commander understood our intention a couple of hours ago," said Keeje bitterly. "He may understand nothing at all soon."

Laughter. "Well, the Mars Base used an emergency radio-channel, so there was no encoding or encryption, and our operator on the Saint Peter could pick it up, from his normal monitoring. They never reached the other ship, but maintained a repeating signal, in case they could re-connect. Then the Saint Peter was identified, because he was on the link, and they knew he was there. So, essentially, we can talk to them. That's it." The radioman was proud of the set-up, which was somewhat of an accident.

"All right," Keeje said. "Go ahead with it."

"Ten minutes, maybe twenty minutes, you will speak to the Mars Base commander, let me work on it," the radio-man said in his thick dialect. He was a big-shouldered, dark-looking man, but with a big smile. He seemed to love his work. The minutes passed. Another tone from the radio-board: *Booooooo-o-oo-oop-booo-o-oop!*

"This is US Mars Base Commander Than, on the Mars-surface. Who is on this line, please? Please identify? Waiting for your response now, please."

Keeje by then had grabbed a hand-set, after more or less pulling himself by the hand-grips (rails), skillfully gliding across the deck, to the radio-kiosk desk, weightless, normal for them all. "This is Commander Prokov Keeje, of the Eastern-Alliance space ship Tolstoy, in orbit above the planet. Greetings, Commander Than."

A pause. The line buzzed with static and warped-distortion for a moment. "What has become of the space-ship known as the Understandable, Commander Keeje? Have you killed our teammates by some means? You may as well tell me, sir." Bojji-Than's voice was steady and fearless on the line. His Asian mother tongue was a feminine counterpart to Keeje's hard-edged tone.

"Mars Base, this is war. You kill us, we kill you back."

"Yes, Keeje, of course, But there's no reason to be, uh, un-civilized about things, now is there?" Bojji-Than responded.

A long pause. Keeje was sizing things up, judging his moment. He knew what he was going to do, but tipping his hand right away, maybe was unwise. Nevertheless, just like his opponent, he knew quite certainly that the battle had gone on far too long. There was no turning back, for his teams. Time was costing them. The US-teams had been able to hang on; they had more resources and could rest. Keeje himself had not set foot on a real-gravity planet surface in 16 months or more, and Earth with nine months or a year's travel back home. It made him and all the men cranky.

"Mars Base, my ship is equipped with a weapon, which we have now applied against the ship, the 'Understandable'. This is an electro-magnetic device, not a bomb, not an explosive. This was applied to our victory, several hours ago. This is war, Commander Than. I am informing you now; we also have regular bombs and missiles. Very large ones, you know?" He paused for effect.

"Go ahead, Tolstoy. Thank you." From the base. Each worded phrase was separated by a brief pause, the static-buzz and humming in the background. Across the command-deck, the video-screen image showing the telescope-camera view of

Berle's ship had not changed (although it was a real-time, 'live' view).

"Yes," said Keeje. "I hope you believe me. I intend to destroy the Mars Base, and use these bombs to do so. Consider this a courtesy-call, so you can get your affairs in order. You understand me then, yes? Please confirm. I don't want to be rude."

A long pause, longer than usual. Bojji-Than finally found what he wanted to say. He was very calm. "That is very bad news for us, Prokov Keeje. Are you willing to negotiate? This war here on Mars has become a madness. My people do not wish to die. This base is a priceless resource. If you destroy us, you have no fuel or food or water or systems-recharge and you will never return to Earth, you will never get home to your own families and your own lives. You are destroying yourself, sir. Surely this is understandable?"

More laughter from the command-deck staff on the Tolstoy. By now, two other navigation-level crewmembers had come up to the helm, with Keeje and others. Not all of them were smiling. They wanted to go home. Keeje was in total command, and rather brutal. None of them would defy him, whatever his decisions were.

"You have 24-hours, Mars Base. I require total surrender. My men will enter the Mars Base within that time frame, or the US Mars Base will be obliterated to less than dust. We can talk about going back to Earth once you have surrendered."

"We've been through this conversation before, Tolstoy. If you send an un-armed group that we can inspect beforehand for weapons or small arms or small bombs, your men can enter the Mars Base within less than 24-hours to negotiate for peace. Just do it, you bloviating fool. This is senseless. Your blustering threats mean nothing. This is our home. Send ten un-armed men, we will talk, we will end all this. I am making a counter-offer. Please respond. I am not powerless, Keeje. Mars Base is on hold for your reply, while there is still time, sir."

Far below where the Tolstoy moved in her orbit, still tracking parallel to the Understandable, which though 'dead', was yet in motion as a result of her momentum before she was disabled, far below, beneath what little atmosphere there was on Mars, the Mars Base was yet in multi-crisis mode. Inside the topmost Command Center, Bojji-Than was sweating it. He didn't tend to show it, but it was obvious. He was waiting for Keeje to respond, hoping the man would be reasonable at this point. The command-center was now of a mode such as a high-tech morgue or off-world funeral parlor. They were so depressed, it was like a fog. They were also exhausted. The computer-screens, tracking and other base-operations were on artificial dark-hours mode. Screens were glowing with powered pixels of perfect data-flow, but none of them on-hand at that hour expected any real change in the scenario. *Hope dies last, Bojji-Than thought.*

Chassidy Katola, the wellness and macrobiotics nutritional base counselor, was beside the base-commander, providing a rare shoulder-and-neck rub, which Bojji-Than enjoyed during the hours of stress. "Stay focused, commander," she said in her sweet Black woman's voice, like a song of his youth. "He'll figure it out. They'll back down. You can do it."

"Thank you, Chassidy. Be sure to stop by my apartment later," he joked, laughing at himself. "Just kidding, dear."

"Hmmmmm," she said. "Why kidding?" Her hands worked the muscles of his shoulders. "Anyway, I heard everything. So, they blow us to hell. Big deal. Shit happens."

Keeje had not replied, by then ten minutes, not unusual, really, given the radio-system protocol. Now, however, the base-security man, Juno Amorrosi, ever loyal and beyond fatigue, walked towards them in the command-center aisle where Bojji-Than was working with a very patient radio-operator. He had a portable computer tablet, which seemed to have some information.

"Commander, please," Juno said. "Something for you now, please, may I interrupt? This is important sir."

"Yes, yes," Bojji-Than. "Go ahead man."

"The signal upload from Vandenberg on the safety-code remote access programming applications for the Russian ships, the thing they stole from the bases down there, when they shut them down, weeks ago, in Mother Russia---the upload was started and confirmed for connection to our restored systems about half an hour ago," Juno said in a hushed tone. "We can start to use this in about five or six hours to remotely screw all five of their ships, if they work." Bojji-Than rested his eyes. The Buddha was with him for an instant's insight. "Once again with that," he said.

"This was what Berle wanted. I know what happened to his ship, sir, word spread around the base within an hour, we heard it from the Saint Peter, or whoever interrupted our emergency call to the Understandable, they were proud of what they've done, you know? Bunch of braggarts. A few more hours of upload from Vandenberg, a quick operations test, we can end this. We know how to use these; they're similar to our own emergency-systems. We shut them down by remote. Emergency safety remote access, we sue them too."

The radio-operator now was alerted. Keeje was back online. Some radio-noises and the headset back to Bojji-Than.

"Here he is now, sir, the Tolstoy" the operator said to Bojii. The Mars Base Commander looked grim, worried.

"All right, he's giving us 24-hours anyway. Juno, move on that as quickly as possible, every resource, top priority. Okay, go!"

"Yes sir," Juno responded, then quickly moved away.

"Tell the Russian Commander I was called away to the dinner-hall," Bojji-Than said to the radio operator. "Delay him. Tell him we have a dinner hour, and the meal is clam chowder with sourdough bread and wine, and I needed to get my portion on schedule, and I will be back on the call in a few moments. Go ahead, do it just that way. You are an actor. Do it now."

The radioman was a bit puzzled, and then seemed to instinctively understand. "Tolstoy, this is Mars Base. The commander has been called to the dinner hall on his schedule, sir, I am very sorry. We have clam chowder tonight with wine, his personal favorite, sir; again, he will be back on your call in just a few moments. Yes, Tolstoy, this is the operator. I am very sorry. I understand. Yes. Well, meals are on a tight schedule, sir, it will only be a few minutes. Yes, uh, I mean no sir, I am not authorized to make any agreements, sir---"

The radio signal grew outwardly from Mars Base, in a circle that reached the Tolstoy in only a few moment's time, there in orbit and then to her radio-receivers, and then of some kind of pain, into Commander Prokov Keeje's stinging pink ears.

## CHAPTER 54: Shut Me Down, Shut You Down

Suddenly, there aboard the 'Understandable', Winton Berle's command-ship in orbit around Mars, it was 'lights-out'. Berle, the pilot and commander of all nine ships that had eventually arrived from Earth to defend the Mars Base, knew better than to panic, but it was scary as hell. He cursed himself privately for not having seen it coming. Apparently it was easier for the Russian-Islamic enemy to position and effectively target their Electro-Magnetic Pulse weapon, than he had thought. They had counted on this task being very laborious, to defend themselves, and hoped for the best. The distances between ships-in-orbit was vast, and the tracking-systems and radar gave them a chance to view any approaching ships. But, here they were, without electric, without computer, no navigation, life-sustain failing quickly, no engines, no communications, drifting in a fractured orbit from only previous momentum, desperately hoping their orbit would not decay into a free-fall friction-burn to certain death below. The effectiveness of the EMP was very similar to what happened to Guy Reisling's ship, in deep space passage, maybe six month's prior to then. They had only a few hours of ambient life-support to stay alive, they had battery-operated back-up's, and they had themselves to restore ship's function and controls, to save their lives. Unlike the 'Penelope', Berle had a considerable advantage, that being the other ships in his 'armada' were by now aware of his emergency, and could fairly easily help or assist; and also, the Mars Base itself had resources.

In Reisling's case, they had lost all power and control, and so on, in deep space, many hundreds-of-thousands of miles from the nearest forms of rescue (that being the Molinari Space Dock). So although 'Kick' could not confirm by standard radio, he felt sure that his teammates were probably already on their way to rescue, or help restore normal operations. So he felt they would survive, they would get through it, it was going to 'be all right', and his anxiety and fears were somewhat relieved.

Immediately after the attack from the 'Tolstoy' which had obliterated his ship's operating-systems, Berle, and his men, had some idea what needed to be done, and began those tasks with determination. In a strange way they could thank Guy Reisling and his crew for this knowledge, and a sort of warning. The EMP did not kill the crew. The first job was to re-set, or re-configure energy-power generation systems and electrical function; then they could re-boot essential life-support (temperature, oxygen, C02 scrub, fluids); after that they needed to quickly re-establish computer-based systems, especially navigation, for normal controls. And all this took time. The ship's orbit was still somewhat stable, but would not last long, without engine-thrusters, so restoring engine-power and maneuvering-ability was urgent. Everything worked from computers and electrical power, so the main batteries and power-generators had to be up-and-running for the same systems to function. Or else it was a friction burn free-fall down to the surface of Mars. Back-up hand-communicators connected all the crewmen aboard the 'Understandable', at their different posts, navigators, communications, engines, cargo, command deck, life-systems, and so on.

"It's a bitch, men," Berle said over his hand-held so all the men could hear. "We can survive this if we don't panic. Electrical power-systems first, life-sustain, and then computer-control, in that order priority. We have about 24-hours. Use whatever operations-manuals, or systems-schematics, that you can find, and get your systems up again, even at a basics-only level. There are enough oxygen-suits for everyone on-board, and extras, so get down there and get yours. They should work fine, you won't need them for a few hours anyway, so obviously, save your air."

Raza, the communications-man, was on the link. "I think someone put cinnamon-pepper from the food-deck in your oxygen suit, Kick. It's the purple one," he joked. Some of the other men on the inter-deck radio-link laughed or commented shortly. Berle laughed too.

"Very funny, Raza," he said, static-on-the-line. "No, my suit is the one in scarlet-red like Santa Claus, right? Ha-ha, very funny. Look, let's get to work. Keep your spirits up. It's not as bad as it seems." And then he added, "Ben, are you there? Number Two?"

After a moment, Ben, his co-pilot, responded. "I'm online, Kick, yeah. Between decks on the way to the battery array deck."

The other men were also in-motion, pulling themselves by the handgrips in the weightless environment, through the dark passageways, and gates between various decks, to places deep within the ship where they could do the work they needed to do. Another long moment between radio-chat. Berle was also moving through the dark tunnels within the ship, searching out the power-generating system-controls, by which the 'Understandable' produced its own electric power, using the ship's engines, and hydrogen fuel, external solar-panel arrays, and other types. He had a flashlight and his hand-held communicator, in his standard flight-suit. He found he was sweating and grunting with the effort, despite the weightless environment. Maybe he really was too old for this. It was dark, and getting cold, and the air seemed to grow stale. Or was it just a projection of his fears? He also had a small portable handheld command-computer pad, a back up for emergencies, operated by batteries, with all sorts of systems information and connections. But it wasn't much use, because there was nothing to connect to. All those systems were down.

"Hey, Ben," he said into his hand-radio, huffing and puffing a bit, without apology. "What was the position of our nearest sister-ship prior to the attack? Let me know immediately, if you remember, I mean. If you have that information."

*What a man knows or may recall, a lifeless or dead computer-machine may forget,* Berle thought. *But the human heart is ever more to the pleasure of any challenge, difficult or not, to somehow ever really forget death. Oh death! Oh philosophy!*

And so it went, with typical courage and gusto, as his crew made to provide for their own survival, there in orbit around Mars, the grinning mockery and laughter of heartless enemies, somewhere off 'that-a-way', a sad truth that had befallen them as well, in their thoughts.

A few hundred miles away (a short distance in space), across the airless expanse of freezing nothingness, over the form and shape of the planet Mars, the Penelope could be observed. Guy's ship, ironically, happened to be nearest to the 'Understandable', at the time the enemy attacked. The 'Penelope' was now under the command of navigation-officer Tom McGee, with Guy 'on vacation' at the Mars Base below. Of course it was no vacation at all, and Guy was recovering from the journey and the transport of the communications-gear to the surface, and the attacks on the way inside to the base itself, and the death of his crewman, young Peter Swain. Beset by common space-travel stresses and strains. McGee, Arron the cargo-chief, and the others operating the Penelope, with Rob Cowan the original co-pilot now gone, and the loss of Peter as well, they were exhausted, growing thin and anemic from the null-gravity lifestyle.

It had been just about six hours from the time the 'Understandable' had taken the EMP hit. Communication back-and-forth from the Mars Base, to the US ships-in-orbit, was still intact, so what had become of Berle's ship was known to those concerned. Within a short time, the 'Penelope' was dispatched for dock-and-rescue operations, to save Berle and his crew. But this was not a simple task at all. The proximity of the Tolstoy, and the Saint Peter, both enemy ships, was an anxiety to be sure. Another EMP attack, or other types of weapons against the 'Penelope', could end things very quickly. But they also knew a second EMP blast was unlikely, given the time involved setting them up and loading such a huge electric charge. McGee was a skilled space-vessel navigator, and understood the dynamics of planetary orbit positions, movement, flight-path, docking, leaving orbit, etc. They could not just turn the 'Penelope' around suddenly, change course quickly, and cruise over to Berle's ship and dock. Both ships were moving at high speed on a trajectory determined by their previous path and momentum. And of course the 'Understandable' could not maneuver at all. So McGee's first task was to figure out the navigations needed, with the help of plotters at the Mars Base, so he could plan out his ship's movement by thruster-changes, slowing or increasing speed, or by horizontal-movements, and elevations, so he could very closely intersect the inevitable and un-changing course of the other ship. And then a docking procedure. The way it worked out, even with their best efforts, it would be 12 hours of work, before they had any chance at all of drawing close enough to the 'Understandable', and bringing out the crew alive. At the same time, his team-mates had to invent docking methods that would safely relieve Berle's endangered crew, including space-walks, air-pressure exchange and seals, and so on. These were 'standard', but far from 'normal'. Nothing was ever really normal in space-travel emergencies. And then there was Keeje.

Down below, Bojji-Than, the Mars Base commander, indeed attempted to enjoy a bowl of hot clam-chowder, with bread and reddish-purple Merlot wine, at one of the base cafeterias. He put his worst enemy out of his mind, and the previous radio-call, and threats. His calm, patience, and quietly reserved determination, were almost eerie. In that hour, other base-regulars distanced themselves from fellowship with Bojji-Than, somehow knowing that he was under a great deal of stress. He took his meal in the third of three base-cafeterias at the base, the smallest one. Five or ten other Mars Base staffers were also there for their own dinner. They came and went as normal, and the food-service provided the best food they possibly could under the circumstances. Communications at the base were stable, and everyone was pleasant. There was no panic. The commander had a small hand-held computer-pad, and was studying the method by which the stolen command-codes now uploading from Vandenberg, back on Earth, via the recently installed radio hardware, also from Earth, could be activated and employed to disable the enemy ships.

It wasn't too terribly complicated, but he really didn't understand it all. Mars Base operators would project a signal, targeted at the specific enemy ships, that would automatically identify and lock on to devices within those ships, coded to certain frequencies and binary-code computer command sequences, like electronic locks and keys. The signals needed to be targeted at each of the ships individually, with exact transmission codes. Linkage or signal-lock to the ship's safety-devices had to be solid and tested, before actual use. Once all that was established, Mars Base operators would use these applications, as if to rescue disabled ships or unconscious crewmen, as the devices were intended. This was their original function and purpose. Now, however, in a delightful turn-about, they would be applied as a weapon-of-choice. Bojji-Than smiled.

The computer information, data-programs and devices that were stolen by the US Marines at the KK/F Region-Six launch-sites in the snowy woodlands of Earth's Ukrainian region, were significantly different than the same on the US side. Transmitted codes and program applications were arriving even at that moment, via radio-transmission from Vandenberg, as Bojji-Than enjoyed his food. The Russians would have typically used dedicated devices and hardware for the same sort of remote transmissions, to assist ships in orbit around Earth, or between the Earth and the moon. So there was no guarantee that their plan would succeed, it just wasn't the same, or anywhere near as reliable, being 'jury-rigged'. What was uploading now was very different. No hardware, no permanent in-place systems, as perhaps had been tested or in-use before, at the Russian launch-sites.

But supposedly, according to his technical experts, the codes would allow for a certain amount of very real manipulation of the enemy ships, by remote, from the Mars Base command-center. They should be able to initiate air lock docking outer-ship entryway or hatches, to open or close, pressurize, and so on, for a presumed rescue-entry. Also, if the crew of a Russian ship were unconscious or disabled, the program-commands would allow for the remote shutdown of engines and thrusters. The reason was obvious. With the crew unconscious, out-of-control engines and navigational thrusters, were a clear danger, and so were included for remote control (shut-down). An obvious safety measure, to avoid the ship speeding into oblivion with no one at the controls. The safety-rescue remote computer applications and codes could also link-up with a ship's radio and flight information, flight logs, position, speed and tracking, so the rescue teams could find them. Yet more, the designers of the remote-rescue applications, invented codes that could shut down or disable, or variously control for operations, any dangerous or volatile systems on any of the ships. Weapons, launch-systems, attack systems, dangerous high-powered fuels and materials, cargo release and gates, descent shuttle drop controls, were all at least somewhat available to the remote operators. Each function had a different remote command, or series of commands. Even the life-sustain systems, although not to be shut down (obviously), but they could be re-booted, or re-started, or measured and monitored, by the same coded-applications.

Bojji-Than smiled to himself again as he studied the materials. The clam chowder was growing cold. He enjoyed clam chowder with red-hot sauce, so he dribbled a bit from a small bottle onto the soupy-white Earth-sea stew before him in a small dark blue bowl. His peaceful smile and confidence were deceptive, he knew. Could they really at last end the war on Mars, with this process? He did not know. At the same time, he and a few others carried the weight of knowing about Keeje's plans to soon annihilate the Mars Base and all its residents, with powerful bombs. As insane as Keeje's threat might seem, with no way out even for his own men and ships, and probably no way for them to return to Earth, it was a dark cloud he wished dearly to remove and push away, for them all. He found it ironic. Their ultimate weapon against a cruel and un-stoppable enemy was a safety-and-rescue method procedure intended originally to save their enemy's lives. He drank again from the reddish-purple Merlot wine, somewhat of a rarity on Mars, this bottle from his personal stock. *Good lord,* he thought to himself. *What an amazing journey it's been, these past two years. I wonder how that asteroid is doing out there?*

He sat quietly for a moment with his thoughts. About that time, Vinces Grant, the base research coordinator, along with Matt Currison, the launch-specialist, entered the cafeteria, and moved over to where the commander was seated. "How many Russian astronauts does it take to screw in a solar-panel?" Grant joked, breaking the ice.

"Please Vinces, no jokes right now. I'm troubled about all of this," Bojji-Than said. Vinces mercifully allowed another one of his vapid one-liners to slip away. Whatever the punch line may have been, it was no doubt sarcastic, dark, and grim, as they all seemed to be.

"Enough already," said Currison. They spoke in hushed tones, as the others in the meals-area allowed them to talk privately. "Commander, the codes from Vandenberg have been successfully uploaded, and we've done simple tests for functionality. We're ready to go for these, with whatever we can do against the enemy ships. We can start immediately."

They left the cafeteria together, like three nuns on their way to a bake-sale, headed back up to the command-center, still talking. Someone else would take away the commander's wine and wineglass, dark blue bowl, spoon and leftover chowder, later, and clean up from his table.

## CHAPTER 55: Never Going Home

*"Allah Akhbar! Mars is ours for the glory of Mohammud!!"*

*--Rezza Alvenu, a space-program worker from India recruited for service under Commander Prokov Keeje, as a surface-fighter against the US Mars Base regulars, 2081.*

The three men stepped inside the elevator lift that would carry them upwards towards the Command Center at the top floor of the Mars Base. The elevator was down the hallway from the cafeteria. The machine rattled a bit going upwards, to where they could control the next phase of the efforts to end the 'war on Mars'. The Mars Base was at that time at least 20-years old, and much of the technology and machinery was less than perfect. But they trusted the things they needed, such as the elevator, and most of the staff at the base were well versed in the operations and even repair of things like air lock gates, doorways and lifts, basic life-support, communications, and the elevators too. It was a requirement of space-travel, in case of emergencies.

"Think in terms of a glorious victory," said Vinces Grant.

Bojji-Than didn't respond with laughter, and it was a source of on-going irritation at times that Vinces insisted on cracking jokes, even during the most dire circumstances. Which, at this point, was clearly at least somewhat the case. However, on their way out of the lift at the proper level, they finally felt they had a workable remedy they could live with. And this was, as they planned it, the enemy's emergency-safety space-ship shut-down radio-transmission applications and codes, that had now successfully arrived from Earth and been installed. They continued to talk, entering the Control-Room area. Bojji-Than wore his usual work-clothes, a sort of all-purpose jump suit, complimented by a long black-woven alpaca over-shirt, that draped down bellow his waist (it was warm, and comforting, and he felt stylish with it). Vinces and Matt Currison dressed in standard 'street clothes', durable pants and shirts, like many at the base who worked inside.

The Command Center was built at the top of all of the other structures. They had a lot to do, and they needed to work as quickly as possible. Despite it all, the mood was somewhat cheerful. There was confidence now that with the right application, they could defeat the Russian-Islamic 'invaders' in their five large inter-planetary ships from Earth, along with the surface-fighters still just beyond the base perimiter, and end the conflict for control of the base. And even very possibly do so without more needless and senseless deaths, even on the part of the enemy. It was a good plan, and it had taken almost a year to put in-place. With the shut down of the Eastern space-program launch-sites on Earth by the Western military, and also the transport of the needed technology aboard Guy Reisling's ship, nearly resulting in his demise and that of his crew, the time had come. But, there was a lot more going on, and it was looking very much like a 'last-ditch attempt' to conclude hostilities, and 'snatch victory from the jaws of defeat'. The fate of the meteor headed towards Earth now wasn't even an issue, there on Mars. It really meant nothing.

"What about Winton Berle's ship," the Commander said. "Have they got him? Is there a rescue docking ship happening? Please inform me quickly."

"It looks like they're going to be fine, Commander," Currison replied. "It's true, the Understandable was blasted with the same kind of electro-magnetic pulse that disabled the transport ship months ago, 30 days from Molinari. That was the transport; the pilot is on the base here now. Berle's ship is disabled, engines, life-support, electric-power, computers---the EMP's are very effective. They're still alive in there. We have radio-contact from a close-by ship, to their hand-helds. It will only be a few hours for a rescue docking to begin, from the other ship. I think it's going to work out. He's in a lot better shape then the transport was in deep-space, that's for sure."

The Command Center at the base was a series of work-spaces and tech-kiosks in small rooms and work-areas, equipped with all kinds of needed technology, computers, controls, monitors, radio-connections, adjusting-meters, tracking-telemetry, life-support. Much like the ships in space, everything the base needed to maintain operations was somehow connected here. But as large as the base was, for the main buildings, there were also many sub-control stations throughout the structures. The center high above it all included at least three large vista-windows, by which they could look out over the Mars-surface below. It was typically a very busy place, but under normal circumstances rather serene and even mundane, given the very desirable normalcy and orderly affairs of a research-and-science outpost. As the three men entered, there was no particular formality, that the Commander was now on-deck, or on the scene. Each of the workers were at their stations, knowing what they had to do, comfortable with the processes, even though the information and data was not always showing things in their favor. Sometimes data-exchange or systems-changes would take hours, and workers would chat or socialize. There were about a dozen people in the room at that time, and with the appearance of Grant, the Commander, and Currison, the others lit up with anticipation, seeming to feel the time had come for some sort of decisive action. The rumors were not without weight. There was no high-level of secrecy or ignorance regarding the shutdown codes, and in general the Mars-program didn't want ignorance or secrecy. It was an hour with a great deal of hope to finally end the bloody struggle and return things to normal.

"Isn't that what we're just about to try to do to the enemy ships, Matt? A rescue?" Vinces asked. They were moving through the room to Bojii's station, an appropriately impressive desk-and-computer kiosk with a nice window view.

"Not exactly," said Bojji-Than. "The rescue docking for Berle and his men is one thing, let's not confuse things. Obviously the Understandable has shutdown and rescue-safety docking external radio-codes, too. Ours, the ones we use. I want tracking stations to follow that, hourly, make sure someone has all data and information until they are safe. We can still relay to the rescue ship, it could be important. At this time we also need to initiate the efforts against the enemy ships. Matthew, you're the launch-specialist, you'll be the front man on this, you know the orbital stuff. But wait for my cue on approval for final decisions to initiate your transmission-sequences. I want the other team leaders, Charley, you Vinces, and the security guy, the tech-people. Create a team for this purpose, the shutdown transmissions. It may take 24-hours, or 72-hours, or more, for success, against all five enemy ships. Don't leave any of them out, ha-ha, not funny. Also make sure that Tutturro is available, the radio-woman, so we can assure that her new installations are suitable."

By now the Commander was seated and comfortable. "Happy to oblige," said Currison. "I do know how to handle the ships in orbit. The available radar-footprint for orbital on this side of Mars is monitored. Once we find their ships, and can keep them on our radar, we target the same way we would a standard radio-transmission. If it works, their ships will shut down at our signal. The trick will be to spot them. Fortunately, the orbit-pathways, inevitably they run their ships somewhere in range. They have to, they have no real choice, and Mars is round. So, I think in a day or so, we can target each one of them. It's going to be a big surprise for those guys, I'll tell you. I've looked at the safety-commands. They don't just pop the airlock hatches, like for a rescue docking. If we did, they'd all die. Those signal-commands are for docking, with unconscious crewmembers. However, we can shut down their basic navigation, shut down the engines and thrusters, link to life-support controls, and main power and computers. That's what the Eastern program had in place for their own safety and crews. So, with any luck, we won't kill them, but the ships will stop cold and lifeless without power or controls, by the time we're through."

"What if the ship's lose orbit? I mean, the orbits decay with power-loss and they fall to Mars in a heat-descent? You know, the orbits decay and they crash??" said Vinces.

There was a moment's pause. "They'd burn up and die, Vinces," said Currison. "What? You're concerned they'll hit you on your head or something? My guess is it's one-in-a-billion chance one of their ships would hit the base and spoil your lunch, so quit worrying."

"All right. As the saying goes, 'just do it'," said Bojji-Than. "As I recall, we had a 24-hour threat from this Russian commander to destroy the base with a powerful bomb or something---what? Six or eight hours ago? Let's try not to let that happen, would also spoil my lunch, too. Just do it, Matt. You have a green-light to go ahead with all such operations."

The other workers in the command-center were keenly interested in what was next. There was still the fighting on the surface, near the base now, and they all knew it. So it was a critical hour. Juno Amorrosi, the security-man, across the hall, was working with life-support specialist Charley Barron, tracking the battles and surface fighting, and providing direction for the leaders on the ground. It was a slow dance, tedious and deadly, with more dead on both sides, the numbers favoring the US Mars teams slightly. There was no real progress other than destruction of equipment, gear, men, and positions. Then, there at his side, Juno heard a pleasant female voice.

"Hi Juno, what's going on?" It was Karen Tutturro, the communications-specialist form Earth on her maiden voyage to Mars. "I was asked to report in up here. How can I help?"

Juno was distracted, watching a radar-screen to follow the progress of a Russian-Islamic squad of about 20 soldiers who were making their way like enemy-ants on his green-screen, towards a stronghold position held by one of the remaining teams in the Survival Igloos. "Huh? Oh hello there," he said. "You want to help? Uh, I think they want you to make sure your radio-stuff works. You can probably report to Grant, over there, cracking his stupid jokes."

He directed her towards where Vinces and the others were gathering to quickly map out procedures for the 'final solution', working by hand on pads of paper.

"This I know," she said. "My installations will function flawlessly."

"Well, that's good then," Juno replied. "Let me ask you, are you in love with me? If you really want to help out, we could get together in a few hours---"

She laughed and so did Charley. "Don't let him bother you," Barron said. "He thinks all the women love him."

"They do," Juno said. "They just don't know it. I have children to prove it. Back home."

More laughter. Within an hour, the process had started. Two of the Russian ships, the Tolstoy and the Saint Peter, were on the base radar-footprint already. Then the Sir Soviet, the Krenika (Captain Zolotny's ship), and the last one, known as the Red Devil. It was quickly decided to target the Tolstoy first, believing it to hold the invading Commander, Prokov Keeje, and also possibly the 'strong bombs' that he had threatened them with only hours before. There would be no warning. The Saint Peter would be targeted at the same time, but was more distant. Keeje seemed to be enjoying his vengeance against the Understandable, watching things proceed; waiting, and he might even try to stop the rescue docking. The ships were hundreds of miles apart, except the Understandable and the Tolstoy, which had necessarily approached within less then ten miles to release the EMP force. Data and telemetry on the other enemy ships was in-flux. But eventually as Currison had said, they would need to pass over, within range.

"It's like duck-hunting," said Vinces. "We wait in the reeds for a fly-by. Once we target the radio-antennas by directional, we send up the signal on each one. Then they go without their microwave ovens for a while."

"Quack, quack," responded one of the radio-tech specialists, working out his antenna positions.

And so it went. High above, Keeje and his men indeed were pleased. His ship was in fine order, humming through the emptiness with trimmed engines, now 100 miles from where the Understandable was drifting powerless, in-view on their own radar and electronic telescopes. "They are stalling us off, Commander," said one of the navigators, now on the helm, there within the Tolstoy's technological machine-like halls and passages. "They blew us off. Eight hours, nothing." The navigator was an Islamic man, with dark features, named Al-Masheesh. He was very skilled and knew his job, and was deeply loyal to their side. "Why should we wait?' he continued. " You gave them a day. 24-hours. The missiles are ready, and we will soon move out of position to strike. Why give them more notice? I say we destroy the base now, start the method, and set your targeting configurations. At least we will have our vengeance."

Keeje was seated now, calmly and peacefully assured of himself, in his pilot's chair before the controls spread before him. He was eating from a Ready-Meal, as were some of the others. "I wanted to save the base, or we can't get home," he said blandly. "You know this."

"It's over, Commander. We are NEVER going home," the other man said grimly.

A long pause. "All right then, you fool. I will give the command after I eat something. The three plutonium-uranium atomic bombs will kill them all very quickly. Maybe there is a way we can still get back to Earth. Maybe Earth will still be there? Ha-ha-ha---too bad. It was a nice base."

The navigator went back to his post, pulling himself by handholds in the null-gravity, with a view of Berle's ship on a view-screen, and their own speed and position.

## CHAPTER 56: 'Give Me Back My Ship!'

*"Never undertake a ground-war against mainland China,"*
*Lao-Tzu, 'The Art of War' (paraphrase, from book)*

There was little or no help from Earth as the Battle at
Tharsis Montes on Mars was now withering away to an ending as
pointless as its beginning, flickering out like a bright light of reds
and blue, furious for a time, then losing power or fuel-source,
rolling out to its conclusion, then to be over at last. Earth had
'other problems'; with asteroid U2753b now a clear shot for a
heavy hit only less than nine months away. As predicted, from
the earliest possible calculations and telescope-work, the huge
rock had tumbled through the nothing, happily wandering in its
ageless and timeless journey, to what was worst-feared soon to
be a random collision with Mother Terra. 'Just one of those
things', the people sighed, but with the armies and military, the
governments and science-community, the rocketry and launches
and bombs, all hot for some sort of remedy, it was a tense and
breathless time. Protests, meetings and conferences, military
combat in reply to previous military combat choices, a few urban
riots, lawless vengeance and chaos, were controlled or organized
as best the leadership could manage. The vast majority of Earth's
population had little or no real information, so it all meant
nothing. The insider view was much more dismal: a rock the size
of the island of Borneo either would, or would not, strike Planet
Earth and very likely create a truly un-livable mess. So any
further effort to win future control of the Mars Base, from the US
side, was a 'lesser issue': they had done all they could, by that
date. Nevertheless, it was still thought by some power brokers,
that Ibrahim Mehudi, who had died unexpectedly during the

troubles, may have been correct after all, and that the under-estimated spin on the asteroid itself, as it moved through space, would recant previous telemetry on the object, enough for a near miss. To help this along, the plan that was put in place was to launch very powerful bombs out to the meteor once it was close enough, and target those to 'push' the spin off slightly more than sufficiently, to protect the Earth-Home, and create the near miss. It was the simplest and most-likely effective option, and was within the manageable available technology, so those good-deeds were in-motion all the same time that the Mars-War was underway: again, a 'What-If War', which had the only purpose of controlling the base, if they were wrong about the meteor, and if they failed to deter a main-hit on Earth.

Hours passing at the Snikta-Basin air-sealed self-sustaining base, so far away on Mars, were now occupied with the following: first they would locate and track the identified enemy ships in orbit; then they needed to wait for those ships to pass into range of the carefully targeted radio-signals with the shut-down command codes, and figure the directional antennas and signal broadcast-transmitters for successful projected linkage to the right ships; the commands to transmit would be issued after quick review, and they could all enjoy what happened to their enemies after that. The two other major tasks were to safely rescue Commander Berle and his crew from the disabled 'Understandable' (already well underway), and somehow maintain or control, or even win, the surface battles with the Mars-Men in their uncomfortable Walker-Suits. As the prophet says, "Everything is happening all at once, all the time."

Of course Bojji-Than knew well that Prokov Keeje, as rude and furious as he was, had only recently issued to him yet another heartless threat of annihilation. So they had no way of knowing when he would launch his deadly bombs, or if his threats were real, or if he would, or could, do as he claimed. A terrorist's deadline doesn't mean much, but at least 12-hours had passed. So, they targeted the Tolstoy first. This was easier than tracking and targeting other Russian-Islamic ships in orbit, because the Tolstoy happened at that time to be 'within their radar footprint range', along with the 'Understandable', as well as the 'Penelope'.

Matthew Currison Von Templar then reported to the base Commander, at the right time. "We can broadcast against the Tolstoy at your order any time now," he said. "Slam-dunk, commander."

"You've identified their ship properly, and you have the signal codes for the shut down perfected to readiness for the process?"

"Yes, sir," said Currison.

"Then go ahead immediately and report back," Bojji-Than said. "And Matthew," he added soberly. "We are not opening the air locks. We are not killing these crews. Is that understood?"

He nodded firmly and went back to his station.

200-miles overhead, and about another 180 miles to one side from the base below, and the foot-soldiers on the surface gasping for air-supply and still hurling their bullets and fire-bombs, a similar scenario with much different intentions was playing out at the command-deck of Keeje's ship.

"What is the circumstance now that delays your bombs? Answer me!" Keeje said to his navigator, also the man on the job to scan or view maps and radar telescopes of the Mars-surface.

"We cannot properly target and launch for another 30-minutes without moving the ship, commander," the navigator said tersely.

"Then give me the thruster-sequence orders you need for the navigation and we move the ship! No more waiting!"

"It is not the best, sir. It would be better to continue the current orbit without----"

"We move the ship, and destroy the base! This war is over! Then we take whatever remains for supplies or fuel or anything we want."

"Yes, sir," the other man said grimly.

Just about then, Keeje and his other officers and technicians found that the Tolstoy was no longer operating properly, for some strange reason. As various thruster-arrangements were set to fire in proper sequence to move the ship laterally, the controls did not function, and the thrusters did not fire, and the ship did not move from its orbit path. At the same time, other main-bus controls were lost, and went dead, the main-engine and power systems dimmed and started their shut-down cycles, control-dashboard lights and meters falling or going blank, one-by-one, at various stations. Fuel sources, the hydrogen-fuel cells and peroxide tank controls and pump valve-systems, were found to be inexplicably frozen or locked, and main-power batteries and solar to electrical that operated large external bay-doors, airlocks, descent-pod release processes, were also out. And then life-support, the ship's air, temperature, potable fluids, inner passageways, lighting, these all suddenly had gone to low-power back up at minimal levels. It took Keeje and his crew a while to realize what was happening.

"Get me control of my ship back online NOW!" Keeje screamed at his crewmen.

His navigators and technicians scrambled, but it was no good. The Tolstoy was now without a working inner-command system from anywhere on the ship, such as normal operations or ship's movements and electrical, as if a Peace-Loving Giant had chosen to end their evil plan in a harmless mystery of failed mechanics. Keeje had lost control of his ship, because the Russian-Islamic-Ukrainian space-program planners had not been so kind as to design the Tolstoy without a fail-safe system to rescue he and his crew if they were somehow disabled or unconscious. Or any other of their space-program workers. And it wasn't long until Keeje, and the other command-deck officers realized that this was the only explanation. The Mars Base had somehow acquired and applied the often-overlooked safety-rescue process command-sequence by radio transmission, and they could do nothing whatever to stop them.

"We are being rescued, Commander Keeje, that is what this is. The rescue-sequence. Our enemies must have somehow sent it from Earth. The radio-transmitter for a dead crew---"

"I know what it is! I know! I know! See if you can over-ride the shutdown on any of my controls. Any controls. Give me back my ship! Hot-wire the damn thing with a pair of pliers! Work on it, all of you! Every system they've locked onto for the safety-systems is for over-ride to temporary controls! Orbit navigational first, then a way to target the bombs."

The men could only marvel at Keeje's determination that they all should die on Mars, instead of possibly going home, or at least ending things some other way, as it now seemed inevitable. Some of the crew were actually laughing, but he did have a horrible temper. Like the so-called 'Son of the Morning Star', the old world US Cavalry's historic Colonel George Armstrong

Custer, leading the doomed attack into Little Big Horn, utterly oblivious to his own insanity, yet in command, the innocent natives and his men, astonished and afraid.

Then the sound of laughter and cheers below, within the Mars Base Command Center. The initial process had taken about 30-minutes, but they still had some remaining potential applications, even for the one ship (Keeje's), that could be transmitted. As the Mars Base task-force assigned to the job now were getting ready for the same procedure against the Saint Peter, and the other three, it was decided to connect to the Tolstoy by voice-radio. The link was set up, and after the usual waiting-period, Bojji-Than was granted the honor.

"Hello, Tolstoy, this is Mars Base. Do you copy?" Bojji-Than said on his handset. A long pause, the radio-link humming.

"Copy, Mars Base. This is Tolstoy. I am Second Commander Pahlevi. Go ahead, please."

A long pause, more radio-beeps and blips. "Greetings, Second Commander. This is Mars Base leader Bojji-Than. Are you having any trouble?"

He had not spoken to Pahlevi before then. "Please wait, Mars Base."

And so it went. Strangely enough, perhaps as a courtesy all well-experienced space-workers habitually extended one-another, for space-travel being so very dangerous and exhausting, the protocol involved was orderly and well established. The US Mars forces never intended to destroy the Tolstoy or its occupants, and that might even have been done, in particular with radio commands to open any external air lock hatches or docking ports, as would be done in a rescue-operation. The functions were confirmed and confirmed again, the stolen codes were working as intended. At that moment, their relief was like a new day, a new hope, an instant of clarity in which the darkness lifted, and the madness seemed to wane, then start to fade or clear, without a hidden malice or sudden violent turnabout, as those who had only research and education in their sights for many long years of service, and dangerous hard labor and training, could see, and feel in their hearts, that a suitable solution was in place.

Currison, Vinces Grant, Charley Baron, Juno Amorrossi, Karen Tutturro, and others working in the Command Center, took a moment to join in with a brief celebration. The Command Center was suddenly awash with cheers and hoots, banging pencils and pens, a paper-airplane flew across the room, with the name 'Tolstoy' written in dark ink on one wing, there was a bell ringing, like a real silver chime, and also compressed air horns like at a ball-game. Word spread quickly by the main dispatch, it was only a short time before the remaining 198 Mars Base regular survivors, and also without a doubt their enemies, numbering some 150 men or so on the surface-battlefield, everyone had soon realized that is was now nearly over.

The conversation with the Tolstoy continued a while more, it wasn't easy to convince Keeje that the choice really was to rescue even them, but Bojji-Than was persuasive. Keeje screamed and raged over the radio-link, certain the base-Commander and his men would never see them to safety at all, making even more threats and quoting space-program doctrine about their rights and the occupational-hazards of his work, and on like this, for quite a bit. Meanwhile, the Saint Peter was moving into the radar-telemetry range in the Tharsis Montes/Snikta Volcanic Ridge Basin region: they had no real choice, because the orbits were consistent to the extent that Mars like any local planet, is round like an orange. And the same process began again, as far as the transmission and use of the rescue-codes. The same job on the Saint Peter took another 10 hours, longer than they had hoped, but was also successful. Apparently the Russian-Eastern Bloc programs were just as interested in their own survival in a dangerous circumstance as anyone else. The computer-applications and connections to the gear and equipment on-board the Saint Peter did their work, and the various systems shut down, with that ship's pilot also screaming his loss. It took another 30-hours for the other three invading ships to likewise be 'disabled'. In the case of the Krenika, and the commander, Zolotny, a descent-shuttle pod was released unexpectedly with men aboard and weapons, as some sort of escape attempt, but it crash-landed on Mars, and they were all killed. There were other complications as well.

Sometime later, details reached Molinari, and then Earth, by relay. Guy Reisling, now recovered and much stronger, still mourning the death of his friend and pod-descent assistant, the dark-haired novice named Peter Swain, was able to connect by radio, mostly because he had no better work-at-hand then to help with some of the standard Command Center work, such as reports to Vandenberg, California. Eventually, he found he was talking with his beloved Lila.

"So, how's your mom?" Lila joked. The radio-link was more stable now, for whatever reason, the delays shorter.

"I dunno'," Guy said. "My mom? She's back home, uh, back home, in Los Angeles, I think. I think she's fine. I think she's buying an RV and wants to go traveling. How's YOUR mom?"

"Well, fine, I guess, but either one of them could be, uh--- gee, what's a good word for it? Uh, ANNIHILATED by the meteor-thing, you know. They have it pegged now at about six or eight months out. No one's quite sure if it will hit, but they think it sure might. Ya, you know. 'No-where-to-go-dot-com'. Heh, oh well." Lila was typically perky and as bright as ever he loved her for.

"Thanks, Lila, I guess. Shit. Does that suck or what?"

"Yeah," she said blandly. "So, they did the thing, with the thing from the bases in Moscow or whatever, and it worked, I guess, is that it? Right?"

"Yeah, it worked. We still have to figure out what to do next, and the soldiers on the surface. I mean, we've got five disabled ships in orbit, and they all have to either somehow land on Mars and make it to the base, as prisoners, or possibly keep fighting until the bitter end as we try to reason with them. They could never have gotten home anyway. But without their ships, it's basically over."

"It's a year or more from Mars back to Earth," Lila pointed out. "Earth may not be very---'homey'---by that time. The meteor would make Vandenberg into an underwater monument. But, Molinari could last maybe a couple of years, without Earth-support, before we all died, too. And I guess the Mars Base, too, maybe about the same, approximately. And the moon bases."

The radio-link buzzed and popped a bit. Guy was laughing to himself now, it was so typical of her. All a big joke. Then a serious tone in her voice, over the half-million miles, the connection wasn't merely electronic at all. "Guy," she said. "Are you, I mean, do you figure you and the Penelope and your crew, are you going to be making the trip home, soon, at all? Has anyone thought of that? And the other men and ships? Has the base-commander even talked to anyone about it--?"

Guy was also hushed for a moment. "We lost a few guys, one of my crew was killed on the surface, enemy fire, I was right there," he said. "His suit lost air. Right now---I mean---the Penelope was the first ship at the scene to help the Command Ship for the US side, when they hit us with another EMP, they took the same hit we did, last time I saw you, lost all power, floating dead. We're needed here. Of course the transports back and forth to Earth are essential---we have tons of casualties from the fighting, and also supplies. The US ships had the base for fresh supplies, including fuel and oxygen and food, like that. But not the Russians. See? They would have taken the base and killed everyone, just to get what they needed to return home, Lila. It's not quite over, that's all. It's a mop-up operation, to be sure. But we're still needed here. One thing for sure, that radio-gal, Tutturro, she's done---I had the cargo gear for her radio-repairs. It's her first time even in space. She hadn't even been in Earth orbit before. She's ready to go home. And a lot of others. It's a mess here, Lila, it's unbelievable."

A long pause, now the radio-delay seemed longer. "Oh well," Lila reassured him. "Hang in there space-cowboy. I like you too much to keep worrying about you. And we're a cute couple, everyone says so."

They talked a while longer, as lovers do.

## CHAPTER 57: Is It Over?

The world powers at work 'back home', by that time in 2081-82, were as-usual ineffective, desperate, corrupt, inclined to extremes, fearful or paranoid, and vastly over-rated. No one really controlled anything, and as a matter of global well-being or the Universal Good for Mankind, it was a real mess as far as any organized, unified, or actual action to repair or properly handle affairs, in this case as-related to both the conflict on Mars, at that time, and whatever could be done to defend the planet from this new 'Wormwood', the disastrous predicted meteor. An observer might conclude that although all the needed tools and technology were at-hand, even easily available, that for whatever reason, leadership and 'will', or clear-thinking and directed effort, were somehow too complex, or too conflicted, or too controversial, or 'too expensive', for realistic remedies to be set in-place, and acted upon. A child might have the answer, world leaders, congresses, assemblies and divination, seemed never to agree, likely by virtue of the enormous ego-pride power-trips involved in the 'real' work of solving a tremendous problem for all. In the Ancient, a king or pharaoh would simply have issued his caveat to move the mountains, build the pyramids or aqua-ducts, or erect the Colossus. In the Modern-Era, even with the science-and-technology, the money-systems, popular-media opinion and uninformed confusion, resulting urban-rebellion, emotional politics, violent wars and small-scale tyranny, and scattered delusion, made co-operation difficult if not impossible. Thus, the doorway to the stars, and local planetary resources and success, was seen precariously to balance on the thresh-hold of eras, or to close that door forever, lost for Mankind to enjoy or benefit. A sad thing, but only a few perceived it.

Among them, in her Texas-style wisdom and busty womanly

charms, was Dr. Lynn Rogers-Smith, Ph.D., US Mars program
manager at Vandenberg, California, accused anti-Christ by some
(Islamic), and variously forlorn and lonely in her laborious task
and leadership. She was no anti-Christ, far from it. But the
science-and-technology crowd, given venomous opposition in
the Islamic Renaissance at that time, circa 2077 and beyond, as
the situation un-folded, and the hyper-conservative US political-
religious-military coalition, could not but envy the knowledge
and freedom essential to the space-program, that Rogers-Smith,
and others, were called upon to perform. She was just a cowgirl
from the ranches and farms of flatland Texas, educated and
introduced to power, by virtue of her loyalty and skill. All her
work was management, and little more. But, as things took place,
it fell to she, and her circle of advisors and experts, to make the
choices and decisions that would, or would not, save them all. It
burdened her heart to be so hated, in public circles, for simple
opinion or form of employment. Tough old gal, and not just a
little attractive, with freckles and large breasts, yellow-blonde
hair-styles, a folky Southern twang, and the position in life to
launch a few space-ships here and there.

What was it, then? All they knew, at the Vandenberg Mars-
program on the California coast, and elsewhere among the
networked stations and US Western military space-program elite,
was what had arrived by radio from the Mars Base. Messages
had been logged and recorded for review, and obviously interest
was quite keen. Bottom line: the intrusive Russian-Islamic space-
ships sent to Mars in secret to overcome and control the US
property Mars Base, had been disabled and rendered more-or-
less powerless. The Rescue-Safety Command Code Applications
retrieved by the US Marines and others from the Eastern-bloc
launch sites had been successfully transmitted against the five
orbiting 'enemy' space ships. Details on the effects and
shutdown computer-transmissions were on going, as Rogers-

Smith and her teams at Vandenberg, and the US White-House, and military commands in the West, rejoiced with cheers. The invaders were now inert, their ships disabled, the crews and men inside still alive, but their weapons and plans against the Mars Base, for every contingency, concluded. "Anything could happen," as military planners with their fears and projections always would tell---but for the most part, aside from the surface battles on Mars with the remaining (and bewildered) soldiers in the Life-Sustain Walker-Suits and their guns and the electric carts, and shoulder-rockets, the superior power represented by the ships-in-orbit for the Eastern-bloc space-program, were defeated. Try though they might, the command crews in the Russian-Islamic-Hindustan ships in orbit above and around Mars, could not easily or quickly re-establish control for navigation, engines, power-systems, docking airlocks, life-sustain monitoring, outer bay release doors or descent pod mechanics, telemetry and viewing. Mars Base could repeat the process if needed, and if threatened, could even 'pop open' the air lock docking ports on the 'enemy' ships, ending all life inside by suffocating air-vent to the vacuum of space-itself. This was not what they wanted, but the enemy knew it was possible. So, it was inevitable. The war on Mars was over, for all practical purposes. Ironically, it was the a-priori technological-science notion of merciful rescue in the event of un-foreseen circumstance and loss of crew-consciousness that defeated them. Had they been more cruel, perhaps, things would not have gone that way. But everyone wants to live.

After a time for sorting out what was really happening, Rogers-Smith was able later to conference on a secure line with President Hazlett, the only woman on the four-person US Presidential Seated Council at that time. The Mars Base conflict was 'an issue', but by no means the center of the political universe, as far as Washington was concerned at that time. If they lost the Mars Base, too bad. If they lost the planet, well, the two women confided together in a way that only the female of the species could understand. Mother Earth---please.

"Just the basic details, Lynn, I don't have time," said President Hazlett. "Is it over? Seriously. What is your view, from what you know?"

The secure phone-line was also yet another electronic connection they relied on, alive with their voices and the in-between relays. "Yes, Madam President," said Rogers-Smith. "I am quite confident the Mars Base is no longer seriously threatened. There is really no way now for the enemy to proceed. They can't even get home, back to Earth. We lost many men, on the surface, and at least one ship, and Commander Berle and his crew are still in serious trouble. But without the orbiting ships, the battle is over, for the enemy. No return home. No re-fueling. No additional troops. No life-sustain or electric re-charge. No command and control, such as battle instructions. No real communications to the men on the surface. No restored oxygen or food, water, for the soldiers. And no significant communication to Earth-based space-program leadership, for them. It may take some time, as a mop-up operation, especially if we don't choose to end their lives. But the US Mars Base is now safely in our hands. Without other ships from Earth or some un-predictable event---it's over. My information is directly from Mars Base and Molinari."

A calculated pause from the White House. President Hazlett was actually on a phone-line from another location in Virginia, a sort of bunker. "Good job, then, Vandenberg," she said. "This took you folks---what? Three or four years? I guess the East felt it might work out. Of course, if the meteor wipes us out, it's all pretty meaningless. But any survivors on Mars, or Molinari, or on the Moon-bases---at least there is some hope, right?"

"Yes, I guess so," said Rogers-Smith. She was herself now back in the ranch-style hospitality-room at the Vandenberg base. It was late in California, later in D.C. The external world with all its problems seemed still and distant, unimportant. "Now your turn. What does Washington really know about the asteroid? Woman-to-woman, I mean."

"Not my area, Vandenberg," Hazlett replied, almost lazily. "Renolds and Bourgalt are saving the world about all that, not me. I'm for all the rest, with Garrido, and the stupid fucking media and international. And the military. That stupid fucking rock is about 500,000-miles out from Earth, about now. They hit it on one side with two big bombs, a month ago, very precise nukes. It was spinning anyway, so now they think it's going to miss. They have two more bombs planned to shove it over just enough to miss. Should be only a few days, or weeks. If they can shove it over just a bit, even a few hundred thousand miles--- problem solved. A miss is as good as a mile, you know?"

"So, is that official? Has there been an announcement?" Rogers-Smith asked her sharply.

"No," said President Hazlett. "Please don't repeat what I've told you. There has been no announcement. Vandenberg and the other sites have all the same telemetry and tracking. You can watch as it hits---or just wait for salvation like the rest of us. I'm not a technician. I only know what they tell me."

"Then---how long?"

Another long pause. "Lynn, look at your charts. If it doesn't hit in the next six months, we're saved. That's all. The precise date and time or location for the predicted impact is Top-Secret. I can tell you, as a friend---six months. Live or die. Six months. Less. Look at your charts."

They talked a bit more, and Lynn Rogers-Smith took note carefully of the tone from D.C. After all, there was only so much any of them could do, or know. To remain calm and make good decisions was also very important. So, she had to let it go. And she did, and likewise President Hazlett. They could even laugh over the phone-line.

"Hey, if I never see you again, I always liked your boobs, okay?" Hazlett joked. "Are all the Texas ladies that way?"

"Standard issue Texas titties," Rogers-Smith responded, also laughing. She never really thought about it. Maybe it was the freckles. The call ended, and they both returned to work.

Far away, somewhere in a Swiss detention-center, Rudolph Terchenko, the KK-F/Region-Six Commander, and Eastern Space-Program leader and global planning insider, had settled into his new routine as a celebrity Prisoner of War. His sense of defeat and powerlessness were palpable. With all they had tried to do, he now was little more than an enemy-property, or ward, along with other Eastern space-program leaders, though not all of them. He had no idea what was really going on, any longer, as far as the battle at Tharsis Montes on Mars, Prokov Keeje and his ships and men, or the asteroid. He felt as if he had been employed to dispatch a high-level mission to nowhere, urgent and critical, with men and ships and crews, and billions of dollars in high-tech resources at his command, and then dropped into a personal darkness of ignorance and confinement, loss and grievance, as one might be lifted up to push a certain button, with a specific knowledge and understanding, and then rushed off to yet another nowhere, his button-pushing finger hot from the effort of an entire lifetime. And hot he was, seething with anger, his dinner-tray in hand with the other prisoners or 'guests', at the prison cafeteria---it was after all a privileged detention. He wondered idly if the food and service was similar in any way to what they served at the US Mars Base, chuckling to himself. Or if the Mars Base even still existed. The secret detention center was an international holding facility for all kinds of war criminals, terrorists, deposed leaders and even several candidates for the next incarnation of the Tibetan Dalai-Llama, at least in theory. It was pleasant enough, but still a prison.

He had been interrogated, of course. A small room, earlier, with white-hot lights. Two men, probably FBI or CIA, or some like that. Everything was recorded. Everything was a lie. He also felt he may have been drugged, by then.

"You fucking American pigs might at least have refrained from murdering an innocent young woman, my assistant Milana," he said bitterly, at one point. "I watched her die, shot in the throat, gasping for air, blood gurgling out as she asked me to help her. Is that how you are saving the world? Huh? You answer me, now."

"It couldn't be helped, Rudolph," said one of the interrogators, a large Caucasian man in a pristine white shirt, with an odor of cologne. "I apologize on behalf of the Western programs."

"Huh! Apology! Apology! Ha!" Terchenko spit back at them.

And so it went. He really didn't know anything of value to them anyway. All the technology was common to both sides. There was some back-and-forth again about international space-exploration agreements and treaties from many years ago, with the Eastern view still to maintain that they had a right to launch ships to Mars without approval or explanation, as free space-exploration and research. And it made sense, at least to the East. Why shouldn't they launch ships? Did the US and the West 'own' Mars, or the travel corridors in-between? But with all that had transpired since then, it was obvious it was more than just science-research. And no one ventured more than a word or two, to inform Terchenko about the meteor, or the out-come of the battle on Mars, or his men and ships, though he pleaded at times to know, even demanding. Late at night, in the dark of his prison cell, trying to sleep and let go of it all, he dreamt of Milana, the young Russian girl rescued from the cold streets, to the halls of power as his devoted and loyal assistant, his lover and desire, friend---and her bitter end, feeling to himself guilty of her death, and the denial that he could not protect her, though he would have, even with all his supposed power. And the images he could not forget, the gun-fire, the bunker, the stray bullet, the blood and gurgling young thing there on the floor, his Milana, and the shouting and panic.

"Fuck you all if the world dies in a hell of some fucking rock from space anyway," he said to himself, bitterly. Yet to him it was as if a prayer.

## CHAPTER 58: Other Types of Tigers

It wasn't supposed to be dark. Normally, there would have been plenty of light to see by, and save their lives. Even by the time the rescue docking had started, to deliver the crew of Berle's Command Ship, the 'Understandable', in orbit around Mars, they had not been able to restore normal power. It always was dark there in space, the deep indigo vacuum of certain death, horrifying to think of at times, even for experienced crews. This will make a good story to tell my grandkids, Kick told himself, there in one of several passageways to the airlock ports used to let people in and out. Maybe when I get back to Earth, I'll write a book about it, he thought. And he laughed, surprised to an extent that he still could.

The Understandable had been drifting powerless, for more than 24 hours. The potent electrical shock wave of the targeted Electro-Magnetic Pulse from the Tolstoy had done its work, much like the Krenika's attack on the Penelope, months previously. Lights-out, power down, controls, operating systems, life-support, navigation, and communications. The damaged ship had a crew of 15 people at that time---the regular crew of ten, and five men working on the descent-shuttle pods and soldiering arrangements, or military. As similar radio-signals of much smaller wattage were disabling the Tolstoy, using the safety-command shut-down programs that had been so cautiously imported from Earth and US military, the Penelope was the nearest ship available to assist with Berle's dilemma. Guy's ship was serviced and repaired when they reached the Molinari space dock, and was in good working condition. So connecting the two ships, and dis-embarking the crew of the Understandable, was their next-best hope.

These were big ships, very large, about the size of four 747 Jumbo-Jets, end-to-end, much thicker and bulky, with many large compartments and engines. The Dupont-Monsanto Condrum 21 Deep-Space Local Planetary Cruiser had to travel many millions of miles to make a single journey to Mars and back to Earth, safely. Fuel, men and crews, all the power systems, food and life-sustaining elements, cargo, engines and external hull features---the ships were probably the first Earth had created for such work, that were dependable, of a consistent design, and were easy enough to operate or control, to do the work. Workhorses for local planetary travel, and very successful, after hundreds of years of technology and science-advances. The Condrum 21 couldn't reach Jupiter, or really even Venus, but was great for journeys to the moon or Mars, and had become regulars, of which they were quite proud.

Docking, however, was a real chore. The Penelope could not just pull up alongside the Understandable. But it was fairly common to exchange crew and staff from ship to ship. This was done using a much smaller inter-ship 'life-raft' vessel, released from the Penelope, and equipped to be connected by an air lock to the other ship. The men opened the inner-gate, and boarded the tiny ship, no bigger than a small private airplane back on Earth, or maybe an RV motor home, seeming in appearance much like a tube, or diving-bell, so that when it was properly docked to the other ship, it stuck out at a long right-angle, like a tunnel, with engines and windows, etc.

Commander Berle, now 'in-defeat', at least for himself personally as a leader, was the last to leave the Understandable, which was 'understandable', and typical pilot's protocol. His crew and staff of 15 men had to make their way through the now cold and dark tunnels inside the ship, from their various stations. Some of the inner doorway passages didn't work well, without normal power, and they had to find other ways. At the same time, after they began to realize that they would not be able to restore Berle's ship to normal on their own, or would not need to, they had to provide some kind of preparation for an 'abandon ship' circumstance. No one wanted the Understandable to lose orbit without power or navigation, and crash onto the surface of Mars in a ball of flame. It would be a horrible loss, and the ship was very much needed for on-going affairs, if only to get many of them home. So, there was essential work there, in the dark, air-supply dwindling, colder than normal, moisture on the walls and machinery, and unfamiliar smells. The rescue docking was handled by the remaining crew of the Penelope, now under command of McGee, the navigator, in concert with what could be done from Mars Base. It took time, and well-planned effort, but they were all eventually safe aboard the Penelope, where they could rest and recover, and think out their next move.

"Welcome aboard, Commander Berle," said one of the air lock docking staffers, from within the Penelope, as Kick was able to crawl through the gates and passages, pulling himself heftily along in the null-gravity, huffing a bit with his age, after the short-flight vessel had docked.

"Sure, sure," said Berle. "Looking good. Good job, you guys. Thanks. I'll need to hook up with your pilot, so let them know."

"Already in process, sir," the man said.

"Damn I hate to fuck up a ship like that. I hope we can save her," Berle added. The narrow passageway was just big enough for about five or six men at a time, and there was a certain amount of loud movement and pressure-gate air-pump action, until they were able to move deeper inside where it was more comfortable.

Viewed perhaps from Phoebos, one of the Mar's moons, one might have seen the sky above Mars, with the reddish surface wide and horizontal, cloudless, below, and the three ships, over an area of hundreds of miles. The Tolstoy was far in a Southerly direction, towards the rising Sun, now herself powerless and tilted off-course. The Penelope and the Understandable were quite close, but following the exercise, the Penelope had moved away safely, the only of the three ships in that area above Mars that could navigate.

At this point, in the 'history of Mars', as Guy Reisling had said to Lila Meetek, things were truly an awful mess. Base-commander Bojji-Than and his teams had really no time at all to celebrate their victory in projecting the shutdown programming at the five enemy ships. It was a 'coup-de-gras' that held the promise of their future release from the war to control the base. However, now, five of the Eastern-program space ships were floating around overhead, with very little control or ability to navigate or change course, or slow, or speed-up, or maintain proper orbital trajectory. Each ship was home to at least 10 men, some less, some more. And, there on the surface, like robots with no operating-instructions, the surviving Russian-Islamic-Hindustan soldiers, hanging on however they could in the essential surface-walker suits, were stuck with a war that was basically over---if they only knew. Low on power, oxygen, or bullets and rockets, the 'enemy' on the ground had only the converted descent-shuttle pods, here and there as they had landed by para-glider chutes, now three months previously, to re-supply or rest. And then there were the US men, such as those in the Rhinoceros Hut. The Russian 'Hit-Man' Prokov Keeje was certainly heartless in his approach to winning the base for their side, as far as these men were concerned. For the teams at work inside the Mars Base Command Center, they well knew it would take weeks to straighten things out. Somehow, using the

American ships-in-space, the Eastern-program ships would need to dis-embark, similar to how Berle's crew was rescued (they were in almost identical condition). In theory, they would be shuttled down to Mars, and placed inside the base as prisoners, along with the 'enemy' soldiers. But of course the Mars Base resources were already strained beyond maximum, and the 'enemy' may well be un-cooperative. And there was the voyage home. It could take months.

"If this is winning, I think I'll pass," said Vinces Grant, the explorer-mapper and planetary-surveyor, now inside the base-cafeteria, along with some others, on a shift-break. Reisling, the Penelope's original pilot, was also there, enjoying what they could of the food and drink.

"No one wins a war," Guy said. "They just finally stop fighting."

"Well, not quite yet, but almost," Vinces answered him. "I can't believe we're still here and more-or-less safe and in-charge of this place. And the lights are still on, too. Amazing."

Guy had been assigned to work back 'up-there', with bringing down the Russian-Islamic-Ukrainian troops, docking and a new series of shuttle-descents---this time as non-combatant 'guests', that is, 'prisoners'. Matt Currison Von-Templar would be launching the 'lifters', or ascending shuttle-pods, because by this time, most of them were sitting idle on the surface. It all had to be organized, and one of two Mars Base launch pads was in ruins.

"You guys rock," said Chassidy Katola, now joining them nearby at a long table, with a cafeteria-tray and food. Chassidy was the popular African-American Wellness Specialist, a beautiful Black woman, who had also found herself out among the troops in a walker-suit with a weapon, at one point, just after the sandstorm. "I mean, thanks, Vinces. You too, fly-boy."

"You're with medical, Chass," Guy said. "What do I do for fatigue? Even in the lighter gravity, I feel like a bag of bologna, you know? Exhausted."

"One word---vacation," she said. "We're all six months or nine months over-time for R-n-R, in the most stressful situation ever. I have people with very serious medical needs, and of course the dead. Some of the men have very bad burns, others lost air long enough to develop psychosis or memory-loss. There are no survivors with bullet-wounds, though. One hit, in the suits, they're gone. Except this one guy, he took a round in the foot, and managed to survive. So, if you got to go back up and work the ships, Guy---check with my office first. I can help. Mostly large fish-oil doses and Vitamin B12. I also give testosterone shots now. We'll keep you together."

They talked more; the cafeteria was busy with about 20 hungry people. The mood was---'wow, enough already'.

Aboard the disabled Tolstoy, Commander Keeje had seethed and raged as long as he was able, but also eventually grew to sullenly accept. He was also able to send messages to Earth, and what remained of his program-managers there, on a limited basis, as Mars Base started to work with him on what they began to call, 'the recovery'. Keeje was a very intense servant of his various masters, 'hired' for his cruel attack-dog mentality, ability as a pilot, and slavish obedience to control. But, he too wanted to live. The Mars Base was now, for him and his staff---very sour grapes. But he could cheer himself somewhat that he would at last be able to actually visit the base and see it for himself. Whatever happened, Keeje was Keeje. He wasn't likely to simply become a US Mars program 'defector' or 'informant', but returning to Earth was beginning to look very appealing. He was softening a bit, but not much.

Keeje was from Russia. The military, and then the space-program, for him, was a lifetime adventure and career, with all the promises of wealth and position, and escape from the powerlessness of corrupt and aging Eastern-bloc cities and conflict. He grew up as a strong, pale-skinned hunk, broad-shouldered, muscled youth, with plenty of energy and enthusiasm for whatever the government had planned. His parents were also in the government, but that was during another time, when things had changed yet again, in that part of the world. Many areas were polluted, some with radiation. Terrorist groups in small enclaves and minor regional factions, were always causing problems, deaths, border-disputes, bloody kidnappings, and Keeje was among those fighting them back for the more secure powers, who were just as evil in their tactics. As a very young child, he did not enjoy much in the way of quality food, or nice homes and cars, and often did hard labor to survive. Keeje increased in strength, weapons, military, and later in the submarine and space-program. In many ways, he was an outstanding officer. He had many women, back home.

"I wish I was home, back in the woods and mountains, even in the snow and cold ice. We still have tigers in Russia, you see them sometimes. I'd love a vodka and hang out at the bars again, the disco was fun, the women are so cheap, I love it. Not much like Mars. I'm not sure I can deal with these Americans," he said. He and his helm-deck co-workers were only now waiting in the dark. They had worked to restore some controls, but the shutdown codes on such a large ship were dedicated to total over-ride, basically working automatically at 200 different computer power-sources to 200 different systems-operations computers, or roughly that. It was not totally uncomfortable, except for their anger and sense of defeat. Soon, they would be prisoners. They had talked about continuing the fight, shooting the first rescue-worker at the air lock docking port, but it just made no sense.

"Mars has other kinds of tigers," his companion said.

## CHAPTER 59: Baby Snikta

In study and research, sociology and so on, wars and conflicts begin with various triggers or goals on each side. How they end, or how hostilities conclude, if they do, is something of maybe lesser interest to academia. Does one side conquer all, taking control and dominating newly, with controls and government, as the other side bows meekly to the superior force, their goods and resources taken as loot? Or does the superior force simply lay waste to the converted lands as vengeance, with no real interest in treasures or rule? On Mars, that year of 2081-82, the conflict had already been called the first real Earth-born 'war in space'. Mars by this time was 'local'; the challenges of travel back-and-forth were simplified, creating an outstanding economy of new wealth at home, in labor, science and high-tech industry. Even without a fundamental profit motivation, as if mountains of gold would be found on Mars, the space-programs created real riches on Earth, as 'something useful', with endless support-business globally. It was but an 'act of God' in the unthinkable form of Asteroid U2753b that even made serious conquest against Mars appealing at all, enough for the Eastern-bloc programs to unite behind the idea. As things ended, the deaths, great cost, losses in very rare and even ground-breaking technology, and venomous hatred, all seen to be pointless.

The meteor threat was deflected more easily than was thought. True, the thing would indeed have hit the planet Earth. The projected impact seemed very much to be in the far Southerly Pacific Ocean, near Australia, north of the Antarctic ice-mass, a vast open region of Big Blue Marble water. Australia, southern Pacific island nations like Indonesia and Malaysia, the Philippines, Viet-Nam, Thailand, Japan, New Zealand, far-East Russia and China, and South America, Chile, Argentina, Bolivia:

these areas would have been utterly obliterated. And the rest of the planet would have suffered a remarkable level of environmental pain, without a doubt, an irreversible Earth-change that would be the end of 'life as we know it'.

But Doctor Ibrahim Mehudi, Ph.D. (deceased) was right, although he didn't have the whole story. In his finer level of science and real-time data, based on the work of others and in-concert with the best efforts of the government, this 'dove', who opposed the militarization of space until his death, had found the key to averting disaster. His funeral was a sad day for those who loved and respected him. A coronary event (heart-attack) took him as he was exercising in the hotel swimming pool, while consulting with the Washington, D.C. big shots about the crisis. The asteroid was of course in motion, but Mehudi was able to carefully map and describe that rate and movement-aspects about how it was also tumbling, like a huge rolling stone, at the same time. By about February, nuclear bombs were carefully targeted and then launched, to explode with enormous power, at a precise point on one side of the meteor. Then two more of the same kinds of bombs were launched with amazing accuracy, in late March. The rocketry and telemetry were fairly standard; after all, Earth was launching very exacting and large rockets all the time by then. The bombs were also understood all too well, the most powerful available.

The result was that the asteroid, as big as Texas, was 'nudged', or deflected to a very slightly new path or trajectory through space. With the great distances involved, by the projected date, and the difference was enough. The world watched with greatest interest then, as Big Baby Bertha thankfully missed the planet. She passed at the outer-circle of the Earth's moon, about 150,000-miles out beyond, or a solid 400,000-miles from Terra herself. Only a hair's-width in the vast

Universe, but more than plenty. It was truly a historic moment. Men of Earth had learned that in such a crisis, we have the knowledge, will, and organized effort of unified purpose, to protect our home world from a very rare Extinction-Level-Event. We had saved ourselves, we wanted to, and we knew how, together. And we felt pretty good about it. The bombs were targeted far enough in advance to make the 'nudge' work 'down-the-road', and true enough; no one really knew for certain it would work. But it did, and the angels sang their hallelujahs. Religious cults and other opportunists eagerly longing for the End of the World were disappointed. "That's how it goes."

Mars Base Commander Bojji-Than would serve for two more years in his role running the much-conflicted outpost on Mars. His rule then ended, and he too would return home, glad indeed. After they learned how to shut down the 'invading' ships-in-orbit by remote, Bojji-Than and his teams began immediately to create and organize what they called the Mars Recovery Program. Two years was hardly enough to clean up the mess. One-by-one, Keeje's five orbiting ships were disembarked down to the surface. These space-men were pleased to set foot on solid ground, though humiliated. They were treated well enough. The Eastern-bloc surface fighters also finally understood, by various means, it was all over. Almost an entire floor of one of two main housing units at the Mars Base was devoted to quarters for the new 'guests' (prisoners). The Russian-Islamic ships were now turned over to qualified pilots to operate normal controls, and piloted for on-going successfully sustained orbit. They were of course very useful later getting them home, as far as those making the trip at that time. Vinces Grant, Juno Amorrossi and Karen Tutturro, found they were to ship out, eventually, aboard the Saint Peter (an 'enemy' ship, pilot and crew handled by US Mars astronauts).

"Well, I guess I really enjoyed my first trip to Mars, despite everything, you know," said Karen, lounging with the other two leisurely together in the Saint Peter's view-port deck, as the Saint Peter left orbit. It was a fine ship. Vinces and Juno had to laugh at Karen's comment.

"How many Russians does it take to invade Mars and take over the only base on the planet?" Vinces joked, tossing Karen a cue for another lame wisecrack.

"All right. I don't know, Vinces. How many?" She rolled her eyes.

"Glad you asked," he said. "Only one, a guy named Terchenko, Commander of the overall Eastern-bloc programs. Because he chose to 'rush in' where angels fear to tread. Get it? Fools rush in?"

"Not funny, not funny," said Juno. They could relax during the orbital-escape main engine ship's-thrusters, that pushed the ship into the Abyss, straight for home, on the view-port deck, with seats and less-than-ideal 'real' windows. All they could see were stars and the distant solar Sun, and far off patches of the reddish globe that was Mars, seeming to drop away behind.

"Was I joking?" Vinces replied sharply.

"He who sits in heaven laughs," said Karen smartly. And they did.

Commander Winton 'Kick' Berle later took command of the 'Penelope'. He worked another three months setting up the Mars Recovery Program. But Santa's 'angels' also had basic needs, and soon scheduled departure back to Molinari, and then Earth. All of his ships would leave Mars on a rotating schedule based on favorable planetary positions, two at a time. The 'Understandable' was restored, and served more years for the US-programs. McGee, Aaron the cargo-specialist in his wheelchair, Raza the radioman, and the other crews, stayed with Berle and his staff, and then left Mars in-turn. Peter, their mate who had died on Mars in the fighting, was cremated, along with all the others who gave their lives on both sides. A special expedition was later undertaken, to carry their ashes nearly five miles high up into the rocky cliffs of the Tharsis Montes Mountains, which took days. The remains of those who died were scattered there, and sometime later a permanent marker-monument was built at the site, to remember the battle, and those who died. Few indeed would ever view the brass marker, even in hundreds of years, so rare a human life either to live, or die, on Mars.

"Sure, sure, so glamorous, so famous, what an adventure," said Berle, to some of the men, now secure again at the flight-deck helm of the Penelope, in orbit and humming. "Fuck that horse-shit, my friends. Freeze the god of war in his hell and fail better next time, to co-exist in peace. War sucks."

Other fates awaited various players. Matthew Currison Van Templar chose to stay on Mars. So he remained on an extended schedule another full year, longer than many of the others. At Vandenberg, the US White House and military bases, and at other US-space centers in Florida, Texas, Puerto Rico, Utah, Colorado and elsewhere, the leadership was relieved every bit as much as a hissing, coiled and deadly snake would turn away his intended bite. Lynn Rogers-Smith spent more time in D.C., along with what remained of her original task force. They weren't reviewing the charts and computer star-maps; they were dancing, and better. It was a new day, a Holy Hour, crisis over, a day of release and freedom. Likewise, the entire elite space-program and military or urban power structure staff and crews, hundreds of not thousands of specialized workers, felt it was a good time to celebrate.

US President Renolds poured a glass of high-quality brandy over ice for his co-President Bourgalt. They had found an hour so to muse reflectively, in one the White House anterooms, quite private and secure. The other two US co-Presidents were fully informed, occupied elsewhere. Renolds also had his own glass.

"Here's to home," Renolds said, lifting his drink.

"Amen," said Bourgalt, and they took a drink each from their glasses. "So, Mister President---what's it all about? Do you know?"

Renolds sighed. "Jesus loves you, Mister President," he said, and they raised their glasses again.

"Well, if the Lord tarries, I wonder if the planet will even be here later when the prophecies are at last satisfied," said Bourgalt. "But until then, must we not care for each other?"

They talked more, about this or that, not so much the details of what had transpired. Planetary politics could be very complex, boring by day's end. Then Bourgalt received a call-tone on his private portable line. When he answered, he was talking to Lynn Rogers-Smith. He hit a button and the call went to speaker, so Renolds could also say hello.

"Congratulations, we've saved the world, I assume you've heard," she said.

"Where are you at, Lynn? Can you come to the White House at all? I can arrange a limousine," said Bourgalt.

"We're down here at the Kennedy Center, with Hazlett and Jesse-Gaurrero, and the team from Vandenberg, and a lot of guys from Houston and Florida, just by accident," she said. "They had a large room set aside for us. Why don't you come down here?"

"Uh---well---I don't like to drink and drive. And this all calls for a drink, you think?" Renolds joked.

"With all due respect, I must say, when they followed the asteroid past the moon, I was in tears. Many of us were. Passover, anyone?" she told them. "You and your staff, all the soldiers and pilots, the science teams and then the bombs launched to budge the big stone off course, and the deaths---I'm done, I'll tell you. You have my respect, sirs."

More talk this way; they all felt pretty much the same.

Away, again, far, far off from there, in Mumbai, India, in the slums, a dark-skinned baby boy was born, crying and sneezing. His mother was named Snikta, a common girl's name in those lands. She was 28 years old, dark and thin, and they lived in extreme poverty. Hardly a handful of rice per week, bad water, hard labor every day grinding steel parts into art-works, mundane, boring, tedious, sweatshop. The boy's father had abandoned them, and later died from alcohol over-use. But Snikta had her friends, and assorted family and relatives, also in the slums.

Not one of them had ever heard of the planet Mars, and some felt the Earth was flat, or might be. They dressed in traditional robes and sashes, the older men wrapping themselves in self-fashioned turbans, stroking their beards under tin-roofs with sheets for shade and insects. Not un-touchable, but 'insignificant people'. The baby was born here, and they knew his life would be hard. Not one of them knew that global powers were launching space ships back-and-forth to Mars and other planets, and had been for years. Not one of them really knew anything about the threat of the meteor, for months or years before the birth. Yet, in other hours, they could gather around television sets, with others from around, when electric was available, and image-audio service, and watch with wide eyes as announcers tried to simplify the passing of the meteor, translated for local communications, and censored, of course. The TV images could only show what the governments were releasing, mostly shots from the space-centers, and blurry satellite images of the meteor in space, and the moon, from which a person could learn very little. Snikta had by then wrapped the baby in some clean rags, holding him in her arms, as they watched the crisis together, with other she knew. It was mid-year, 2081. Just another day. But a sense or spirit moved from Earth to Earth, a global moment, and even the poorest and most under-educated could sense it.

Saved. Delivered. Rescued.

## CHAPTER 60: Home At Last

One thing that both Guy, and lovely Lila, and every other long-lived space-worker in the US Mars or US Moon programs, had learned: 'when on Earth, do as the Earthlings do'. In other words, cheer up, you made it, welcome home, life is good, you've crossed over the river. Like a tiny flickering spark of awareness, almost nothing at all in the seemingly endless vast unity of deep-space, as experienced from ships and orbiting space docks, any human life in-transit between worlds or floating around somewhere, wondering what it's all about, was certainly a fragile thing. If, as the space-program planners and science-guys said, 'the Universe is actively hostile to intelligent life', which was maybe debatable---but given as much, for Guy and Lila, there was really no way to be 'that tough', or 'that strong', and it eventually became clear to each and every one of them. They did what they did out there, the technology and machines worked or they didn't, they could breath life-sustaining oxygen moment-by-moment or they could not, and the space-ships and shuttles landed successfully, or lifted off successfully, or entered orbit successfully, or navigated from planet to planet successfully---or they did not. So, it was a quick study---'be happy now', or you might not see another chance, when things went wrong. And there was no happier time for two like these, than to finally be back home together for real, on solid ground, in California, about 2082.

Lila was the beautiful one, it was true. But her appeal was also intellectual; she was bright as any, smart, easily quoting Shakespeare or trivial facts about the geography of Russian/Balkan sedimentary rock formations. And, as a space-worker, she was also athletic. They had to be, the work was rigorous and physically demanding. So she had the body of an

Olympic swimmer or runner, though perhaps not perfect. A perfect woman? Not so, as Guy could attest, and he wouldn't have it any other way. Her heart was a journey in itself, to gaze into her eyes a longer and even more demanding adventure than any travel through space. Alive, aware---at times his love for her was overwhelming, and he was a simple man. And it was only ever the thought of losing her.

As the Mars-Recovery became a more organized program, ships were once again on their way back-and-forth from Earth on a more regular schedule. The Mars Base was in urgent need of all sorts of supplies, now. Many repairs and re-stocked fuels or substances, technology, medical-evacuations, prisoner-evacuations, leadership and common transport needs, were essential. Without a truly Earth-like surface ecology, the base was extremely fragile, and given the 'war', Mars Base management was ever on-edge over things like air-seals, power-supplies, oxygen levels and water levels and important technological failures, at least for a time. The invading 'enemy' had done all the damage they could. Over now, for hostilities, but not for survival. So, some of the orbiting US ships under Winton Berle headed home, with personnel aboard, including Guy. As it worked out, he did not pilot the Penelope back to Earth as her triumphant pilot/commander. Medical revealed that Guy was suffering from an unspecified fatigue-syndrome of some kind; he had unusual weakness, difficulty sleeping, poor concentration and saggy reflexes or reaction time, and also minor heart-palpation. Many of them did, it wasn't unusual at all, and he had been more than a year in deep space doing very stressful work. So he was finally assigned to return to Earth aboard a ship under Berle's command, with as many Mars-regulars as they could berth along with him, those whose 'work-shifts' had become dangerously long. And of course the journey was equally lengthy heading home, and needed to be plotted for the planetary alignments and so on. It was more than six months before he saw Lila, as they docked at Molinari, and then they were together for the rest of the journey home, like a cranky old married couple on a Greyhound bus out of Saint Louis, only more so. And then the disembark and shuttle down to Earth-base as the ship entered Earth-orbit, the medical and de-briefing, a week's hospital-rest and health-tests as standard for that length of service in space, and also some simple matters of personal grooming. Guy's beard had grown quite scraggly and matted, he needed a haircut, and he had a rash. And for all her beauty, Lila complained of a tummy-ache---which Vandenberg medical soon learned was due to the fact that she was pregnant.

"I wouldn't say I blame you," Lila told Guy, now together with him in the pleasant afternoon sunshine in Santa Barbara, California, where they found themselves walking together on the beach. The blue Pacific seemed every bit as ageless as Mars, her cold waves lapping towards the dirty-looking sand, sea-gulls calling to one another, a mist in the clouds that lingered as a marine-layer that cooled from the jealous Sun. "I mean, me being pregnant."

"Blame me? How about---'thanks'?" said Guy. "Race-horse quality sperm, lady-baby-girl. Our kid will fly high and far in his lifetime. I'm looking forward to it."

"What if it's not yours?"

Guy was by now a lot stronger, the journey had been harder on him than he realized. Four years older, ruddy looking, somewhat taller than Lila but not much, of European ancestry, with blonde-brown-reddish hair, and freckles, He was trim, strong in the chest and arms, with blue-gray eyes. For walks on the beach and around downtown Santa Barbara, he wore khaki-jeans and tennis shoes, with a burgundy long-sleeved polo shirt, sunglasses and a floppy sun-hat. Lila was in a very modern blue spandex sort of dancing-gear outfit, tightly showing her figure, with a draped-over colorful pink and gray sash-robe, and sandals. She also liked to wear jewelry, and had shaved her hair to almost bald, which was somehow sexy.

"Heck, I dunno'," Guy responded. "It's not like we've been that active. What are the odds? You could answer that better than me."

She took her time to reply. There was the guy at Molinari, the hunky external-hull worker, but any pregnancy would have been obvious since those romps. And she usually used protection to avoid a pregnancy. As near as they could place it, the baby must have been conceived when Guy 'docked' at Molinari on his way back home. She really wasn't all that promiscuous, and DNA tests would give them the real info. Maybe it wasn't important.

"It's probably yours Guy, okay? You fucking dick!" Then she looked at him ferociously, turned and ran away from him down the beach towards where they had parked his car, suddenly very emotional. He waited, letting her go.

"Hey! Hey!" he called. Like the sea gulls, their cries to one another back-and-forth, different and so very human, yet meaning much. Lila was fast, and then 100 yards away, up some rocks, by a trail through the dogwood bushes and ice plant and grasses. Guy started after her slowly, knowing. Things had changed.

Lila had private living quarters provided by Vandenberg, when she was Earthside, and then usually lived in San Francisco at a family home, one of the older brownstones there on the hilly streets, still preserved or restored at that time. She and Guy had chosen to spend time together at his place on the Central Coast, the same private house in a quiet residential neighborhood, single-story but with plenty of the usual comforts. By the time they were back at his place, she had calmed down and was easier with him. But it was plain, they had some choices to make, neither of them sure what was ahead.

"Does anyone even know how many babies have been conceived off-planet? Are there even records? Does that mean our baby is even an American?" Lila was somewhat flustered. She could easily handle tracking dangerous meteor showers that even threatened her own life, and those of friends and co-workers, or solar flares, or planet-level massively destructive asteroids hurtling towards Earth. She could deal with space-ship docking and passenger transfer procedures flawlessly, she knew how to deal with long-range radar and heat-maps by computer, and in an emergency, Lila was prepared to work outside a ship or the space dock (Molinari), in a life-suit, or even pilot one of the ships, for basic operations. But making a baby, giving birth to a new life, and cementing her relationship with Guy Reisling for the rest of her life, was somehow too much, for that evening anyway. They had enjoyed some wine earlier, so perhaps it was telling in her mood, as she was getting herself ready for a hot shower before bed.

"Well, he's not a Martian, I guess," Guy answered, laughing. "Maybe sort of a Citizen of the Galaxy. A galaxy baby. An interplanetary local? Seriously, Lila, there are almost certainly records of some kind, and I can tell you from what I've looked at, there have been hundreds of babies conceived in space before yours. Lighten up."

He followed her into the shower-stall area in his bathroom, complete with twin-nozzle 120-degree hot water, ruby-colored stone tiles accented with white marble, double-occupancy bathing, a separate tub big enough for three, and the usual toilet features and grooming. Lila was dropping her street clothes as she started the hot water, with steam beginning to fill the room, fogging the mirror. Guy was still in his khaki-jeans and polo shirt. Lila never failed to arouse him, and watching her undress, he faced another choice, but decided to wait. She grabbed a towel, stepping lightly out of her skin-tight blue spandex. Her legs went on forever.

"I have a career, Guy! I have friends at Molinari! I wanted to advance to management and planning! I could do it, too! I've put 12-years into my work at space dock!!"

"You think I want to lose my credentials as a pilot?" Guy responded. "Why is that different? I'll never fly again! They said something medical was going on with me, fatigue syndrome or something. I was out for 18 months! Just to deliver their radio-gear! So don't feel so sorry for yourself, Lila, you're just afraid it will hurt when it comes out, that's all. A woman your age!"

Now she stepped into the flowing hot water, the plexi-glass sliding doors deftly shutting behind her with sincerity. All he could see was her naked body now as a blurry form somewhere in the steam as she began to soap. "It's no mystery to me that you love me, Guy. I'm a prize race-horse myself, not like I don't know it, the space-program isn't for old ladies, is it?"

"We can make a different kind of life for ourselves now, Lila, that's all," Guy went on. "This whole thing, these last few years. We didn't realize it, but---they don't need us any more, you know?"

"Speak for yourself," she said. "I love my work!"

"Plenty of women, or men for that matter, can run those monitors out there," Guy said. "Look---woman---damn it---

marry me! Man-and-wife---you know—a wedding and a house and kids and all that crap. What's wrong with that?"

It took a moment for the two of them to recognize what had happened, with Guy's obvious and at least somewhat passionate proposal. The streaming hot water accented the silence like a pattering rain. Lila was standing naked to the world behind the glass, doing her nails. She seemed distracted, absently moving the ideas around about her future, and the new life inside her. For Guy as well, he suddenly realized that he would have to live up to those words, and follow through, perhaps for the rest of his life. But he did love her, more than she knew.

"I'll think about it," she said finally, like a shoe dropping. "Just take your pants off and get in here and make love to me. I feel sad."

He was happy to oblige.

## CHAPTER 61: Biscuits and Gravy

*"Star-crossed lovers really do exist, they tend to forget each other over long periods apart, but it makes a great story"-Lucille Ball to Desi Arnaz, concerning an early 1968 episode of Star Trek, a Desilu Studios production.*

Rob Cowan, the original co-pilot for that cargo trip across the Earth-Mars travel corridor aboard the Penelope, had invited Guy, and Lila Meetek, to spend time with himself and his family at their considerable Idaho Ranch. After the news regarding the 'near-miss' status on the meteor that had started the past four years of panic, things had settled down for many of the US Mars program workers. They could at last take care of medical and health issues, relationships, financial or home-matters such as sons-and-daughters, investments, schooling, and so on. And not incidentally, take part in any planning or future-vision meetings and background, as far as the US Mars program, based on hard-earned experience.

Rob had inherited the family cattle ranch, and it was a working ranch, even when he was absent. In that part of the country, very large-tall stony mountains and hills or small rocky creeks, many trees such as oak or pine, also birch and cottonwood trees. In a typical season, he would run 500 to 1,000 head of prime-beef cattle, heifers or steers, and some Brahma-bulls, for breeding. For an astronaut, or many types of demanding or high-paying professions, a ranch or farm, or often a wine vineyard, was a way to escape the many demands and pressures of these careers. And Rob had grown up here, it truly was his home, so he suffered none of the 'out-of-towner' local-yokel pretense or rejection. For many of his neighbors, folks who had lived in the area five generations or more, the fact that Rob worked in the space-program meant less than how his cattle were handling a particularly deep snow that year, or what he had chosen to do about a certain variety of livestock fungus that was going around.

Lila and Guy arrived later towards summer, as it began to get warm and the bright sunshine over and beyond the clouds and blue-blue sky, was so very unlike the thin-silvery-distant light from the same Sun, back at Mars Base. It was true, and Lila agreed, along with other medical professionals and experts, that the adjustment from deep-space work, to regular gravity Earthside home-life, was at least somewhat real, to the senses. After a year in space, or half a year, walking again on Earth was sort of dreamlike, even Heavenly, with so many sensory delights and potent, fertile memories, so 'real' as be even overwhelming, by stark contrast to the empty, dark, airless void of space and the sterile-technical, machinelike chamber-passageways, airlock doors, and rooms and suits and controls, in the ships and the space dock. So Idaho looked wonderfully natural and pleasant.

The couple traveled by road-car, from Guy's place near Santa Barbara, in California. Guy was driving a 2080 Corvette Electron, one of the new-generation hydrogen-fuel cell 'electrics', in this case designed in the basic mold of the classic Chevy Corvette. It was an incredible car. He could easily cruise at 120 mph, on quality highways, and also virtually silent-running. The journey was about 1,280-miles, somewhat more, and Lila made a point for Guy to stop frequently, so they could enjoy the sights and towns and small cafes. It was especially pleasant, because no one had any idea they were space-workers, and they could be easy with themselves among the public. Not that Guy was a celebrity. But with all that had happened regarding the meteor, and the so-called 'space war', of which they had been a part, either one of them would sometimes be asked for autographs or interviews, or photos with the family. The Corvette-Electron whispered through the mountains near Santa Fe, New Mexico. Lila had the top down, it was quite warm.

"So, was Rob really crazy, or what?" Lila asked Guy, as he drove. She raised her voice above the wind, lifting her sunglasses to glare at him from the side. "I know what the doctors said, Guy. I just wanted your opinion."

"I think without a doubt he developed psychosis," Guy told her. "It's pretty common in space-travel, and we had the stress of figuring out how to survive the EMP from the Krenika. He was also taking experimental gene-trigger antibiotics, which later they said was a factor. But he's okay, and he knows how to deal with his emotions, now. He's s smart guy."

"That's why he was relieved of duty at Molinari, when you finally docked," Lila responded, then turning. "Oh, look---cows!"

She pointed to a grassy field across from them, where a herd of 100 or so large reddish heifers were grazing peacefully. The Electron whizzed past.

"The steaks out this way are to die for," Guy said. "The quality ranches do corn-fed or grains-only, best organic. Makes a mighty-fine hamburger."

Guy drove on, cruising now at about 80 mph on one of the open highways through the reserve-lands. "So, Rob's okay now, then, right?" Lila said.

"Yeah, he's fine."

"Crazy is---crazy," Lila said seriously. "I never cared for it in responsible people, or friends."

They met at the main ranch house, and said hello. Rob's wife and two kids were there, although the teens were distracted by some needed daily chores, both schoolwork and farm-work. Rob looked a lot better, and indeed, they laughed nervously and smiled at all they had shared and suffered together, with all the tension-and-release of knowing 'it had all worked out'. By this time it was late afternoon, a breeze was blowing through the cottonwood trees, with a characteristic sound unique to that variety. The ended up sitting around the dinner table inside. Rob's wife, Carol, had prepared a nice plate of fresh biscuits-and-gravy, with fruit and wine, brandy, or hot coffee.

"We got into more trouble in the bars in Vegas than we ever did on the way to Mars!" Rob joked. "That was nothing!"

They laughed. "Depends what you mean by 'trouble'," Guy responded. "At least in Vegas you got air and gravity!"

"And hookers," said Lila. Carol joined her with a smile.

"The abyss is a great whore, it's true," Guy added. "Vegas is a lot like the Molinari anyway, I always thought. Always a gamble."

"These are fresh-baked farm biscuits, with ranch-gravy," Carol said. "Rob's favorite in the morning. You make it like corn bread, but use stone-ground local flower, eggs and milk. The gravy is with flour, milk, oil, pepper, and baking soda, and beef-stock."

"Real home cooking," said Rob. They all ate together silently for a moment. "So, you two---no more work in ships for a while? Rumor was you were getting hitched? About time. Shit,

Guy, take responsibility for a change. What'd you do, make a baby or something? Science has proven what causes that, these days, you know."

Then Guy shared that Lila's ultra-sound had shown the baby was girl, and they were going to name her Snikta. This got at least a serious chuckle or two, as dinner ended. Marriage was another matter, but it was fairly clear that both Guy and Lila were in a significant transition from space-work, to something else. The men went their own way after a while into Rob's 'man-cave', just to drink and shoot-the-breeze. Carol, Rob's farm-wife for more than 12 years, took Lila aside, out into a garden she kept on an open-patio, where she had plants and ornamentals, including lavender, statice-sinuata, mertle-bush, and other somewhat odd perennials, which seemed to like the moonlight. Girl-talk.

"Okay, so how does a woman keep her love alive when her man is working on another planet a million miles away?" Carol mentioned slyly, working a little with the plants, plucking a few stray flowers and stems that had grown too long. "Know what I mean? A million miles away for six months or a year at a time, and only radio or video chats to say hello now and then. It's awful. Good money. But hard on a marriage. Rob has been eight years with the program, as co-pilot, I guess five years with Guy. But---what do you and Guy do?"

Lila was ready for her. "I cheat on him," she said. "And he cheats one me. Does that answer your question?"

Carol huffed a snickering farm-girl's breath into the night. "You mean you have sex with the other astronauts on the Molinari? Oh great. You naughty thing."

"It's healthier than fooling myself that I'll even ever see him again," Lila said. "Work in space is dangerous."

"I wouldn't know," Carol said. "Here, some fresh lavender for you and Guy. You can make a scented gift bag. Or just put it around somewhere."

Late into the evening, they retired to separate bedrooms for the two couples. The Cowan teens also had rooms, staying up late to watch favorite TV-shows. The ranch-life and Idaho mountains seemed to quiet, with only the wind through the cottonwood trees, and frogs in water-paths, or sometimes a night bird. Lila and Guy were exhausted, it was only a few weeks since they had returned to Earthside, gone through disembark and medical de-briefing, and returned to 'normal' life. Rob's guest-room had a huge, sweetly scented bed with thick blankets of wool-and-cotton, with cowboy patterns and art. Lila had placed Carol's lavender on an end table. They listened to the local radio a while, then trimmed the night-lamp, and got cozy together, naked under the sheets, a bliss of rare opportunity for them.

"Rob had some news, girl," Guy said. "He wants to give us a piece of land, down from here about 80 miles, with some small cabins and a barn, part of his rancho. Real easy terms, like, nothing, or next-to-nothing, just cover the taxes. He showed me pictures, it's pretty sweet."

Lila turned on one side like a long lazy cat. She wasn't 'showing' yet, or not much. But a woman can tell, and it was her first. "Might be good," she said. "Good place to raise a kid."

"Yeah, or even have some more," Guy joked. They cuddled and talked more, then drifted off in a lover's embrace, this time with the subtle help of real gravity. Rob's Idaho land and small 'rancho' was worth millions. Lila had some misgivings about living so near Rob and his family, thinking it may grow uncomfortable over a few years. But Guy reassured her; the spot was twenty miles away, with plenty of hills and small creeks or cattle-grazing fields in-between. Under other circumstances, had the meteor hit the planet, even their Idaho homestead might have been seriously impacted. A strike by the rock in the Atlantic ocean, maybe North such as Canada, or mainland continental America somewhere, and they would have faced very harsh consequences to the environment---huge dust and particle-debris clouds, bizarre weather patterns, maybe wandering groups of displaced homeless Americans looking for food or shelter, lawlessness, loss of normal power-grid and services, or much worse. Even deep inland in the US, a large asteroid like that would not have been without its effect, even complete disaster.

But it wasn't really the military or space-programs, or scientists who had been their salvation. The meteor had missed, wasn't that enough? The battle on Mars was not much more than a political opportunity for the East, the meteor panic was only an excuse, for what they wanted all along. And all the deaths and tremendous effort and expense, meaningless. So, three highly trained space workers for the US programs on Mars and elsewhere, and their families, easily and painlessly ended their careers on active-duty, that year. With Lila ready to bear a child, and Guy's fatigue-syndrome, as well as Rob's now medically demonstrated illness, though past, the powers that had lead them off world, with all the glamorous attention and adventure, now released them. It made sense. If it had been 1,000-years there for them and their kin and offspring, it couldn't have been sweeter. Work as astronautics trainers themselves, or in consulting, or deep-background research and experienced 'retired' space-experts, may yet be ahead---or some other crisis, maybe a child who didn't make it, in the cold snow, one day yet ahead, wandering after a lost calf or lost dog he loved, and following it into the wild, then himself lost, and the parents to weep. But that's life, that's pain, that's birth, that's death. And between the dark hours, they had each other. And it was enough.

The planets still turned and danced around us all, then, ageless, smiling it seemed at neighbor Earth, and her many disgraceful days.

THE END

Cover art "Mars Outpost" (by BookBaby.com for Luong Films)

Cover art "Tharsis Montes" by David Schleinkofer | artist20@verizon.net

## ABOUT THE AUTHORS

Tom Luong | Film Producer | Director
tvluong1@hotmail.com | 951-660-6010

Tom Luong is a Vietnamese American born in his native country just after the Vietnam War ended and raised mainly in California. Tom was born in November 11, 1976. His parents were refugees of the war and fled Communist Vietnam in the late 70's and was awarded sponsorship with a relative to live in Orange County in Southern California in late 1981. Like many Vietnamese that fled the Communism, his Dad (Mike Manh Van Luong born in 1949; Mom, Nancy Ngat Thi Le born 1954) was a POW during the war and this affected Tom in many ways. To understand about how his Dad felt during the war, Tom joined the US Army at one point and underwent basic training and was later deployed to South Korea. Tom later went to film school to hopefully make films someday about the war. After working with Julian Phillips on the first movie script, he expanded to writing books. Tom went to many colleges to gain a thorough understanding of the physical world and has a BS degree in Aerospace Engineering from Cal Poly Pomona in 2001. He completed film studies in 2008 and directed his first feature film "The Grounded" in 2011. Tom is a futurist and likes to make movies about the future of Human existence.

Michael Julian Phillips | Author | Film Screenwriter | Producer
pog777@inbox.com | www.kumaskitchenfreelancemedia.webs.com

Michael Julian Phillips is a 56-year old retired journalist-artist, a Fifth-generation Californian whose career has included many projects for film-video, stage, numerous children's books, newspapers and magazines, and music. Julian was born in 1957, at Seaside Hospital in Long Beach, California, the son of John William Phillips, an Army Radio Man who served in World War 2 and the Korean Conflict, and his young German born bride, Brigitte, an orphan girl whose childhood included witness of Hitler's murderous rule, and Russian military incursion into East Germany. Julian's grandfather, Edward Julian Phillips, known as 'Jules', was a Salinas area farmer and crop scientist, who was schoolyard pals with famous California writer John Steinbeck. Julian has worked in freelance and small newspapers since about 1973, and was editor of a countywide weekly tabloid newspaper, year 2000 to 2003, in San Luis Obispo County, California, where he spent most of his life, and grew up on ranches and farms. The writer earned a BA degree in Journalism/Communication from San Jose State University in 1981. Julian has been married for 25 years to his wife, Carol Lynn, and his son Preston Laverne Phillips is now an art student in San Francisco.